The Effects of the COVID-19 Pandemic on the Digital Competence of Educators

The Effects of the COVID-19 Pandemic on the Digital Competence of Educators

Editors

Boni García
Carlos Alario-Hoyos
Mar Pérez-Sanagustín
Miguel Morales
Oscar Jerez

MDPI • Basel • Beijing • Wuhan • Barcelona • Belgrade • Manchester • Tokyo • Cluj • Tianjin

Editors
Boni García
Universidad Carlos III de
Madrid (UC3M)
Madrid
Spain

Carlos Alario-Hoyos
Universidad Carlos III de
Madrid (UC3M)
Madrid
Spain

Mar Pérez-Sanagustín
Paul Sabatier University -
Toulouse III
Toulouse
France

Miguel Morales
Universidad Galileo
Ciudad de Guatemala
Guatemala

Oscar Jerez
Universidad de Chile
Santiago
Chile

Editorial Office
MDPI
St. Alban-Anlage 66
4052 Basel, Switzerland

This is a reprint of articles from the Special Issue published online in the open access journal *Electronics* (ISSN 2079-9292) (available at: https://www.mdpi.com/journal/electronics/special_issues/COVID_Digital_Competence).

For citation purposes, cite each article independently as indicated on the article page online and as indicated below:

LastName, A.A.; LastName, B.B.; LastName, C.C. Article Title. *Journal Name* **Year**, *Volume Number*, Page Range.

ISBN 978-3-0365-6460-9 (Hbk)
ISBN 978-3-0365-6461-6 (PDF)

Contents

About the Editors vii

Boni García, Carlos Alario-Hoyos, Mar Pérez-Sanagustín, Miguel Morales and Oscar Jerez
The Effects of the COVID-19 Pandemic on the Digital Competence of Educators
Reprinted from: *Electronics* **2023**, *12*, 82, doi:10.3390/electronics12010082 1

Rosa Estriegana, Jose-Amelio Medina-Merodio, Rafael Robina-Ramírez and Roberto Barchino
Analysis of Cooperative Skills Development through Relational Coordination in a Gamified Online Learning Environment
Reprinted from: *Electronics* **2021**, *10*, 2032, doi:10.3390/electronics10162032 5

Ahmed Al Mulhem and Mohammed Amin Almaiah
A Conceptual Model to Investigate the Role of Mobile Game Applications in Education during the COVID-19 Pandemic
Reprinted from: *Electronics* **2021**, *10*, 2106, doi:10.3390/electronics10172106 27

Jain-Shing Wu, Ting-Hsuan Chien, Li-Ren Chien and Chin-Yi Yang
Using Artificial Intelligence to Predict Class Loyalty and Plagiarism in Students in an Online Blended Programming Course during the COVID-19 Pandemic
Reprinted from: *Electronics* **2021**, *10*, 2203, doi:10.3390/electronics10182203 41

Pierpaolo Limone, Giusi Antonia Toto and Barbara Cafarelli
The Decision-Making Process and the Construction of Online Sociality through the Digital Storytelling Methodology
Reprinted from: *Electronics* **2021**, *10*, 2465, doi:10.3390/electronics10202465 61

Álvaro Antón-Sancho, Diego Vergara and Pablo Fernández-Arias
Self-Assessment of Soft Skills of University Teachers from Countries with a Low Level of Digital Competence
Reprinted from: *Electronics* **2021**, *10*, 2532, doi:10.3390/electronics10202532 79

Iman Oraif and Tariq Elyas
Applicability of Collaborative Work in the COVID-19 Era: Use of Breakout Groups in Teaching L2 Translation
Reprinted from: *Electronics* **2021**, *10*, 2846, doi:10.3390/electronics10222846 99

Mar Pérez-Sanagustín, Iouri Kotorov, António Teixeira, Fernanda Mansilla, Julien Broisin, Carlos Alario-Hoyos, et al.
A Competency Framework for Teaching and Learning Innovation Centers for the 21st Century: Anticipating the Post-COVID-19 Age
Reprinted from: *Electronics* **2022**, *11*, 413, doi:10.3390/electronics11030413 115

Živilė Sederevičiūtė-Pačiauskienė, Ilona Valantinaitė and Vaida Asakavičiūtė
'Should I Turn On My Video Camera?' The Students' Perceptions of the Use of Video Cameras in Synchronous Distant Learning
Reprinted from: *Electronics* **2022**, *11*, 813, doi:10.3390/electronics11050813 149

Mohammed Amin Almaiah, Sarra Ayouni, Fahima Hajjej, Abdalwali Lutfi, Omar Almomani and Ali Bani Awad
Smart Mobile Learning Success Model for Higher Educational Institutions in the Context of the COVID-19 Pandemic
Reprinted from: *Electronics* **2022**, *11*, 1278, doi:10.3390/electronics11081278 161

Esen Sucuoğlu and Azubike Umunze Andrew
Administrators and Students on E-Learning: The Benefits and Impacts of Proper Implementation in Nigeria
Reprinted from: *Electronics* **2022**, *11*, 1650, doi:10.3390/electronics11101650 . . . . **175**

Talal Musaed Alghizzi and Tariq Elyas
The Effect of Graded-Reading Websites/Applications on EFL Undergraduates' Reading Comprehension during COVID-19 Pandemic
Reprinted from: *Electronics* **2022**, *11*, 1751, doi:10.3390/electronics11111751 . . . . **187**

Flora Woltran, Katharina-Theresa Lindner, Tanja Dzojic and Susanne Schwab
Will–Skill–Tool Components as Key Factors for Digital Media Implementation in Education: Austrian Teachers' Experiences with Digital Forms of Instruction during the COVID-19 Pandemic
Reprinted from: *Electronics* **2022**, *11*, 1805, doi:10.3390/electronics11121805 . . . . **209**

Tommy Tanu Wijaya, Jianlan Tang, Yiming Cao, Robert Weinhandl and Zsolt Lavicza
Factors Influencing Students' Continuous Intentions for Using Micro-Lectures in the Post-COVID-19 Period: A Modification of the UTAUT-2 Approach
Reprinted from: *Electronics* **2022**, *11*, 1924, doi:10.3390/electronics11131924 . . . . **225**

Maura A. E. Pilotti, Emaan Nazeeruddin, Mohammad Nazeeruddin, Ibtisam Daqqa, Hanadi Abdelsalam and Maryam Abdullah
Is Initial Performance in a Course Informative? Machine Learning Algorithms as Aids for the Early Detection of At-Risk Students
Reprinted from: *Electronics* **2022**, *11*, 2057, doi:10.3390/electronics11132057 . . . . **245**

Eda Birsa, Matjaž Kljun and Barbara Kopačin
ICT Usage for Cross-Curricular Connections in Music and Visual Arts during Emergency Remote Teaching in Slovenia
Reprinted from: *Electronics* **2022**, *11*, 2090, doi:10.3390/electronics11132090 . . . . **261**

About the Editors

Boni García

Boni García (Associate Professor) works at Universidad Carlos III de Madrid (UC3M) in Spain. His main research interest is software engineering, with a special focus on automated testing. Also, he is Staff Software Engineer at Sauce Labs, working in the Open Source Program Office. He is a committer at Selenium, and the creator and maintainer of several projects belonging its ecosystem, such as WebDriverManager, Selenium-Jupiter, and BrowserWatcher. He wrote the books Mastering Software Testing with JUnit 5 (Packt Publishing, 2017) and Hands-On Selenium WebDriver with Java (O'Reilly Media, 2022). Moreover, he is the author of more than 45 research papers in different journals, magazines, and conferences.

Carlos Alario-Hoyos

Carlos Alario-Hoyos (Associate Professor) received the M.S. and Ph.D. degrees in information and communication technologies from the Universidad of Valladolid, Valladolid, Spain, in 2007 and 2012, respectively. He is currently an Associate Professor with the Department of Telematics Engineering, Universidad Carlos III de Madrid, Getafe, Spain. His skills and experience include research and development in MOOCs, social networks, collaborative learning, or evaluation of learning experiences.

Mar Pérez-Sanagustín

Mar Pérez-Sanagustín (Associate Professor) works at the Université Paul Sabatier Tolouse III (France), researcher at the Institute de Recherche Informatique de Tolouse (IRIT), and associate researcher at the Pontificia Universidad Católica de Chile (PUC). Mar also founded the team T4DLab at PUC, and she continuous collaborating with it. Mar worked from 2011 until 2014 as a researcher at the Gradient Lab of the Gradient del Grupo de Aplicaciones y Servicios Telemáticos (GAST) at the Universidad Carlos III de Madrid (UC3M) and as a teacher of the department of Telematics Engineering, where she was with a Post-Doctoral Fellow Alianza 4 Universidades. She is a Doctor in Information and Communication Technologies since July 2011 by the University Pompeu Fabra, where she obtained the cum laude grade with European Mention. During her PhD, Mar completed her studies with a moths stage at the LTRI group from the London Metropolitan University in London.

Miguel Morales

Miguel Morales (Associate Professor) is Ph.D. Information and Knowledge Engineering from Universidad de Alcalá (Spain). He has a Master's Degree in Digital Education, from the Universidad Carlos III de Madrid (Spain) and Electronic Engineer from Universidad Galileo (Guatemala). He is Sub-Director of IVN (Institute Von Neumann), Teaching and Learning Center coordinator and responsible for management of virtual innovative academic programs and MOOCs in Galileo University. He has been involved in international projects related to MOOCs and e-Learning, among the projects are the following: ESVI-AL (ALFA III), MOOC-Maker and PROFXXI (Erasmus+).

Oscar Jerez

Oscar Jerez (Associate Professor) is a Doctor in Psychology and Education, and with a Diploma of Advanced Studies in Psychology and Education, from the University of Granada in Spain. He has a Master's Degree in Education, with a mention in Educational Informatics from the University of Chile, a Bachelor's Degree in Philosophy from the Università Pontificia Salesiana di Roma and a Bachelor's Degree in Philosophy from the University of Chile. Double Degree in Education and Professor of Philosophy, from the Pontificia Universidad Católica de Chile. Consultant and advisor for various government entities and higher education Latin America, Europe and Africa. He has been an undergraduate and postgraduate teacher in various programs on Quality Management, Teaching in higher education and co-creator of Teaching and Learning Centers in more than 37 universities around the world.

electronics

Editorial

The Effects of the COVID-19 Pandemic on the Digital Competence of Educators

Boni García [1,*], Carlos Alario-Hoyos [1], Mar Pérez-Sanagustín [2,3], Miguel Morales [4] and Oscar Jerez [5,6]

1 Telematics Department, Campus Leganés, Universidad Carlos III de Madrid, 28911 Leganés, Spain
2 Institut de Recherche en Informatique de Toulouse (IRIT), Université de Toulouse, 31062 Toulouse, France
3 Department of Computer Science, Pontificia Universidad Católica de Chile, Santiago 7810000, Chile
4 Galileo Educational System Department, Universidad Galileo, Ciudad de Guatemala 01010, Guatemala
5 Economics and Business School, Universidad de Chile, Santiago 8320000, Chile
6 Laspau Affiliated with Harvard University, Cambridge, MA 02138-6095, USA
* Correspondence: boni.garcia@uc3m.es

1. Introduction

The COVID-19 pandemic is having an undeniable impact on all aspects of society. Regarding teaching and learning activities, most educational institutions suspended in-person instruction and moved to remote emergency teaching during the lockdown of March and April 2020. Although many countries progressively re-opened their educational systems, online and hybrid education became a common practice aimed at reducing the spread of the COVID-19 disease. This disruption has caused an unprecedented acceleration in the digitalization of teaching and learning. Teaching professionals have been forced to develop their digital competence quickly, achieving mastery in the management of information, creation of audiovisual content, and use of technology to keep their students engaged. This Special Issue (SI) presents contributions regarding adopting distance learning strategies, experiences, or lessons learned in this domain.

2. The Present Issue

A total of 15 papers are presented in this SI. Birsa et al. [1] conducted a qualitative case study by interviewing teachers from Slovene primary schools focusing on implementing cross-curricular connections in music and visual arts content during emergency remote teaching. Pilotti et al. [2] proposed machine learning algorithms to predict students' final grades based on early performance. Wijaya and Weinhandl [3] presented a study to explain and predict the factors influencing students' continuous intention to use micro-lectures to learn mathematics after the pandemic based on the unified theory of acceptance and use of technology (UTAUT-2). Woltran et al. [4] studied Christensen and Knezek's theoretical Will–Skill–Tool model to examine various factors that influenced teachers' use of digital technologies during remote emergency education due to the materialization of COVID-19 in the spring of 2020. Alghizzi and Elyas [5] conducted a study to analyze the effect of learning environments (blended and online) on the reading comprehension in English of Saudi undergraduates during the COVID-19 pandemic. Sucuoğlu and Andrew [6] presented a paper investigating the impact of adopting e-learning to fix the dilapidated Nigerian educational structure. Almaiah et al. [7] proposed an article examining the most critical drivers influencing the adoption of smart mobile learning (M-learning) using the Technology Acceptance Model (TAM). Sederevičiūtė-Pačiauskienė et al. [8] studied how students perceived the quality studies related to the use of video cameras during synchronous remote learning. Pérez-Sanagustín et al. [9] evaluated the PROF-XXI framework, which proposes a holistic set of competencies that Higher Education Institutions (HEIs) can take as a reference to develop the actions and strategies of their Teaching and Learning Centers (TLCs). Oraif and Elyas [10] investigated the learners' acceptance of breakout

Citation: García, B.; Alario-Hoyos, C.; Pérez-Sanagustín, M.; Morales, M.; Jerez, O. The Effects of the COVID-19 Pandemic on the Digital Competence of Educators. *Electronics* **2023**, *12*, 82. https://doi.org/10.3390/electronics12010082

Received: 12 December 2022
Accepted: 13 December 2022
Published: 25 December 2022

groups on the Learning Management System (LMS) Blackboard in a second language (L2) English learners. Antón-Sancho et al. [11] analyzed the degree of acquisition of soft skills in Latin American university teachers using hybrid educational models. Limone et al. [12] illustrated the experimentation conducted in the initial training of teachers to observe the processes of negotiating content, making decisions, and group building through the use of Digital storytelling (DST) in an educational context. Wu et al. [13] proposed a method combined with the parsing trees of program codes and the fuzzy membership function to detect plagiarism in an online, blended programming course during the COVID-19 pandemic. Al Mulhem and Almaiah [14] presented an experimental design to examine the role of scaffolding learning strategy in students' use of mobile educational games. Estriegana et al. [15] presented an empirical study of an online learning environment using a set of web-based resources (such as virtual laboratories, interactive activities, or educational videos) that examined the influence of the combination of such resources with active and collaborative learning.

3. Conclusions

The COVID-19 pandemic initiated a sudden and profound digital transformation. By April 2020, around half of the world's population was under lockdown. This situation forced society to take an extraordinary digital leap. In the high education space, about 186 countries worldwide closed their institutions due to this lockdown, switching from on-site teaching to remote emergency teaching using online resources. But as happened before, a crisis also means an opportunity. This way, educators and students were forced to develop their digital competence quickly. Nowadays, online and hybrid education is still essential in the post-COVID world, and this digital transformation has changed the education landscape forever.

Acknowledgments: The authors acknowledge PROF-XXI, which is an Erasmus+ Capacity Building in the Field of Higher Education project funded by the European Commission (609767-EPP-1-2019-1-ES-EPPKA2-CBHE-JP).

Conflicts of Interest: The authors declare no conflict of interest.

References

1. Birsa, E.; Kljun, M.; Kopačin, B. ICT Usage for Cross-Curricular Connections in Music and Visual Arts during Emergency Remote Teaching in Slovenia. *Electronics* **2022**, *11*, 2090. [CrossRef]
2. Pilotti, M.A.E.; Nazeeruddin, E.; Nazeeruddin, M.; Daqqa, I.; Abdelsalam, H.; Abdullah, M. Is Initial Performance in a Course Informative? Machine Learning Algorithms as Aids for the Early Detection of At-Risk Students. *Electronics* **2022**, *11*, 2057. [CrossRef]
3. Wijaya, T.T.; Weinhandl, R. Factors Influencing Students' Continuous Intentions for Using Micro-Lectures in the Post-COVID-19 Period: A Modification of the UTAUT-2 Approach. *Electronics* **2022**, *11*, 1924. [CrossRef]
4. Woltran, F.; Lindner, K.T.; Dzojic, T.; Schwab, S. Will–Skill–Tool Components as Key Factors for Digital Media Implementation in Education: Austrian Teachers' Experiences with Digital Forms of Instruction during the COVID-19 Pandemic. *Electronics* **2022**, *11*, 1805. [CrossRef]
5. Alghizzi, T.M.; Elyas, T. The Effect of Graded-Reading Websites/Applications on EFL Undergraduates' Reading Comprehension during COVID-19 Pandemic. *Electronics* **2022**, *11*, 1751. [CrossRef]
6. Sucuoğlu, E.; Andrew, A.U. Administrators and Students on E-Learning: The Benefits and Impacts of Proper Implementation in Nigeria. *Electronics* **2022**, *11*, 1650. [CrossRef]
7. Almaiah, M.A.; Ayouni, S.; Hajjej, F.; Lutfi, A.; Almomani, O.; Awad, A.B. Smart Mobile Learning Success Model for Higher Educational Institutions in the Context of the COVID-19 Pandemic. *Electronics* **2022**, *11*, 1278. [CrossRef]
8. Sederevičiūtė-Pačiauskienė, Ž.; Valantinaitė, I.; Asakavičiūtė, V. 'Should I Turn on My Video Camera?' The Students' Perceptions of the use of Video Cameras in Synchronous Distant Learning. *Electronics* **2022**, *11*, 813. [CrossRef]
9. Pérez-Sanagustín, M.; Kotorov, I.; Teixeira, A.; Mansilla, F.; Broisin, J.; Alario-Hoyos, C.; Jerez, Ó.; Teixeira Pinto, M.d.C.; García, B.; Delgado Kloos, C.; et al. A Competency Framework for Teaching and Learning Innovation Centers for the 21st Century: Anticipating the Post-COVID-19 Age. *Electronics* **2022**, *11*, 413. [CrossRef]
10. Oraif, I.; Elyas, T. Applicability of Collaborative Work in the COVID-19 Era: Use of Breakout Groups in Teaching L2 Translation. *Electronics* **2021**, *10*, 2846. [CrossRef]

11. Antón-Sancho, Á.; Vergara, D.; Fernández-Arias, P. Self-Assessment of Soft Skills of University Teachers from Countries with a Low Level of Digital Competence. *Electronics* **2021**, *10*, 2532. [CrossRef]
12. Limone, P.; Toto, G.A.; Cafarelli, B. The Decision-Making Process and the Construction of Online Sociality through the Digital Storytelling Methodology. *Electronics* **2021**, *10*, 2465. [CrossRef]
13. Wu, J.S.; Chien, T.H.; Chien, L.R.; Yang, C.Y. Using Artificial Intelligence to Predict Class Loyalty and Plagiarism in Students in an Online Blended Programming Course during the COVID-19 Pandemic. *Electronics* **2021**, *10*, 2203. [CrossRef]
14. Al Mulhem, A.; Almaiah, M.A. A Conceptual Model to Investigate the Role of Mobile Game Applications in Education during the COVID-19 Pandemic. *Electronics* **2021**, *10*, 2106. [CrossRef]
15. Estriegana, R.; Medina-Merodio, J.A.; Robina-Ramírez, R.; Barchino, R. Analysis of Cooperative Skills Development through Relational Coordination in a Gamified Online Learning Environment. *Electronics* **2021**, *10*, 2032. [CrossRef]

electronics

MDPI

Article

Analysis of Cooperative Skills Development through Relational Coordination in a Gamified Online Learning Environment

Rosa Estriegana [1,*], Jose-Amelio Medina-Merodio [2,*], Rafael Robina-Ramírez [3] and Roberto Barchino [2]

1 Departamento de Automática, Universidad de Alcalá, 28871 Madrid, Spain
2 Departamento de Ciencias de la Computación, Universidad de Alcalá, 28871 Madrid, Spain; roberto.barchino@uah.es
3 Business and Sociology Departamento, University of Extremadura, 10002 Cáceres, Spain; rrobina@unex.es
* Correspondence: rosa.estriegana@uah.es (R.E.); josea.medina@uah.es (J.-A.M.-M.); Tel.: +34-918856669 (R.E.)

Abstract: One of the main problems of the sudden digital transition to online education due to the COVID-19 pandemic is the increased isolation of students. On the other hand, one of the main goals of higher education is to develop students' cooperative competence. This experimental study presents an online learning environment, consisting of a set of web-based resources such as virtual laboratories, interactive activities, educational videos and a game-based learning methodology. The study also examines the influence of the combination of such resources with active and collaborative learning on the improvement of students' relationships and the development of cooperative competence. To this end, an analysis was conducted based on the data collected from a core subject of the Computer Engineering and Computer Science Engineering degree courses. The answers of an online survey ($n = 289$) were examined by using the structural equation modeling technique (SEM). The results suggest that the proposed learning environment has a significant and positive impact on the two dimensions of relational coordination; communication and relationships, and plays a key role in the acquisition and development of cooperative competence. Findings also indicate that effective, accurate, frequent and timely communication, positively influences on students' relationships. Additionally, this study addresses other important issues with significant theoretical and practical implications for higher education.

Keywords: collaborative learning; cooperative competence; virtual learning; educational technology; game-based learning; relational coordination

Citation: Estriegana, R.; Medina-Merodio, J.-A.; Robina-Ramírez, R.; Barchino, R. Analysis of Cooperative Skills Development through Relational Coordination in a Gamified Online Learning Environment. *Electronics* **2021**, *10*, 2032. https://doi.org/10.3390/electronics10162032

Academic Editors: Boni García, Carlos Alario-Hoyos, Mar Pérez-Sanagustín, Miguel Morales and Oscar Jerez

Received: 23 July 2021
Accepted: 20 August 2021
Published: 22 August 2021

1. Introduction

The integration of technology in educational systems has been a key driver to enhance teaching and learning for many years. However, it has been in the COVID-19 situation that the use of technology and online learning has grown exponentially, highlighting the countless opportunities and advantages it can offer, but also emphasizing some deficiencies and limitations. One of the main problems of online learning is the feeling of social isolation of the learner [1], which is the reason of many failures and dropouts. According to [2] the feeling of connectedness and the perception of belonging to a community is crucial for learners.

Aligned with that sense of inclusion, companies are increasingly proclaiming the importance of strong communication and cooperation skills among professionals. The computing professional no longer fits the old stereotype of an isolated computer programmer working alone; instead, most computer jobs require professionals to interact and cooperate with others.

For these reasons, one of the main goals of higher education is to develop students' collaborative or cooperative ability. Consequently, many institutions promote the integration of collaborative learning strategies. In this line, the European Higher Education Area (EHEA) and Bologna process urged European higher education institutions to redefine their degree programs on competency-based achievements that better prepare graduates for their future role in society [3]. However, although the acquisition and development of these essential competences is highly demanded nowadays and so important that they are described in degree programs and teaching guides, there are no clear guidelines on how to improve these skills in addition to knowledge. Technology and appropriate instructional strategies can play a key role in addressing this problem [4,5]. Therefore, by combining learning strategies and technology resources, it is possible to create meaningful learning scenarios that increase communicative and collaborative skills and also improve in-depth learning.

In order to increase students' motivation, encourage participation and improve the learning process, we have created an online learning environment (OLE) from scratch, which integrates interactive exercises, virtual laboratories video teaching and game-based learning [6,7]. The use of this OLE has been crucial to increase students' interaction and involvement and to allow active and collaborative learning methods in the classroom. This increase in participation and interactions has had an impact on enhancing student relationships [7] which is more necessary that ever, especially due to the COVID-19 pandemic.

The main goal of this study is to analyze the effect of this web-based online learning environment. To achieve this, we make contributions in three aspects. Firstly, by evaluating whether the online learning environment has had a significant impact on students' relationships. Secondly, by analyzing whether the learning environment has led to a significant impact on cooperative competence acquisition. Thirdly, by studying the impact of relational coordination on students' cooperative competence.

This paper is structured as follows: Section 2 contains a literature review and the theoretical framework. This section explains what the online learning environment consists of and what it is based on, as well as the learning methodology used which has been facilitated by the online learning environment. Section 3 presents the model and its components; also in this section the hypotheses are presented. Section 4 describes the research methodology, the instrument employed, the participants, and data collection, while the data analysis and results are presented in Section 5. A discussion follows in Section 6. Finally, the paper ends with the conclusions drawn from the study.

2. Literature Review and Theoretical Framework

During the global pandemic sparked by COVID-19, the educational system has experienced a radical shift and online learning tools have become more essential than ever before. This shift is a great opportunity for universities to better adapt and enrich their educational practices by means of innovative learning technologies [8]. The advantages offered by online learning and technological learning tools have been extensively studied and documented over recent years; for example, to enhance learning accessibility and flexibility [9,10], improve students' achievement [6,11], and give highlighted support for learning in practice and experimentation [8,12] and for active learning approaches [13]. However, despite the numerous studies on the use of technology in education, there are limited studies to evaluate the impact of relationships and the acquisition of competences in active, participative and collaborative learning environments. Hence, this study addresses this gap in the literature by applying and testing a model to gain insights into the impact of an online learning tool and an active and collaborative environment on the improvement of relationships and the development of cooperative competence.

2.1. Active Learning

In an increasingly digitized society, the mere transmission of information no longer constitutes efficient teaching, so the traditional classroom learning process does not seem to be the most suitable learning model to achieve the new challenges facing higher education.

Active learning is a range of teaching strategies that focus on the effective participation and involvement of learners. Thus, in active learning, the main responsibility for learning shifts from the teacher to the learner, who takes on a central role. According to Prince [14] to put active learning into practice, students need to participate in meaningful learning activities and reflect on what they are doing.

There is empirical evidence that proves the benefits and the effectiveness of active learning compared to traditional lecturing [14,15]. Active learning involves students in the learning process, adapts to the learner's style and provides spatial and temporal flexibility [16]. In addition, active learning promotes the acquisition and development of key competences [17,18].

In line with [13], technology is fundamental to support the implementation of active methodologies inside and outside the classroom. Thus, thanks to pre-study using online learning tools, class time can be spent using cooperative and participative learning strategies which can also be supported by technology. Teachers can foster active learning, propose challenges and collaborative projects for developing students' skills, and get students to feel more involved with their own learning.

In a study analyzing the different types of active methodology, ref [19] found six main types of active methodology, namely: Flipped classroom, Game-based learning, Problem-Based Learning, Project-Based Learning, Peer instruction, and Team-Based Learning. The last four would fall under collaborative learning.

Flipped classroom [20] is probably the best known and most widely implemented active learning methodology, which is why the two terms are sometimes used interchangeably. In this approach, teachers provide students with learning resources, usually video lessons; the students can work, watch or listen at home at their own pace. Later, in the classroom, students have time to apply acquired knowledge, and collaborative and participative activities can be carried out.

Below we discuss collaborative learning and game-based learning.

2.2. Collaborative Learning

The abrupt adoption of virtual learning tools and e-learning experimented with in face to face universities due to the COVID-19 pandemic has been a great challenge for many students and lecturers. Although students have a high level of training and ease of adaptation to technological tools, this has been particularly difficult because of the feeling of social isolation. On the other hand, cognitive development depends to a large extent on social interaction and collaboration with other people. Thus, according to [21], collaborative learning is the educational use of small groups of students to work together in order to maximize their own and others' learning in a well-defined period of time. Similarly, ref [22] defines collaborative learning as an instructional approach in which a small number of learners interact together and share their knowledge and skills in order to reach a specific learning goal.

Collaborative learning structures allow more interaction and dialogue among learners and help to create a feeling of connectedness and belonging among learners, which, according to [2], is crucial for learners. This perception of belonging to a group, a team, or a community avoids the perception of isolation of learners, increasing enthusiasm and motivation and decreasing failure and drop-out rates.

Collaborative learning can be an effective teaching method in both traditional and online learning environments, and may be organized in a wide variety of ways, including as a means to complement more traditional activities. It can also be useful with quite a wide range of students.

Different experiences applying collaborative methodologies have produced successful results showing numerous benefits, including increasing knowledge retention [23], enhancing creativity [24], helping to prepare students professionally [25], enhancing results in terms of competences acquired compared to other teaching methods [26], or improving students' theoretical and practical ability, cooperative ability and autonomous learning ability [27].

Although collaborative learning has been seen as a difficult pedagogical strategy, in agreement with [28], technology can make online collaborative learning more effective and ubiquitous.

2.3. Game-Based Learning

Higher education institutions have a growing interest in recent years in harnessing the potential of game-based learning for integration into both traditional learning and e-learning, and as a complement to both active and collaborative learning.

Game-based learning is defined as the use of game elements in non-game contexts to promote participation and engagement [29]. Including this strategy in education attracts students and helps them pay attention to their studies plus challenges and involves them in an active learning process. Game elements such as badges, content unlocking, avatars, collections, gifting, level progressions, quests, social graphs, and virtual goods are used in learning activities to achieve educational objectives [30].

Several authors have investigated the positive effect of games to motivate and engage students [31,32], and to improve performance and results in practical work [33]. It also develops their experience in terms of students' learning attitudes, interest in learning, and acceptance of technology [34].

An increase in students' engagement results in improved learning. Furthermore, according to [31], increased engagement also means increased cooperative competence.

In addition, exceptional situations such as those experienced with the COVID-19 pandemic require creative and critical thinking to find new answers and solutions. Incorporating elements such as participation or competition motivates students and creates engaging learning environments that help students to continue their studies. Thus, game-based learning has been widely used to enhance the learning process in different environments since the emergence of COVID-19.

2.4. Virtual Learning Tools

According to Adel and Dayan [35] virtual learning is the concept of digitalizing learning material and placing it online, so both students and instructors can manage the educational process effectively. The use of virtual learning tools provides new opportunities and a wide range of benefits to improve the learning process that have been widely studied and documented; for example, increase students' motivation [36], provide autonomy, flexibility and accessibility to the learning contents enabling study from anywhere [9,10], develop students' autonomous learning ability [37], increase the efficiency of teaching and improve students' achievement [4,11], acquire and develop competences [5,38] and support active learning approach [13]. Furthermore, virtual learning could bring other important benefits, such as eliminating unnecessary travel, saving costs and waste of time and also providing other indirect benefits such as the decentralization of urban centers or environmental improvement.

However, especially with the global pandemic sparked by COVID-19, virtual learning has become unquestionably indispensable. Universities around the world have converted all or part of their activities to an online mode to ensure the safety of students and lecturers, and all the work that has been conducted should be useful for the future. According to several experts, the influence of COVID-19 will impact higher education for much longer after the epidemic has been finally controlled [39]. Therefore, it is foreseeable that the increase in the use of e-learning and virtual learning tools will be very notable in the future. In this line, ref [40] affirm that finding the right balance between online and on-site will be the real strategic challenge for the universities in the future. Other authors, e.g., ref [10], also argue that it will be necessary to develop a blended approach to education that flexibly combines virtual and face-to-face learning. Thus, this emergency caused by COVID-19 could be an opportunity for education to accelerate the adoption of other forms of learning supported by virtual learning tools.

The technological development and wide availability of virtual learning tools have made the adoption of online learning relatively easy. However, the real challenge for universities today is to integrate and combine these tools with an effective learning approach and to also consider the need for hands-on experience as well as skills development [41].

Many examples of recent research show that the combination of virtual tools and real components employed collaboratively has proven to be a success, especially in times of COVID-19. For example, Adel and Dayan [35] propose a blended learning model where digital technology is utilized to virtually engage students. The blended learning model proposed integrates managed virtual support and traditional learning environments, and includes a web server that hosts all applications and tasks and provides a virtual platform that simulates real labs. Additionally, Loukatos et al. [8] highlight educational activities for agricultural engineers that challenge them to use problem solving techniques and team collaboration skills through a blended learning activity by means of a robotic arm platform to assist the students to tackle the lack of physical presence in the classroom due to COVID-19. Similarly, Smigelski, Movassaghi and Smal [42] describe the impact of the COVID-19 pandemic on urology trainees, with a focus on virtual learning initiatives. Smigelski, Movassaghi and Smal [42] conclude that despite the undeniable devastation of COVID-19, some good things may be found in the innovative solutions that have arisen in education.

COVID-19 has accelerated the use of virtual learning tools, and everything learned and all educational resources developed during this time of the pandemic will have a lasting effect on education and continue to be useful in the years to come. Universities should combine these virtual learning tools with learning methods of collaboration and participation that engage students and foster efficient learning, experience, communication, and competence development inside and outside the classroom.

2.5. Learning by Doing and Hands on Activities

Learning activities with hands-on exercises in which students play an active role engage and motivate them more effectively than learning activities where they are passive.

Learning by doing is essential in scientific and technological fields such as engineering, which involve laboratory practical work in addition to theoretical lectures, and where students' hands-on exercises and laboratory work form crucial aspects of their learning. In these fields, virtual laboratories, simulators, and interactive tools are essential to provide students with virtual practical work [41].

Several authors have suggested that a successful strategy for developing practical learning activities is the creation of web-based simulators and virtual laboratories. Some examples include a virtual electrical machine laboratory for electrical engineering courses [43], a remote laboratory applied to control engineering learning [44], or a laboratory framework for a Networked Control System design to provide plug-in free online experiments [45].

There are many advantages of such simulators and virtual laboratories designed for practical activities. Thus, ref [46] found that students using a virtual laboratory acquire a better conceptual understanding and develop better procedural skills than students using a traditional laboratory. Likewise, ref [47] suggested that learning outcomes are equal or even better using virtual or remote versus traditional laboratories. In a similar study, ref [45] verified the pedagogical effectiveness of their virtual control system against a physical one. Additionally, ref [48] claim that a virtual interactive laboratory in engineering mathematics enhances students' learning.

In addition to all of the above, face-to-face labs are expensive, both because of the instruments used and because the groups of students have to be reduced; therefore, according to [49], virtual labs also provide remarkable cost advantages.

2.6. Teaching Videos

According to [50], video is one of the educational technology resources more widely used and with greater potential. The results of many studies find videos to be an effective and successful learning tool that reported significant knowledge gains [51], grade improvement and student satisfaction [52], or higher practical skills acquired [53].

Videos are used in a wide variety of ways; e.g., the massive, online, and open courses (MOOC), the infinite number of courses and tutorials that can be found on YouTube, and other alternatives such as Vimeo, Yahoo Video, Viddler, or Screencast. Some are expressly created for educational purposes, such as School Tube, Teacher Tube or Teacher TV and, of course, videos are used in distance education, but also as the linchpin in hybrid education environments as an effective tool of autonomous learning.

There is no doubt that teaching videos are extraordinarily useful as learning support, especially in a pandemic situation such as COVID-19; nevertheless, media such as video are not effective on their own—according to [54], they must be embedded in appropriate instructional contexts to bring out their full educational potential.

2.7. Online Learning Environment (OLE)

Following previous research, both in the search for effective virtual learning tools and in the search for the most appropriate learning approaches to apply them, an online learning environment (OLE) integrating web-based resources was developed [6].

The OLE is based on appropriate instructional strategies, learned over several years of teaching. Taking into account the main needs and deficiencies of the students, through tests, evaluations and based on feedback from students and teachers, the application has been improved year after year.

The platform integrates a range of randomly generated interactive and graphical activities such as computer simulation, virtual laboratories or explanatory teaching videos with exercises to evaluate understanding. The activities in OLE cover the contents of the subject Fundamentals of Computer Technology. from exercises with different numbering systems, truth tables, circuit simplification exercises, Karnaugh maps (min terms, and max terms) of different numbers of variables, combinational circuit simulation exercises, circuit analysis and synthesis, logic gate and integrated circuit exercises, sequential circuits, bistables truth and transition tables, counters, graphs of finite-state machines to memory system activities, operations with address, data, and control buses, to a simulator of an EPROM memory recorder.

The OLE also includes game-based learning that aims to encourage learners to work with the application. Learners can choose the type of activity, watch videos, or see their performance and the badges earned, encouraging a level of competence that stimulates learners. The purpose of the badges is to increase student motivation, since, according to [55], this is a primary component that positively affects learning.

In previous studies [6], we evaluated and confirmed the effectiveness of OLE in motivating students, increasing learner satisfaction, as well as improving the learning process and learning outcomes.

One of the most useful options of OLE is to be used as a self-study tool. With the time saved with this previous study, active learning activities can be carried out. Thus, teachers can devote more time to student diversity and students have more time for problem solving.

Furthermore, OLE allows activities to be carried out from any mobile device, which has made it easier to carry out individual and group activities in the classroom.

In addition, OLE has been helpful as an assessment tool, as students' successes or failures can be registered. This reduces the risk of impersonation or cheating as the exercises are randomly generated.

However, also in the situation experienced due to the COVID-19 pandemic, OLE has proved to be a useful and efficient tool to use in online classes, to maintain interest and to make lessons more dynamic, which is really very necessary in synchronous lessons of two or more hours.

Figure 1 shows some examples of interactives exercises that are randomly generated. Although students can attempt an unlimited number of exercises, they have only three opportunities to solve each activity before the correct result is displayed, another activity is randomly generated, and the sequence begins.

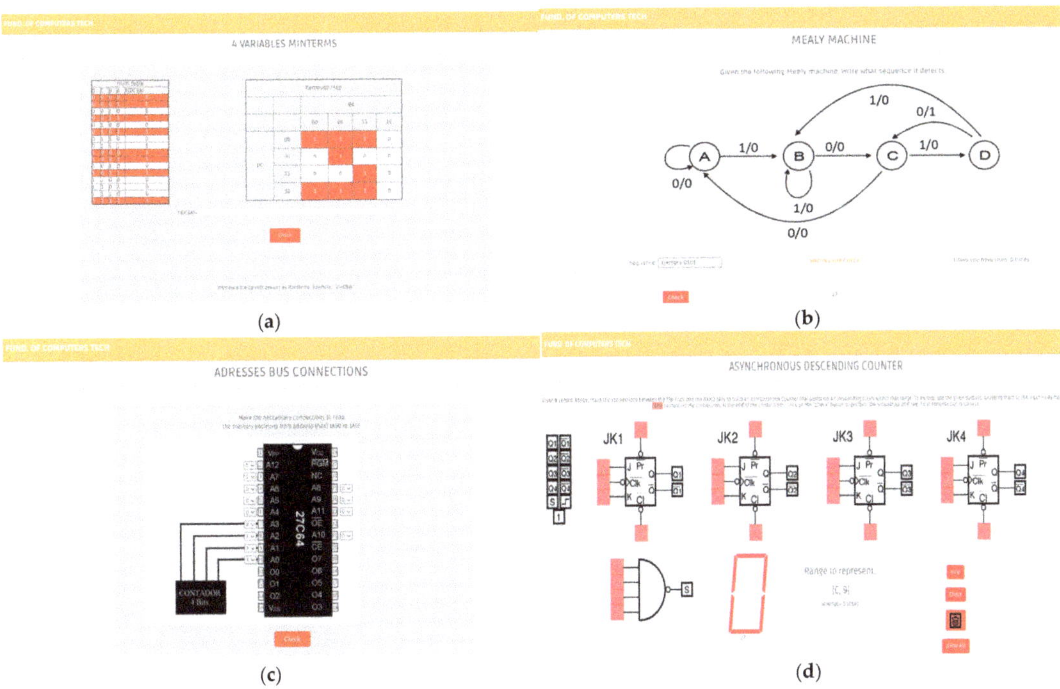

Figure 1. Example of OLE interactive activities randomly generated. (**a**) Karnaugh Maps, min terms, (**b**) example of sequential circuit: asynchronous counter with JK bistables, (**c**) example of memory system circuit: addresses bus connections, and (**d**) example of finite state machine graph: Mealy.

3. Research Model and Hypotheses

Based on the previous research, a theoretical model was developed to evaluate the impact of the OLE on students' relationship and on cooperative competence acquisition and also to understand the influence of relational coordination on cooperative development.

Each of the hypotheses presented below correspond to each path in the SEM and forms part of the aforementioned objective.

3.1. The Learning Environment (LE)

Just as active learning is supported by technology [13], technological resources do not provide a learning solution by themselves [56]. Hence, combining the OLE resources with Learning Methodologies (LM) such as active and collaborative learning strategies in the Learning Environment (LE) has been created to increase classroom interaction and encourage participative and cooperative work. Communication and interaction is achieved through questions and answers between teacher and students. Similarly, cooperative work is reached when two or more students learn or attempt to learn something together, for example by using problem solving, discussion and agreement. These types of activities are essential for students, especially in the first year of university [57], not only due to the connection to like-minded or same-age peers but also to obtain social and academic integration and achievement. In addition, student involvement is increased with challenges and discussions, which enhances student learning in a cognitive and affective way.

By means of the previous autonomous study, OLE supports active learning. Thus, participatory and collaborative activities using game-based learning have been employed, e.g., using mobile-based learning applications that allow activities for rapid development, with immediate visible results such as Socrative, Kahoot or Mentimeter that are easy, intuitive applications that perfectly match the stage of class attendance.

On the other hand, as suggested by [58], is essential to provide online or virtual learners with multiple channels, both synchronous and asynchronous, in order to accommodate their preferences for different communication styles. Therefore, OLE has been used not only to motivate and involve students in synchronous classes, but also for students to do individual and collaborative activities asynchronously. An example of collaborative activity implemented is the jigsaw method [59]. Each member of a group takes responsibility for a piece of content and then shares it with the other participants to create a possible final joint representation. However, there are many possible variations of this methodology, as many as we can imagine.

In view of the above, we hypothesized that the LE created through the combination of the OLE resources and the active learning methodologies would have a positive impact on communication (H1) and on student's relationships (H2), and also that this LE would be a significant factor that positively influenced cooperative competence development (H3).

3.2. The Relational Coordination Model

The model of relational coordination (RC) has proved to be a power driver for quality and efficiency outcomes. Relational coordination model [60] emphasizes the need to understand the coordination in relationships and the dynamic of communication at organizations to reach best organizational results. So, different research has applied the RC model in different sectors such as airlines, healthcare, cloud computing, or education.

In the field of health services, the effect of relational coordination on the nurses' job satisfaction and on the quality of life of nursing homes residents was evaluated, and the results indicated that there was a positive relationship between the level of relational coordination and both results [61].

In addition, ref [62] analyzed a case of a medical service for the organ donation and transplantation process. The results showed that service interaction emerges as the atomic unit of productivity analysis, since it is significant at the process level and exhibits a strong cause–effect relationship, where service interactions identify who is responsible for what in complex services.

The relational coordination model has also been studied in other areas such as education, thus [63] indicated that ICTs are tools that facilitate the online educational framework and that have been extended to practically all business areas.

On the other hand, ref [64] analyzed the application of relational coordination in three Public Universities, the Rey Juan Carlos University (URJC), the Complutense University (UCM) and the Carlos III University (UC3M). The results showed that UC3M presents higher percentages of relational coordination, since it has a strategic plan focused on good coordination.

Besides, ref [65] observed that the communication channel has an impact on the frequency of communication and that quality relationships increase the degree of teacher satisfaction with their work and also increase the degree of satisfaction with the learning management system, and thus improve student satisfaction.

Relational coordination was also applied to some online courses given in Spanish universities and private companies, and it was found that organizations obtain better results in terms of satisfaction by providing shared knowledge, shared goals and mechanisms of mutual respect [66].

Therefore, according to the previous research, quality relationships and quality communication increase teacher and student satisfaction by providing shared knowledge, shared goals and mechanisms of mutual respect, supported by frequent, timely and problem-solving communication, leading to better results. Despite this, there is a lack of literature exploring the impact of relationships and communication on learning environments and competence development.

According to the RC model, the coordination process takes place through a network of relationship and communication dimensions. Gittel [61] indicated the RC model is relatively unique and includes relational coordination dimensions: Relationships and Communication.

Relationships are based on the relational dimensions included in the model as a shared goal, shared knowledge and mutual respect. They enable students and lecturers to coordinate more effectively the work processes they are involved in [60,66]. These relational ties are mutually reinforced through communication links that allow effective coordination of work.

Communication is based on the communication dimensions included in the model such as frequency, timeliness, accuracy and resolving problems. Thus, accurate information makes it easier for teachers and students to teach, learn, help and share at the same time. In addition, frequent and timely communication means that students do not have to wait long time to receive information from teachers, which would create uncertainty about their learning tasks [67].

Based on the previous studies, we hypothesized that communication positively influenced acquisition and development of cooperative competence (H4) and, on the other hand, communication would positively impact relationships (H5). In turn, it seems logical to assume that good and fluid relationships positively influence cooperative competence improvement (H6).

3.3. Cooperative Competence

Universities are often criticized for being disconnected from the world of work. Thus, while companies are increasingly looking for professionals with communication and cooperative competences, leadership and teamwork skills, many universities continue to use the same old teaching methodology, focused only on learning content.

According to [68], competence-based education helps to bridge the gap between university and companies' requirements.

Cooperative competence can be defined as the ability of working with others effectively. It means the capacity to exchange information and to establish, develop and maintain social relationships. Along with communication competence comes social skills. Cooperative competence includes interpersonal skills, critical and self-critical capacities, the ability to communicate with others and the ability to work in a team.

Social skills in general and cooperative competence in particular are undoubtedly key for higher education. Hence, it is necessary to develop tools and to implement efficient learning methodologies that help in the development of these competences. Regarding tools for the development of cooperative competence, following [5], online learning environment have proven to be very successful to promote the students' acquisition of competences as well as social skills. As for the appropriate methodology, some authors such as Schaeper [18] claim that active learning environments enhance the acquisition of social competences (cooperative and communicative).

That is why, based on the interactive and collaborative activities and learner involvement offered by the online learning environment and the relationships that are created in the active and collaborative learning approach, our study suggests that this learning scenario enhances cooperative competence.

The conceptual framework of this research model and its different elements can be seen in Figure 2.

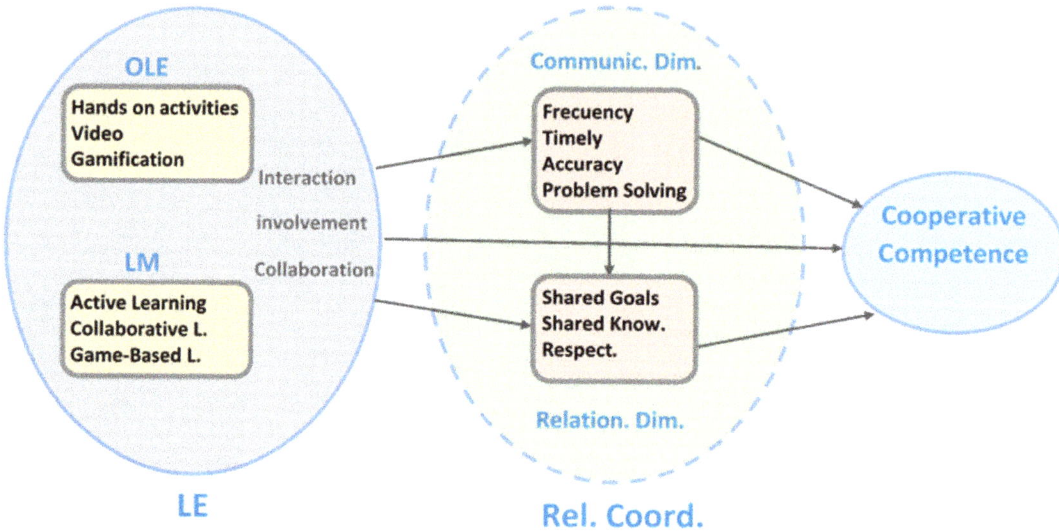

Figure 2. Conceptual research framework.

4. Research Method

Based on other reviewed and validated models and following several criteria as guidelines [69], an online questionnaire was designed to test the hypotheses.

4.1. Participants and Data Collection

Data were collected from students by means of voluntary and confidential online questionnaires. The full sample consisted of 289 students (253 males and 36 females) aged mostly from 18 to 20 years old. These students were taking the course Fundamentals of Computer Technology, a core subject in the first year of the Computer Engineering and Computer Science degrees at the University of Alcalá, Spain.

The main objective of the subject Fundamentals of Computer Technology is to understand the functioning of a computer at different levels of abstraction, from logic gates to basic electronic devices, combinational and sequential circuits, with an introduction to functional units at the architecture level.

There was a first survey conducted at the beginning of the course, and its purpose was simply to determine the initial knowledge of students and their willingness to use OLE tools (virtual interactive exercises, teaching videos, gamification) and active learning activities (active learning, mobile-app activities, gamification). In this questionnaire, students were asked if they had ever used it, what their experience was and what their expectations were about the efficiency of each of the LE tools.

The answers from this initial questionnaire were contrasted with the answers to the same questions at the end of the course. In this questionnaire, other questions to determine the degree of satisfaction with the OLE tools were included. These questions are based on a validated questionnaire used by other investigators [33,36].

Nevertheless, the central survey of this study aimed to analyze the research model (Figure 2) and to evaluate the hypotheses presented. This questionnaire was conducted upon completion of the course and is addressed in the following section.

4.2. Instrument

Questionnaires are highly accurate and widely used for all types of research, including those related to learning and competence acquisition [70,71]. The questionnaire follows a 5-point Likert bipolar scale [72], with answers ranging from 1 "strongly disagree" to 5 "strongly agree", adopting the usual method for measuring variables that are not directly quantifiable [73].

Items for each variable were adapted from scales that have been validated in previous studies. Hence, the LE construct use collaboration, involvement and interaction. Questions on Interaction, and Collaboration were adapted from the Distance Education Learning Environments Surveys (DELES) [74]. This instrument assesses students' perceptions of virtual learning environments and has been used in numerous studies with strong reliability and validity [75,76]. Questions on involvement were based on the classroom environment instrument: What is happening in this class? (WIHIC) [77], that examine students' perceptions of the classroom by combining some relevant scales from existing questionnaires and is validated in several studies [78–80]. Questions on relational coordination are based on an adaptation of the original questionnaire provided by Gitell [61]. It has also been adapted by [63,64,81] in previous research applied to education. Items to evaluate cooperative competence follow the "Evaluation in Higher Education: Self-Assessed Competences" (HEsaCom) [70,71].

The questionnaires used simple questions and easily understood language to minimize errors in items related to variance. No research intentions or hypotheses were mentioned, items were clearly formulated, abstract questions or terms were avoided, and there were no double-barreled items. Several experts revised the questionnaires to determine whether the questions were appropriate, unambiguous and easy to understand; a few modifications were made following the feedback.

5. Data Analysis

Data have been processed by using SmartPLS 3.2.6, through the PLS multivariate technique based on structural equation modeling to visually examine the relationships between unobservable or latent variables, such as those conveyed in the hypotheses [73]. PLS does not require any parametric conditions and is recommended for small samples [82] due to is predictive accuracy, which means that the model could be replicated in other scenarios [83].

5.1. Justification of Sample Size

The sample size issue is one of the main features of PLS [84] (p. 198). The segmentation process used by the PLS algorithm renders divide complex models into subsets in order to calculate sample size based on the highest number of structural paths directed at a particular dependent latent variable. According to [85], increasing sample size to 100 is recommended in order to reach acceptable levels of statistical power. Roldán and Cepeda [86] proposed specifying the size effect for each regression while consulting the power tables developed by Cohen [87] to obtain a more precise assessment.

Instead of using the old heuristic rule of 10 cases per predictor proposed by Barclay, Higgins and Thompson [88], Hair et al. [89] suggested using programs such as G*Power or G*Power 3.0 for a specific power analysis according to model specifications [90,91].

The expected effect size (ES) and the significant alpha (α) and power (β) values were used to calculate the minimum necessary sample size. For this purpose, a multiple regression study was conducted with four predictors, an average effect size (ES) of 0.15, an alpha of 0.05, and a power of 0.95, in accordance with Cohen [87]. In this case, the minimum sample result was N = 129 participants. Since the study sample was comprised of 289 valid cases, the sample comfortably exceeded all criteria for the analysis by using the structural equation modeling technique.

5.2. Measurement Model Evaluation

According to Carmines and Zeller [92], individual reliability should be analyzed at the beginning of the process of ascertaining the proper indicators. As shown in Table 1, loads (λ) were greater than 0.707, indicating that individual item reliability was acceptable [92].

Table 1. Outer model loadings.

	Coop-Comp	Comunic.	LE	Relations
c-coo-1	0.822			
c-coo-2	0.800			
c-coo-3	0.880			
c-coo-4	0.798			
Com-1		0.788		
Com-2		0.767		
Exa-2		0.707		
Frec-1		0.782		
Rsp-2		0.721		
int-1			0.843	
int-2			0.866	
int-3			0.816	
inv-1			0.851	
collab-1			0.862	
collab-2			0.898	
collab-3			0.844	
Con-1				0.842
Obj-1				0.816
Res-1				0.830
Res-2				0.710

Cronbach's alpha and rho A test values are above 0.7 [93]. The effectiveness of each construct was also validated using the composite reliability (CR); each value was also above the level of acceptance of 0.7 in all the cases [94]. Convergent validity is assessed via the average variance extracted (AVE). AVE values should be greater than 0.50 [95]. Consistent with this suggestion, AVE measures for all variables exceed 0.568 (Table 2).

Table 2. Reliability and validity of the constructs and criterion by Fornell–Larcker.

	Cronbach's Alpha	rho_A	CR	AVE	Fornell-Larcker Criterion			
					Coop-Comp	Comunic.	LE	Relations
Coop-comp.	0.845	0.852	0.895	0.682	0.826			
Communic.	0.817	0.833	0.868	0.568	0.625	0.754		
LE	0.938	0.939	0.950	0.730	0.765	0.689	0.855	
Relations	0.815	0.833	0.877	0.642	0.562	0.703	0.588	0.801

In accordance with Fornell and Larcker [95], the discriminant validity was analyzed through the comparison between the square root of each AVE on the diagonal and the correlation coefficients. As shown in Table 2, The values of the diagonal were higher than rows and columns, indicating acceptable discriminant validity of the measurements.

In addition, the Heterotrait–Monotrait method (HTMT) was also applied [96]. The values obtained should be less than 0.90, although a value higher than 0.85 [96] should be interpreted as a weak correlation between constructs. Table 3 shows that the HTMT ratio was valid because to their values were all less than 0.85. In our case, this requirement was fulfilled.

Table 3. Discriminant validity matrix (HTMT), R^2 y Q^2 results.

	Coop-Comp	Comunic.	LE	Relations	R^2	Q^2
Coop-comp.					0.610	0.396
Communic.	0.710				0.475	0.251
LE	0.846	0.754				
Relations	0.657	0.813	0.659		0.514	0.317

5.3. Structural Model Analysis

In response to the literature reviewed, a research model (Figure 3) was developed to investigate the influence of the proposed learning environment on the improvement of students' relationships and the development of cooperative competence.

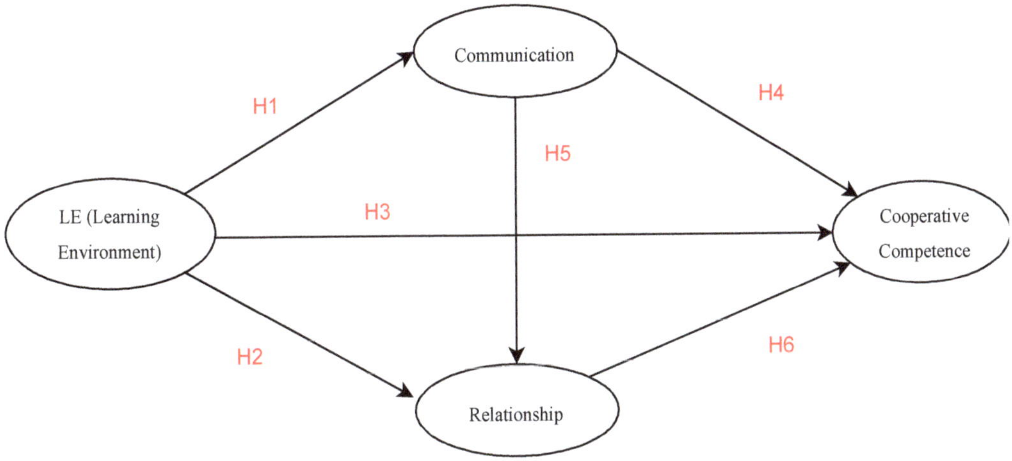

Figure 3. Structural model (baseline model).

The PLS program can generate t statistics for significance testing of both the inner and outer model, using the procedure called bootstrapping [97]. In this procedure, a large number of subsamples (5000) are taken from the original sample with replacement to give bootstrap standard errors, which in turn give approximate T-values for significance testing of the structural path.

The results of the bootstrapping procedure were as follows:

Coefficient of determination (R^2) defines the prediction of the model. It means the amount of variance explained by the construct within the model. R^2 is strong, moderate, and weak when the values are higher than 0.67, 0.33, and 0.19, respectively [98].

Table 4 shows the variance explained (R^2). The model explained 61% of the total variance.

Table 4. Structural model results.

	R^2	Sample Mean (SM)	Standard Deviation (STDEV)	T Statistics (|O/STDEV|)	*p*-Values	Q^2
Coop-comp.	0.610	0.616	0.042	14.590	0.000	0.396
Communic.	0.475	0.476	0.050	9.458	0.000	0.251
Relations	0.514	0.518	0.055	9.376	0.000	0.317

On the other hand, the predictive relevance (Q^2) is calculated for the Stone–Geisser test [99] and for the latent variables that predict "cross-validated redundancy" [97]. There is a relevance prediction in the cases where $Q^2 > 0$. Table 4 shows that all the constructs Coop-comp, Communic, and Relations fulfil this requirement.

Standardized regression coefficients show estimates of structural model relationships, that is, the hypothesized relationships between constructs. Hence, the algebraic sign is analyzed if there is a change in sign; the magnitude and statistical significance (T statistics) was greater by 1.64 (t (4999), one-tailed test). Next, the hypotheses were checked and validated, and the relationships were positive, mostly with high significance, as shown in Table 5.

Table 5. Structural model results. Path significance using percentile bootstrap 95% confidence interval (n = 5000 subsamples).

Hyp.	Results	Influence	SPC	Sample Mean (SM)	Standard Deviation (STDEV)	T Statistics | O/STDEV |	p-Values	± Change
H4	Accepted (*)	Communic → Coop-comp	0.122	0.121	0.069	1.768	0.039	No
H5	Accepted (***)	Communic → Relations	0.567	0.567	0.056	10.128	0.000	No
H3	Accepted (***)	LE → Coop-comp	0.612	0.612	0.055	11.144	0.000	No
H1	Accepted (***)	LE → Communic	0.689	0.689	0.037	18.769	0.000	No
H2	Accepted (**)	LE → Relations	0.197	0.198	0.063	3.154	0.001	No
H6	Accepted (*)	Relations → Coop-comp	0.116	0.119	0.066	1.752	0.040	No

Note: t (0.05, 4999) = 1.645158499, t (4999, 0.01) = 2.327094067, t (0.001, 4999) = 3.091863446, * $p < 0.05$, ** $p < 0.01$, *** $p < 0.001$, ns. No significant based on t (4999), one-tailed test.

When a percentile bootstrap was applied to generate a 95% confidence interval using 5000 resamples, Hypotheses H1 to H6 were supported because their confidence interval did not include zero (Table 5). Thus, all hypotheses were confirmed. These results complete a basic analysis of PLS-SEM in our research. The result for PLS-SEM is shown in Figure 4.

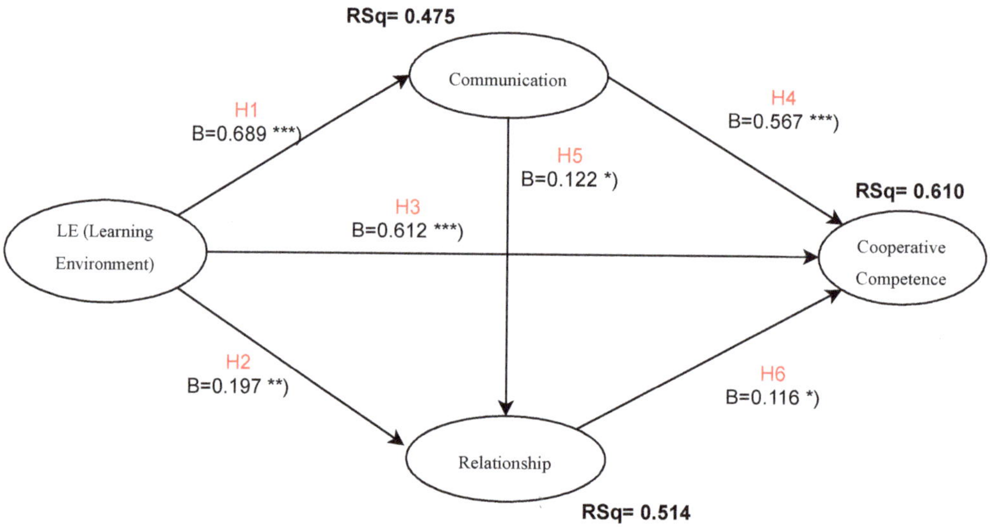

Figure 4. Results of testing the model significance * $p < 0.05$, ** $p < 0.01$, *** $p < 0.001$.

Therefore, Figure 4 and Table 6 show the variance explained by R^2 in the dependent constructs and the path coefficients for the model and the effects on endogenous variables.

Table 6. Effects on endogenous variables (extended model).

Dependent Variable	R^2	Q^2	Antecedents	Path Coeff.	Correlation	Explained Variance (%)
Coop-comp.	0.610	0.396				61.0
			H4: Communic	0.122	0.625	7.62
			H6: Relations	0.116	0.562	6.51
			H3: LE	0.612	0.765	46.81
Communic.	0.475	0.251				47.5
			H1: LE	0.689	0.689	47.5
Relations	0.514	0.317				51.4
			H5: Communic	0.567	0.703	39.86
			H2: LE	0.197	0.588	11.58

5.4. Analysis of Efficiency and Satisfaction

Although the main objective of this study was to analyze the influence of the combination of OLE resources with active learning on the improvement of student relationships and cooperative competence, students' perceptions and opinions on the different learning resources and methodologies used were also gathered, as students' feedback is a key aspect in defining the quality and effectiveness of these tools and methodologies. Thus, in order to find out the efficiency of the OLE tools (virtual interactive exercises, teaching videos, gamification) and the learning activities (active learning, mobile-app activities, gamification), the students' opinion of these resources before the course was compared with their opinion at the end of the course. The results are shown in Table 7.

Table 7. Comparative efficiency of learning tools.

	Had Already Used (%)	Pre-Efficency	Post-Efficiency
Videos	77.01	4.498	4.574
Exercices/VLab	64.37	4.464	4.609
Mobile app.	62.45	4.138	3.737
Activ. Learning	55.17	4.188	4.266
Game in class	34.87	4.268	4.339
Game OLE	26.44	4.291	4.547

Regarding the results on students' degree of satisfaction with the OLE globally, Table 8 shows the questions and a summary of the students' answers. These values show that experience and appreciation of the OLE were certainly positive.

Table 8. Questions and results of the ole satisfaction survey.

OLE	Mean	STDev	STDErr
It was presented effectively	4.326	0.692	0.041
It was simple	4.269	0.734	0.043
I found it useful	4.574	0.561	0.033
The N° of exercises was appropriate	3.906	0.887	0.052
I found it motivating to use	3.996	0.897	0.053
I found it interesting	4.157	0.773	0.045
It enhanced my experience	4.369	0.692	0.041
It helped to pass the course	3.934	0.918	0.054
It helped me use my time better	3.819	0.950	0.056

6. Discussion

Structural equation modeling (SEM) was used to examine the impact of a web-based online learning environment, which includes interactive activities, virtual laboratories, teaching videos, and a game-based learning methodology, on student relationships and the development of cooperative competence.

According to the results, the proposed model to evaluate the hypotheses is completely satisfactory. Thus, simple and compose reliability were acceptable. Additionally, there were high levels of internal consistency reliability among latent variables. The independent explanatory variables, the values for validity and discriminant validity of the measurements were also acceptable.

All the hypotheses were tested and validated, and the relationships were positive, mostly with a high level of significance. Therefore, the results confirm the hypotheses.

As shown in Table 6, the LE created by combining the OLE resources and the learning approach (active, collaborative learning and gamification), clearly affects communication (H1), explaining 47.5%. Furthermore, the LE affects relationships (H2), explaining 11.58%. Likewise, as Gittel [61] states, communication plays a fundamental role in improving relationships (H5), accounting for 39.86% in our analysis. This indicates that communication fosters the relationships dimension, creating a climate that facilitates learning and cooperation.

As suggested by some authors, the acquisition and development of key skills such as cooperative competence can be improved through active learning strategies [17,18], collaborative learning [26,27], game-based learning [32] and also with online learning resources [5,38]. Therefore, in line with these authors, and as was hypothesized, the LE created by combining OLE resources and active learning strategies, collaborative learning and gamification, greatly affects the development and enhancement of cooperative competence (H3), explaining 47%.

The communication and relationship dimensions of the relational coordination model also affect the acquisition of cooperative competence, although with much fewer significant values than those of LE. Thus, the communication dimension explains (H4) 7.62% and the relationship dimension explains (H6) 6.51% of the acquisition and development of cooperative competence.

From the results, we can see that the OLE tools and the learning environment facilitated by them have a direct and significant influence on the acquisition and improvement of cooperative competence, and that they also have an indirect influence through communication and relationships, in line with the results of De Pablos et al. [64], Margalina et al. [65,66] and Gallego et al. [81].

On the other hand, with regard to the analysis of efficiency of the different learning tools and the level of satisfaction with OLE, Table 7 shows that the students' perception of effectiveness with each of the tools at the end of the course is slightly higher than their initial expectations, with the exception of the activities with mobile applications.

Moreover, it is noteworthy that there are no major changes between the pre and post values, due to students already having quite high expectations when they answered the questionnaire at the beginning of the course.

Additionally, in Table 7 it can be seen that the learning tool most previously known by the students was the use of videos, with 77% of students having already used them, and the least known was gamification both inside and outside the classroom, with 34.87% and 26.44%, respectively.

In any case, the rating assigned to each of the learning tools was quite positive. The satisfaction with OLE was also very positive, as can be seen in Table 8. These results show that the experience and appreciation of the learning process was very satisfactory and that the LE motivated and encouraged the students. Thus, in line with [32], increasing student engagement translates into enhancing learning and also cooperation skills.

7. Conclusions

Higher education institutions are required to do much more than just provide students with content knowledge, they must create quality educational processes in which students acquire skills and competences as well as knowledge, through appropriate learning resources and modes in order to converge with the needs of today's society.

This study contributes to the existing literature on the use of new educational technologies as it presents an online environment devised from scratch that integrates computer simulations, virtual laboratories, interactive activities and educational videos in a game-based approach.

The main objective of this study is to analyze the impact of this online learning environment combined with active learning methodologies on the improvement of relationships and the development of cooperative competence.

Our results indicate the following: first, the structural model developed in this research has proven to be a useful theoretical instrument to test and validate the proposed hypotheses. Therefore, this study constitutes a contribution to the literature supporting the effectiveness of structural equation modelling (SEM) to analyze the correlation of variables.

Second, the online learning environments combined with active and collaborative learning methodologies had a significant impact on students' relationship. Technology, with the right instructional approach, can make online learning more participative and collaborative. This learning environment supports and improves students' relationships both inside and outside the classroom.

Third, this learning environment leads to a strong and significant impact on cooperative competence acquisition. The participative and collaborative activities and the use of game based learning allow more interaction and dialogue between learners, which has contributed to creating a feeling of connectedness that promotes teamwork and the development of cooperative skills

Fourth, as expected, cooperative competence is also affected by the relational coordination model (communication and relationships). Improved relationships between students, but especially effective, frequent and timely communication and sharing of knowledge and goals, has a significant impact on the development of cooperative competence.

Students' feedback also revealed a high level of satisfaction with the OLE resources and a very positive attitude towards the active teaching method.

According to the results, virtual activities based on game-based learning and appropriate instructional strategies are key to engaging learners in the learning process and to promote commitment and involvement. Therefore, these kinds of activities should be appropriately designed to enhance learners' relationships and promote the development of cooperative competence.

Many of the changes experienced in higher education during the global pandemic sparked by COVID-19 will remain for a long time. It is foreseeable that the percentage of time spent on online learning will be higher than before COVID-19, and we must make sure that we do not fail in this educational transition. This study addresses important issues, such as improvement of relationships and the development of essential competences such as communicative and cooperative. Given the problem of the isolation of students in online and blended learning, these issues are especially relevant in the COVID-19 pandemic and also post-pandemic education. As the results show, technology and appropriate instructional strategies can play a key role in promoting the development of relationships and social competences. Therefore, the results of this study are valuable for researchers, academic institutions and educators to help them make the best decisions about which learning environments (learning tools and strategies) are most useful and efficient, to improve the learning process, but also to develop important skills for graduates' future careers.

This study addresses important issues in higher education, namely relationships and the development of essential competences such as communication and cooperation. As the results show, technology and appropriate instructional strategies can play a key role in promoting the development of these important skills for graduates' future careers. Hence, the findings of this study are valuable for academic institutions and educators to create efficient learning environments for competence development, by applying active and participatory learning strategies and designing the online learning tools that support them.

Some limitations should be noted. Firstly, with regard to the development of the OLE system, it has been created ad hoc from scratch, which implies certain limitations in terms of adaptability for other courses. The lack of appropriate virtual tools and the need to create web resources adds complexity, so further research is needed to achieve greater standardization. Secondly, with regard to the model used to carry out the research, despite the total variance explained for the dependent variables being quite high, it is possible that other predictors were excluded from the study.

In the future, we hope to add other modules to the OLE that will provide greater communication between students and facilitate the scalability and flexibility. We also intend to carry out longitudinal research using similar OLE in other subjects. The suggested study will evaluate how other factors such as gender or experience with the OLE may affect the model. This was not possible to assess in this present study due to the small sample size in both cases.

Author Contributions: Conceptualization, R.E. and J.-A.M.-M.; methodology, R.E., R.R.-R. and J.-A.M.-M.; software, R.R.-R.; validation, R.E. and R.B.; formal analysis, J.-A.M.-M. and R.E.resources, R.E. and R.R.-R.data curation, R.E. and J.-A.M.-M.; writing—original draft preparation, R.E., J.-A.M.-M., R.R.-R. and R.B.; writing—review and editing, R.E., J.-A.M.-M., R.R.-R. and R.B.; visualization, R.R.-R. and R.B.; supervision, R.E.; project administration, J.-A.M.-M.; funding acquisition, R.B. All authors have read and agreed to the published version of the manuscript.

Funding: This research received no external funding.

Data Availability Statement: The data that support the findings of this study are available from the corresponding author upon reasonable request.

Conflicts of Interest: The authors declare no conflict of interest.

References

1. Wieser, D.; Seeler, J. Online, Not Distance Education: The Merits of Collaborative Learning in Online Education. In *The Disruptive Power of Online Education*; Emerald Publishing Limited: Bingley, UK, 2018.
2. Hara, N.; Bonk, C.J.; Angeli, C. Content analysis of online discussion in an applied educational psychology course. *Instr. Sci.* **2000**, *28*, 115–152. [CrossRef]
3. EHEA and Bologna Process [Online]. Available online: http://www.ehea.info/ (accessed on 12 July 2021).
4. Livingstone, S. Critical reflections on the benefits of ICT in education. *Oxf. Rev. Educ.* **2012**, *38*, 9–24. [CrossRef]
5. Diaz Redondo, R.P.; Fernandez Vilas, A.; Pazos Arias, J.J.; Gil Solla, A. Collaborative and Role-Play Strategies in Software Engineering Learning with Web 2.0 Tools. *Comput. Appl. Eng. Educ.* **2014**, *22*, 658–668. [CrossRef]
6. Estriegana-Valdehita, R.; Barchino Plata, R.; Medina-Merodio, J. Educational Technology in Flipped Course Design. *Int. J. Eng. Educ.* **2017**, *33*, 1199–1212.
7. Estriegana, R.; Medina, J.; Robina, R.; Barchino, R. Virtual learning environment to encourage students' relationships and cooperative competence acquisition. In Proceedings of the 26th ACM Conference on Innovation and Technology in Computer Science Education V. 1, Paderborn, Germany, 26 June–1 July 2021; pp. 53–59.
8. Loukatos, D.; Zoulias, E.; Chondrogiannis, E.; Arvanitis, K.G. A Mixed Reality Approach Enriching the Agricultural Engineering Education Paradigm, against the COVID-19 Constraints. In Proceedings of the 2021 IEEE Global Engineering Education Conference (EDUCON), Vienna, Austria, 21–23 April 2021; pp. 1587–1592.
9. Wu, J.; Tennyson, R.D.; Hsia, T. A study of student satisfaction in a blended e-learning system environment. *Comput. Educ.* **2010**, *55*, 155–164. [CrossRef]
10. Eringfeld, S. Higher education and its post-coronial future: Utopian hopes and dystopian fears at Cambridge University during COVID-19. *Stud. High. Educ.* **2021**, *46*, 146–157. [CrossRef]
11. Lai, K.; Hong, K. Technology use and learning characteristics of students in higher education: Do generational differences exist? *Br. J. Educ. Technol.* **2015**, *46*, 725–738. [CrossRef]
12. Sell, R.; Seiler, S. Improvements of multi-disciplinary engineering study by exploiting design-centric approach, supported by remote and virtual labs. *Int. J. Eng. Educ.* **2012**, *28*, 759–766.
13. Balzotti, J.M.; McCool, L.B. Using Digital Learning Platforms to Extend the Flipped Classroom. *Bus. Prof. Commun. Q.* **2016**, *79*, 68–80. [CrossRef]
14. Prince, M. Does active learning work? A review of the research. *J. Eng. Educ.* **2004**, *93*, 223–231. [CrossRef]
15. Battaglia, D.M.; Kaya, T. How Flipping Your First-Year Digital Circuits Course Positively Affects Student Perceptions and Learning. *Int. J. Eng. Educ.* **2015**, *31*, 1126–1138.
16. Rosenberg, M.J. *E-Learning: Strategies for Delivering Knowledge in the Digital Age*; McGraw-Hill: New York, NY, USA, 2001; Volume 9.
17. Hurtubise, L.; Hall, E.; Sheridan, L.; Han, H. The Flipped Classroom in Medical Education: Engaging Students to Build Competency. *J. Med Educ. Curric. Dev.* **2015**, *2*, 35–43. [CrossRef]
18. Schaeper, H. Development of competencies and teaching-learning arrangements in higher education: Findings from Germany. *Stud. High. Educ.* **2009**, *34*, 677–697. [CrossRef]
19. Ribeiro, M.I.C.; Passos, O.M. A Study on the Active Methodologies Applied to Teaching and Learning Process in the Computing Area. *IEEE Access* **2020**, *8*, 219083–219097. [CrossRef]
20. Sams, A.; Bergmann, J. *Flip Your Classroom: Reach Every Student in Every Class Every Day*; ISTE: St, Arlington, VA, USA, 2012.
21. Johnson, D.; Johnson, R.T.; Smith, K. *Active Learning: Cooperation in the College Classroom*; Interaction Book Company: Edina, MN, USA, 1998.
22. So, H.; Brush, T.A. Student perceptions of collaborative learning, social presence and satisfaction in a blended learning environment: Relationships and critical factors. *Comput. Educ.* **2008**, *51*, 318–336. [CrossRef]
23. Bauerle, T.L.; Park, T.D. Experiential learning enhances student knowledge retention in the plant sciences. *HortTechnology* **2012**, *22*, 715–718. [CrossRef]

24. Ayob, A.; Majid, R.A.; Hussain, A.; Mustaffa, M.M. Creativity enhancement through experiential learning. *Adv. Nat. Appl. Sci.* **2012**, *6*, 94–99.
25. Straub, J.; Kerlin, S.; Whalen, D. Teaching Software Project Management Using Project Based Learning (PBL) and Group Projects. In Proceedings of the 2017 IEEE International Conference on Electro Information Technology (EIT), Lincoln, NE, USA, 14–17 May 2017; pp. 16–21.
26. Blanco, M.; Gonzalez, C.; Sanchez-Lite, A.; Sebastian, M.A. A practical evaluation of a collaborative learning method for engineering project subjects. *IEEE Access* **2017**, *5*, 19363–19372. [CrossRef]
27. Wang, J.; Chen, X.; Gao, X. Economic Management Teaching Mode Based on Mobile Learning and Collaborative Learning. *IEEE Access* **2020**, *8*, 200589–200596. [CrossRef]
28. Koschmann, T.; Hall, R.P.; Miyake, N. *CSCL 2*; Routledge: England, UK, 2002. Available online: https://books.google.co.jp/books?id=SaHMPo7cx2YC&dq=Koscmann,+T.%3B+Hall,+R.P.%3B+Miyake,+N.+CSCL+2%3B2002&lr= (accessed on 1 August 2021).
29. Werbach, K.; Hunter, D. *For the Win: How Game Thinking Can Revolutionize Your Business*; Wharton Digital Press: Philadelphia, PA, USA, 2012.
30. Buckley, P.; Doyle, E. Individualising gamification: An investigation of the impact of learning styles and personality traits on the efficacy of gamification using a prediction market. *Comput. Educ.* **2017**, *106*, 43–55. [CrossRef]
31. Auvinen, T.; Hakulinen, L.; Malmi, L. Increasing students' awareness of their behavior in online learning environments with visualizations and achievement badges. *IEEE Trans. Learn. Technol.* **2015**, *8*, 261–273. [CrossRef]
32. Hasan, H.F.; Nat, M.; Vanduhe, V.Z. Gamified collaborative environment in Moodle. *IEEE Access* **2019**, *7*, 89833–89844. [CrossRef]
33. Domínguez, A.; Saenz-de-Navarrete, J.; De-Marcos, L.; Fernández-Sanz, L.; Pagés, C.; Martínez-Herráiz, J. Gamifying learning experiences: Practical implications and outcomes. *Comput. Educ.* **2013**, *63*, 380–392. [CrossRef]
34. Hwang, G.; Wu, P.; Chen, C. An online game approach for improving students' learning performance in web-based problem-solving activities. *Comput. Educ.* **2012**, *59*, 1246–1256. [CrossRef]
35. Adel, A.; Dayan, J. Towards an intelligent blended system of learning activities model for New Zealand institutions: An investigative approach. *Humanit. Soc. Sci. Commun.* **2021**, *8*, 1–14. [CrossRef]
36. Chu, E.T.; Fang, C. CALEE: A computer-assisted learning system for embedded OS laboratory exercises. *Comput. Educ.* **2015**, *84*, 36–48. [CrossRef]
37. Feng, M. *A Study on Teaching Strategy of Autonomous Learning for English Majors in Higher Vocational Colleges*; Amer Scholars Press: Marietta, GA, USA, 2011; p. 268.
38. Dlouha, J.; Henderson, L.M. e-Learning as an Opportunity for Virtual Mobility and Competence Development within European Universities. In *The Proceedings of the 7th European Conference on e-Learning (ECEL)*; Academic Publishing Limited: Reading, UK, 2008; Volume 1, pp. 297–307.
39. De Oliveira Araújo, F.J.; de Lima, L.S.A.; Cidade, P.I.M.; Nobre, C.B.; Neto, M.L.R. Impact of SARS-Cov-2 and its reverberation in global higher education and mental health. *Psychiatry Res.* **2020**, *288*, 112977. [CrossRef]
40. Agasisti, T.; Soncin, M. Higher education in troubled times: On the impact of COVID-19 in Italy. *Stud. High. Educ.* **2021**, *46*, 86–95. [CrossRef]
41. Potkonjak, V.; Gardner, M.; Callaghan, V.; Mattila, P.; Guetl, C.; Petrović, V.M.; Jovanović, K. Virtual laboratories for education in science, technology, and engineering: A review. *Comput. Educ.* **2016**, *95*, 309–327. [CrossRef]
42. Tanyildizi, E.; Orhan, A. A virtual electric machine laboratory for synchronous machine application. *Comput. Appl. Eng. Educ.* **2009**, *17*, 187–195. [CrossRef]
43. Vargas, H.; Sánchez, J.; Jara, C.A.; Candelas, F.A.; Torres, F.; Dormido, S. A network of automatic control web-based laboratories. *IEEE Trans. Learn. Technol.* **2010**, *4*, 197–208. [CrossRef]
44. Lei, Z.; Zhou, H.; Hu, W.; Deng, Q.; Zhou, D.; Liu, Z.; Lai, J. Modular web-based interactive hybrid laboratory framework for research and education. *IEEE Access* **2018**, *6*, 20152–20163. [CrossRef]
45. Smigelski, M.; Movassaghi, M.; Small, A. Urology virtual education programs during the COVID-19 pandemic. *Curr. Urol. Rep.* **2020**, *21*, 1–4. [CrossRef]
46. Kollöffel, B.; de Jong, T. Conceptual understanding of electrical circuits in secondary vocational engineering education: Combining traditional instruction with inquiry learning in a virtual lab. *J. Eng. Educ.* **2013**, *102*, 375–393. [CrossRef]
47. Brinson, J.R. Learning outcome achievement in non-traditional (virtual and remote) versus traditional (hands-on) laboratories: A review of the empirical research. *Comput. Educ.* **2015**, *87*, 218–237. [CrossRef]
48. Cheong, K.H.; Koh, J.M. Integrated virtual laboratory in engineering mathematics education: Fourier theory. *IEEE Access* **2018**, *6*, 58231–58243. [CrossRef]
49. Bose, R. Virtual labs project: A paradigm shift in internet-based remote experimentation. *IEEE Access* **2013**, *1*, 718–725. [CrossRef]
50. Giannakos, M.N. Exploring the video—based learning research: A review of the literature. *Br. J. Educ. Technol.* **2013**, *44*, E191–E195. [CrossRef]
51. Kay, R.; Kletskin, I. Evaluating the use of problem-based video podcasts to teach mathematics in higher education. *Comput. Educ.* **2012**, *59*, 619–627. [CrossRef]
52. Wells, J.; Barry, R.M.; Spence, A. Using Video Tutorials as a Carrot-and-Stick Approach to Learning. *IEEE Trans. Educ.* **2012**, *25*, 453–458. [CrossRef]

53. Donkor, F. The Comparative Instructional Effectiveness of Print-Based and Video-Based Instructional Materials for Teaching Practical Skills at a Distance. *Int. Rev. Res. Open Distance Learn.* **2010**, *11*, 96–115. [CrossRef]
54. Blomberg, G.; Sherin, M.G.; Renkl, A.; Glogger, I.; Seidel, T. Understanding video as a tool for teacher education: Investigating instructional strategies to promote reflection. *Instr. Sci.* **2014**, *42*, 443–463. [CrossRef]
55. Weiss, R.E.; Knowlton, D.S.; Morrison, G.R. Principles for using animation in computer-based instruction: Theoretical heuristics for effective design. *Comput. Hum. Behav.* **2002**, *18*, 465–477. [CrossRef]
56. López-Pérez, M.V.; Pérez-López, M.C.; Rodríguez-Ariza, L. Blended learning in higher education: Students' perceptions and their relation to outcomes. *Comput. Educ.* **2011**, *56*, 818–826. [CrossRef]
57. Byl, E.; Struyven, K.; Meurs, P.; Bieke, A.; Tom, V.; Nadine, E.; Koen, L. The Value of Peer Learning for First-year Postgraduate University Students' Social and Academic Integration. *Procedia-Soc. Behav. Sci.* **2016**, *228*, 299–304. [CrossRef]
58. Curtis, D.D.; Lawson, M.J. Exploring collaborative online learning. *J. Asynchronous Learn. Netw.* **2001**, *5*, 21–34. [CrossRef]
59. Aronson, E. *The Jigsaw Classroom*; Sage: Beverly Hills, CA, USA, 1978.
60. Gittell, J. Coordinating mechanisms in care provider groups: Relational coordination as a mediator and input uncertainty as a moderator of performance effects. *Manag. Sci.* **2002**, *48*, 1408–1426. [CrossRef]
61. Gittell, J.H. *High Performance Healthcare: Using the Power of Relationships to Achieve Quality, Efficiency and Resilience*; McGraw Hill Professional: New York, NY, USA, 2009.
62. López, D.; de Pablos, C.; de la Puerta, E.; Fernández, C. Productivity in service systems: Towards a managerial framework. *Serv. Sci.* **2011**, *3*, 223–238. [CrossRef]
63. De Pablos, C.; Montes, J.; Soret, I. Relational coordination and business outcomes: Application to the practices of cloud computing. *Interciencia* **2014**, *39*, 76–84.
64. De Pablos Heredero, C.; Ramos Gomez, L.; Garcia Martinez, A.R.; Angon Sanchez de Pedro, E.; Perea Munoz, J.M. *How Students Value a Good Coordination at University*; ICERI Conference: Sevilla, Spain, 2013; p. 2154.
65. Margalina, V.M.; de Pablos Heredero, C.; Botella, J.L.M.; Martinez, A.R.G. The role of relational coordination in final teacher satisfaction in e-learning. *Procedia Technol.* **2014**, *16*, 365–375. [CrossRef]
66. Margalina, V.M.; De-Pablos-Heredero, C.; Montes-Botella, J.L. Achieving quality in e-Learning through relational coordination. *Stud. High. Educ.* **2017**, *42*, 1655–1670. [CrossRef]
67. Song, L.; Singleton, E.S.; Hill, J.R.; Koh, M.H. Improving online learning: Student perceptions of useful and challenging characteristics. *Internet High. Educ.* **2004**, *7*, 59–70. [CrossRef]
68. Gasmi, H.; Bouras, A. Ontology-based education/industry collaboration system. *IEEE Access* **2017**, *6*, 1362–1371. [CrossRef]
69. O'Leary, Z. *The Essential Guide to Doing Your Research Project*; Sage: London, UK, 2017.
70. Braun, E.; Leidner, B. Academic Course Evaluation Theoretical and Empirical Distinctions Between Self-Rated Gain in Competences and Satisfaction with Teaching Behavior. *Eur. Psychol.* **2009**, *14*, 297–306. [CrossRef]
71. Braun, E.; Woodley, A.; Richardson, J.T.E.; Leidner, B. Self-rated competences questionnaires from a design perspective. *Educ. Res. Rev.* **2012**, *7*, 1–18. [CrossRef]
72. Likert, R. A technique for the measurement of attitudes. *Arch. Psychol.* **1932**, *22*, 5–55.
73. Hair, J.F., Jr.; Ringle, C.M.; Sarstedt, M. Partial Least Squares Structural Equation Modeling: Rigorous Applications, Better Results and Higher Acceptance. *Long Range Plann.* **2013**, *46*, 1–12. [CrossRef]
74. Walker, S.L.; Fraser, B.J. Development and Validation of an Instrument for Assessing Distance Education Learning Environments in Higher Education: The Distance Education Learning Environments Survey (DELES). *Learn. Environ. Res.* **2005**, *8*, 289–308. [CrossRef]
75. Fernández-Pascual, M.D.; Ferrer-Cascales, R.; Reig-Ferrer, A.; Albaladejo-Blázquez, N.; Walker, S.L. Validation of a Spanish version of the Distance Education Learning Environments Survey (DELES) in Spain. *Learn. Environ. Res.* **2015**, *18*, 179–196. [CrossRef]
76. Ferrer-Cascales, R.; Walker, S.L.; Reig-Ferrer, A.; Dolores Fernandez-Pascual, M.; Albaladejo-Blazquez, N. Evaluation of hybrid and distance education learning environments in Spain. *Australas. J. Educ. Technol.* **2011**, *27*, 1100–1110. [CrossRef]
77. Fraser, B.J. Classroom Environment Instruments: Development, Validity and Applications. *Learn. Environ. Res.* **1998**, *1*, 7–34. [CrossRef]
78. Aldridge, J.; Fraser, B.; Huang, T. Investigating classroom environments in Taiwan and Australia with multiple research methods. *J. Educ. Res.* **1999**, *93*, 48–62. [CrossRef]
79. Dorman, J.P. Cross-National Validation of the What is Happening in this Class? (WIHIC) Questionnaire using Confirmatory Factor Analysis. *Learn. Environ. Res.* **2003**, *6*, 231–245. [CrossRef]
80. Margianti, E.S. Learning environment research in Indonesia. *Stud. Educ. Learn. Environ.* **2002**, 153–168.
81. Gallego Sanchez, M.d.C.; De Pablos Heredero, C.; Medina Merodio, J.A. Relational Coordination in Online Education. *Interciencia* **2015**, *40*, 869–874.
82. Hulland, J. Use of partial least squares (PLS) in strategic management research: A review of four recent studies. *Strateg. Manag. J.* **1999**, *20*, 195–204. [CrossRef]
83. Wong, K.K.K. Partial least squares structural equation modeling (PLS-SEM) techniques using SmartPLS. *Mark. Bull.* **2013**, *24*, 1–32.

84. Roldán, J.L.; Sánchez-Franco, M.J. Variance-based structural equation modeling: Guidelines for using partial least squares in information systems research. In *Research Methodologies, Innovations and Philosophies in Software Systems Engineering and Information Systems*; IGI Global: Pennsylvania, PA, USA, 2012; pp. 193–221.
85. Reinartz, W.; Haenlein, M.; Henseler, J. An empirical comparison of the efficacy of covariance-based and variance-based SEM. *Int. J. Res. Mark.* **2009**, *26*, 332–344. [CrossRef]
86. Roldán, J.; Cepeda, G. *Modelos de Ecuaciones Estructurales Basados en la Varianza: Partial Least Squares (PLS) para Investigadores en Ciencias Sociales*; Universidad de Sevilla: Sevilla, Spain, 2016.
87. Cohen, J. The Earth is Round (P-Less-Than.05). *Am. Psychol.* **1994**, *49*, 997–1003. [CrossRef]
88. Barclay, D.; Higgins, C.; Thompson, R. The Partial Least Squares (pls) Approach to Casual Modeling: Personal Computer Adoption Ans Use as an Illustration. *Technol. Stud.* **1995**, *2*, 85–309.
89. Hair Jr, J.F.; Hult, G.T.M.; Ringle, C.; Sarstedt, M. *A Primer on Partial Least Squares Structural Equation Modeling (PLS-SEM)*; Sage Publications: Newbury Park, CA, USA, 2016.
90. Cohen, J.; Rothstein, H.; Borenstein, M. *Power and Precision*; Taylor & Francis: Englewood, NJ, USA, 2001; Volume 1.
91. Faul, F.; Erdfelder, E.; Lang, A.; Buchner, A. G*Power 3: A flexible statistical power analysis program for the social, behavioral, and biomedical sciences. *Behav. Res. Methods* **2007**, *39*, 175–191. [CrossRef]
92. Carmines, E.G.; Zeller, R.A. *Reliability and Validity Assessment*; Sage Publications: Newbury Park, CA, USA, 1979; Volume 17.
93. Nunnally, J.C.; Bernstein, I. *Psychometric Theory (McGraw-Hill Series in Psychology)*; McGraw-Hill: New York, NY, USA, 1994; Volume 3.
94. Werts, C.E.; Linn, R.L.; Jöreskog, K.G. Intraclass reliability estimates: Testing structural assumptions. *Educ. Psychol. Meas.* **1974**, *34*, 25–33. [CrossRef]
95. Fornell, C.; Larcker, D. Evaluating Structural Equation Models with Unobservable Variables and Measurement Error. *J. Market. Res.* **1981**, *18*, 39–50. [CrossRef]
96. Henseler, J.; Ringle, C.M.; Sarstedt, M. A new criterion for assessing discriminant validity in variance-based structural equation modeling. *J. Acad. Mark. Sci.* **2015**, *43*, 115–135. [CrossRef]
97. Chin, W.W. The partial least squares approach to structural equation modeling. *Mod. Methods Bus. Res.* **1998**, *295*, 295–336.
98. Falk, R.F.; Miller, N.B. *A Primer for Soft Modeling*; University of Akron Press: Akron, OH, USA, 1992.
99. Geisser, S. A predictive approach to the random effect model. *Biometrika* **1974**, *61*, 101–107. [CrossRef]

electronics

MDPI

Article

A Conceptual Model to Investigate the Role of Mobile Game Applications in Education during the COVID-19 Pandemic

Ahmed Al Mulhem [1] and Mohammed Amin Almaiah [1,2,*]

[1] College of Education, King Faisal University, Al-Ahsa 31982, Saudi Arabia; ahamulhem@kfu.edu.sa
[2] College of Computer Sciences and Information Technology, King Faisal University, Al-Ahsa 31982, Saudi Arabia
* Correspondence: malmaiah@kfu.edu.sa

Abstract: During the COVID-19 pandemic, educational mobile games may play a significant role to facilitate students' learning. Several studies have indicated that these games using mobile phones may improve students' learning motivation and effectiveness when they are equipped with appropriate learning strategies. However, investigating the impact of learning strategies in students' utilization of educational mobile games has received little scholarly attention during the COVID-19 pandemic. Hence, this research proposed two learning games scenarios to fill this gap. In the first scenario, students were offered an educational mobile game with a learning strategy called 'scaffolding strategy'; while in the second scenario, the same game was offered without the strategy. To achieve this objective, an experimental design with a research model was developed to examine the role of scaffolding learning strategy in students' use of educational mobile games. In this experimental study, 43 students from two classes participated in the two learning scenarios. The results indicate that educational mobile gaming with the scaffolding learning strategy significantly influenced students' utilization of the mobile game. In addition, the adoption of the learning strategy significantly affected students' perceived enjoyment, perceived usefulness, perceived ease of use, and behavioural intention to use, compared with the same game without the learning strategy. The results also indicate that the introduction of the scaffolding learning strategy into the educational mobile game will increase students' learning effectiveness and motivation.

Keywords: mobile games; scaffolding strategy; students' utilization; sustainability; TAM and TUT; COVID-19

Citation: Al Mulhem, A.; Almaiah, M.A. A Conceptual Model to Investigate the Role of Mobile Game Applications in Education during the COVID-19 Pandemic. *Electronics* **2021**, *10*, 2106. https://doi.org/10.3390/electronics10172106

Academic Editors: Carlos Alario-Hoyos, Mar Pérez-Sanagustín, Miguel Morales, Oscar Jerez and Boni García

Received: 11 August 2021
Accepted: 26 August 2021
Published: 30 August 2021

1. Introduction

Recently, the emergence of the COVID-19 has brought about an acceleration towards the use of mobile games in learning and education [1]. Educational mobile games have been increasingly regarded as a powerful tool to enhance students' learning motivation. They offer a learning environment in which students acquire knowledge from playing by using their mobile devices [2,3]. They also enable students to play an active role in learning, which is why they are regarded as more effective than educational computer games in terms of improving students' learning motivation [4]. Moreover, this kind of game requires students to accomplish a challenging task by which students can also develop their problem-solving capability [5], or acquire knowledge in particular subject through the drill-and-practice process [6]. In this case, students will become not only receivers of knowledge from their instructors but active knowledge constructers [7], and thus, they achieve meaningful learning. Hence, educational mobile games have attracted many researchers' attention and have been introduced into many disciplines.

Several studies have further indicated that educational mobile games will improve students' learning effectiveness when they are equipped with appropriate learning strategies [8,9]. They differ from educational computer games as educational mobile games must be designed to be entertaining, informative, and effective [10]; otherwise, they will have negative impacts on students' learning. For example, [11] stated that educational mobile games need to be designed carefully, so as to avoid disappointing learning effectiveness caused by the conflict between gaming missions and learning tasks. Accordingly, integrating appropriate learning strategies into the educational mobile game context has become a crucial issue, for such integration can provide students with proper learning guidance or hints to ensure balance between gaming missions and learning tasks [12]. Researchers have thus incorporated learning strategies into the design of educational mobile games and explored their influence on students' learning effectiveness. For instance, [13] designed an educational mobile game application for math learning. The study adopted a digital role-playing game for students to learn number and calculation concepts in the mobile game application. This approach improved students' learning effectiveness on how to solve the equations during the game playing. [14] found that a mobile game improved students' learning effectiveness and awareness about important healthcare issues in teaching microbiology. [15] also confirmed that the combination of a mobile-based assessment strategy and educational mobile games can significantly improve students' learning effectiveness because it stimulates students thinking. This kind of game also encourages students to enjoy the learning process, and thus improves their motivation to learn effectively.

However, the important question as to whether the embedding of learning strategies will influence students' acceptance of educational mobile games has not been specifically addressed. Several studies have clearly indicated that a successful educational digital game should be accepted by students wholeheartedly in addition to improving students' learning effectiveness; otherwise, the effect will be negative [16]. Accordingly, investigation into students' acceptance of educational mobile games has been regarded as a vital issue and even as one of the criteria for evaluating the quality of educational mobile technologies [17,18]. More importantly, this kind of investigation not only helps designers or developers optimise educational technologies in a more effective manner, but also enables teachers to exploit the full potential of the technology [19]. While a majority of the literature concentrates on the factors that influence students' intention to use educational digital games [20], the role of learning strategies in students' acceptance of educational mobile games has received surprisingly little scholarly attention. As a result, this research tried to fill the gap by introducing learning strategies into the exploration of the factors behind students' acceptance of educational mobile games, thereby helping researchers design and develop more effective and acceptable educational mobile games.

To achieve this objective, we conducted an experimental study with two experimental design-based mobile games, one of which was supported by a learning strategy called the "scaffolding strategy" and the other was not. Then, we developed a research model by extending the technology acceptance model (TAM) [21], with technology utilization theory (TUT) to examine the effect of learning strategies on students' acceptance of educational mobile games. Accordingly, we designed a questionnaire to collect the students' opinions on the two mobile games, from which the role of learning strategies in students' acceptance of educational mobile games was assessed. Then, we analysed the collected data using structural equation modelling (SEM). Finally, this research provides useful suggestions for designers, developers and educational providers in order to help promote the further development and usage of educational mobile games effectively.

2. Literature Review

Educational mobile games are a new approach to learning and teaching through utilizing mobile devices. According to [22], introducing mobile games in learning and teaching has become one of the most interesting topics in education. In fact, there is an increasing research trend from many researchers towards investigating the usage of mobile games among students in the learning process [23]. Another study focused on studying the influence of these games on students' learning effectiveness [24]. Educational mobile games have been increasingly regarded as a powerful tool to enhance students' learning motivation. They offer a learning environment in which students acquire knowledge from playing by using their mobile devices [25]. It offers not only a mobile virtual learning space and enables them to play an active role in learning, which is why they are regarded as more effective than educational computer games in terms of improving students' learning motivation [26]. Educational mobile games must be designed as entertaining, informative, and effective [27]; otherwise, they will have negative impacts on students' learning.

Several studies have investigated the role of digital and mobile games in enhancing the students' learning effectiveness. For example, [28] examined how mobile game application influenced math learning. They demonstrated that this approach improved students' learning effectiveness in learning number and calculation concepts during game play. [29] demonstrated that students could achieve better learning outcomes and motivation when using mobile games than with a non-gaming learning approach. A study conducted by [30] found that playing by educational mobile games has a positive effect on students' knowledge acquisition and engagement in learning. Since they can learn facts, acquire new knowledge, and engage with content very well by mobile game playing. In another study, [31] indicated that mobile gaming could show students the learning tasks and guide them to find the target; thus, this strategy will help to improve students' learning motivations, achievements and develop their skills in problem solving. [32] confirmed that there is need more research on the investigation about the crucial factors that enable students to use mobile games in effectively way. [19] mentioned that students' acceptance is the crucial significant factor in order to ensure the success and usage of mobile games in the learning environment among students.

In order to address this problem, many scholars have focused on understanding the main aspects that influence learners' usage and acceptance of digital games. For instance, [33] employed the TAM model to understand the main determinants that affect students' intention to use digital and mobile games. The study found that learners' usage of digital games was significantly impacted by self-efficacy and perceived ease of use. The researchers also recommended that the TAM model was a powerful in explaining the adoption of mobile games among learners. Another study conducted in China by [34] extended the TAM model to investigate the adoption and usage of mobile games in classroom among students. The study findings indicated that students' intention to use mobile games was positively affected by perceived ease of use and perceived usefulness. [35] also employed the TAM model to explore the main factors that affect students' usage of computer games. The results also confirmed that perceived ease of use and perceived usefulness had significant effect on students' intention to use computer games. Overall, previous studies did not pay attention to the critical role of learning strategies behind students' acceptance of educational mobile games. Thus, in our study we employed the TAM model to examine the effect of learning strategies on students' acceptance of educational mobile games.

The Role of Mobile Games in the Learning Context

Several mobile game applications are being developed and used in the learning and teaching context. Using mobile games in learning focuses on enhancing students learning skills in particular subjects such as English learning. In addition, these games also help in improving students critical thinking skills. Some other mobile games can be applied to improving knowledge in a specific subject such as math games. In contrast, some other games like chess cannot be classified as educational games because they do not deliver content or relay curriculum material, but only improve logic and thinking skills. Mobile games that contain curriculum contents or other educational material are referred to as educational games. In this study, we will focus on mobile games that can help students to improve their English knowledge and understanding. In addition, mobile games with the scaffolding strategy offer language practice, helping learners to enhance their skills in reading, writing, speaking and listening. They encourage students to interact and communicate. They create a meaningful context for language use. Mobile games are games that are developed to help students learn about specific subjects, reinforce development and help them in improving learning skills through they play. Through this paper, we will provide evidence that the use of mobile games could support and increase English language learning outcomes.

3. Research Design

3.1. Technology Acceptance Model (TAM)

In general, one of the critical concerns that plays an important role in the literature of mobile games is assuring learners usage [14]. Many works have attempted to determine which factors influence students' acceptance of mobile games. Most of these papers have used the TAM model to explore adoption of mobile games in education [36,37]. In other words, researchers have employed the main constructs of TAM such as actual use and intention to use to study the acceptance.

The TAM model developed by [22], which has been widely used to study users' acceptance of educational technologies [38–40]. In fact, investigating the critical factors behind learners' choices of educational technologies has proven helpful in providing users with a more acceptable educational technology, and therefore has been widely regarded as a vital issue [41–44]. [22] developed the TAM based on five main constructs, namely, perceived usefulness (PU), perceived ease of use (PEU), attitude toward to use (ATU), behavioural intention (BI) and actual use (AU). PU means "the degree to which a person believes that using a particular system would enhance his or her job performance", PEU is "the degree to which a person believes that using a particular system would be free from effort," [22]. ATU is defined as the degree to which a user holds positive or negative feelings about using a particular technology, and BI is defined as the degree to which a user is willing to use a particular technology [22].

Due to the success the TAM model in exploring the user acceptance of technology, many researchers have used it extensively to clarify the factors that affect students' acceptance of educational digital games. For instance, [45] employed the TAM model to understand the main determinants that affect students' intention to use digital and mobile games. The study found that learners' usage of digital games was significantly impacted by self-efficacy and perceived ease of use. The researchers also recommended that TAM model was a powerful model in explaining the adoption of mobile games among learners. Another study conducted in China by [46] extended the TAM model to investigate the adoption and usage of mobile games in the classroom among students. The study findings indicated that students' intention to use mobile games was positively affected by perceived ease of use and perceived usefulness. [47] also employed the TAM model to explore the main factors that affect students' usage of computer games. The results also confirmed that perceived ease of use and perceived usefulness had significant effects on students' intention to use computer games. Overall, previous studies did not pay attention to the critical role of learning strategies behind students' acceptance of educational mobile games. Thus, in our study we employed the TAM model to examine the effect of learning strategies on students' acceptance of educational mobile games as shown in Figure 1.

3.2. Technology Utilization Theory (TUT)

In the TUT, utilization of technology is determined by two main factors, namely, predictive effectiveness and the predictive efficiency, as illustrated in Figure 2. Effectiveness means capability of producing an effect. In other words, effectiveness means getting the right things done [48]. Predictive effectiveness means the expected effect or impact of the specific technology. On the other hand, Efficiency simply refers to the extent to which we create output out of particular amount of input [49]. In other words, efficiency means doing things in the most economical way. Predictive efficiency means the expected output created out of particular amount of input (eg. cost, time) for the specific technology.

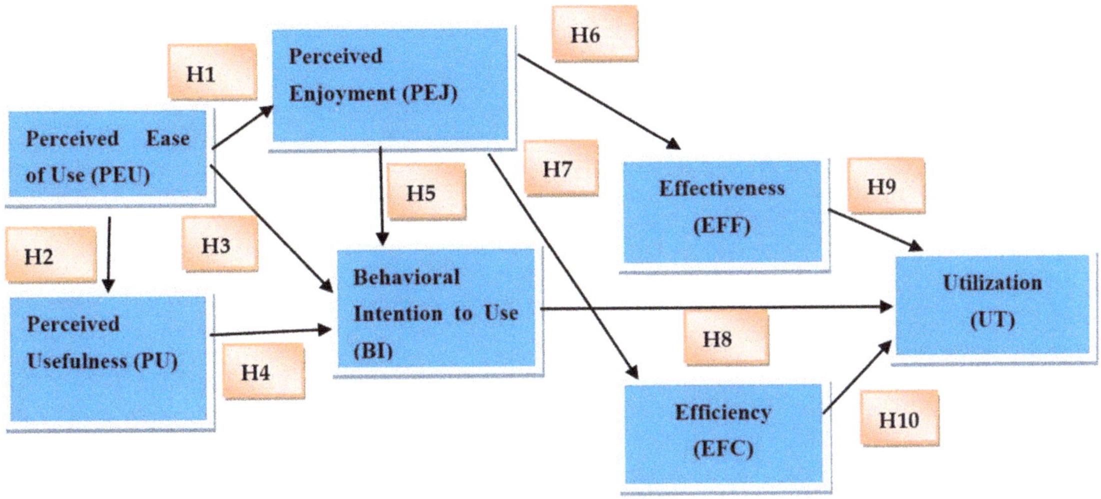

Figure 1. Research model.

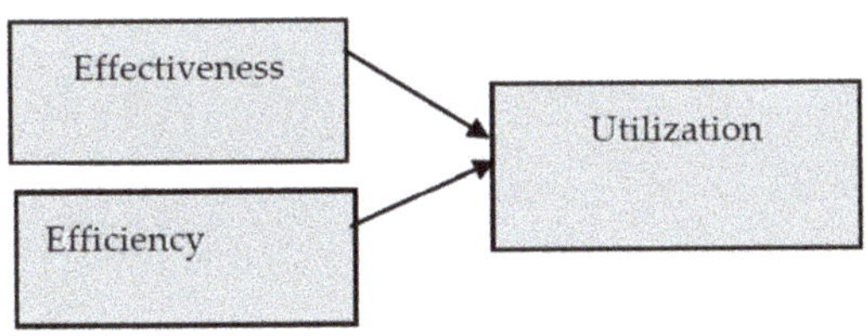

Figure 2. Technology Utilization Theory (TUT).

4. Methodology

4.1. Participants and Measurement

The participants in this experimental study were 43 students from the Educational Technologies Department enrolled in two classes of English language at King Faisal University, Saudi Arabia. Their average age was between 20 and 22 (SD = 0.91), with an average of 5 years (SD = 4.22) of experience in playing digital games and 3 years (SD = 3.67) of experience in playing mobile games. This suggested that all the students in this study had considerable experience in playing digital and mobile games.

In the experimental design for this study, the participants were divided into an experimental group and a control group. The experimental group comprised 23 students from an English language class, while the control group comprised 20 students from another class. The participants in both groups were required to take a pre-test and fill in a pre-questionnaire concerning their familiarity with digital and mobile games. After that, the participants in the experimental group played the educational mobile game supported by a scaffold strategy to learn a vocabulary of fruits, while those in the control group played the same game without the assistance of the scaffolding strategy. At the end of the learning activity, the participants in both groups were requested to fill in a post-questionnaire concerning their views on the educational mobile games.

4.2. Scenario of the Experimental Design

In this study, the experimental design applied two educational mobile game applications in order to enhance students' skills in learning vocabulary of fruits. One of the applications was designed based on a scaffolding strategy, while the other application was not. According to [50], scaffolding strategy is one of the most effective learning strategies in improving students' learning performance. Scaffolding refers to providing assistance according to students' requirements; such assistance varies with the increase in students' knowledge and ability [51,52] noticed when students first begin to use the educational computer game, they may lack the competence to tackle the challenges in the game, and therefore give up learning out of frustration. Accordingly, to avoid that, we designed an educational mobile game based on a scaffolding strategy that enables students to meet the challenges in the game and thereby immerse themselves in vocabulary learning. The game was developed using HTML5, so students could install the game on different mobile devices such as Android phones and iPhones.

Simply, the game requires students (players) to break through different barricades in different missions, through which the players developed a wider vocabulary of fruits. The main menu of the game based on a scaffolding strategy consists of three options: Play, Help, and Play More, as illustrated in Figure 3. The Play option leads the players directly into the game. The Help option offers the rules of the game to the players. The scenario of the game consists of two missions: the first is about a monkey try to collect the fruits by jumping to reach the outstanding fruits on the trees. After accomplishing the first mission, the player can guess the type of the fruit based on the alphabetical letters (A, B, C....). In this way, the game equips the players with the vocabulary of fruits required for accomplishing the second mission, in which they can learn the spelling and pronunciation of the vocabulary as well as the wholesomeness of the fruits. Figures 4 and 5 illustrate the scenes of the game based on a scaffolding strategy. In the game, students needed to control the monkey who jumps to collect the fruits from trees. To help students accomplish this task, the scaffolding strategy was adopted to provide them with timely tips. Then they must select the correct fruit based on the alphabetical letter within a limited period. On the other hand, Figure 3 shows the game that does not offer any tips to accomplish the task, that is, the scaffolding strategy is removed from the game. Accordingly, students who play the game without the assistance of a scaffolding strategy will face difficulty in the beginning because they obtain no tips from the game. In contrast, when introducing the scaffolding strategy, students need not face difficult challenges at the beginning of the game but incrementally adapt to the challenges in the game.

Figure 3. Mobile game without applying scaffolding strategy.

Figure 4. Mobile game applying the scaffolding strategy.

Figure 5. Mobile game applying the scaffolding strategy.

4.3. Experimental Procedure

In order to conduct the experimental design, the participants were divided into an experimental group and a control group. The experimental group comprised 23 students from anadvanced programming class, while the control group comprised 20 students from another class. The participants in both groups were required to take a pre-test and fill in a pre-questionnaire concerning their familiarity with digital and mobile games. After that, the participants in the experimental group played the educational mobile game supported by a scaffold strategy to learn a vocabulary of fruits, while those in the control group played the same game without the assistance of the scaffolding strategy. At the end of the learning activity, the participants in both groups were requested to fill in a post-questionnaire concerning their views on the educational mobile games. Figure 6 illustrates the steps of the experimental design.

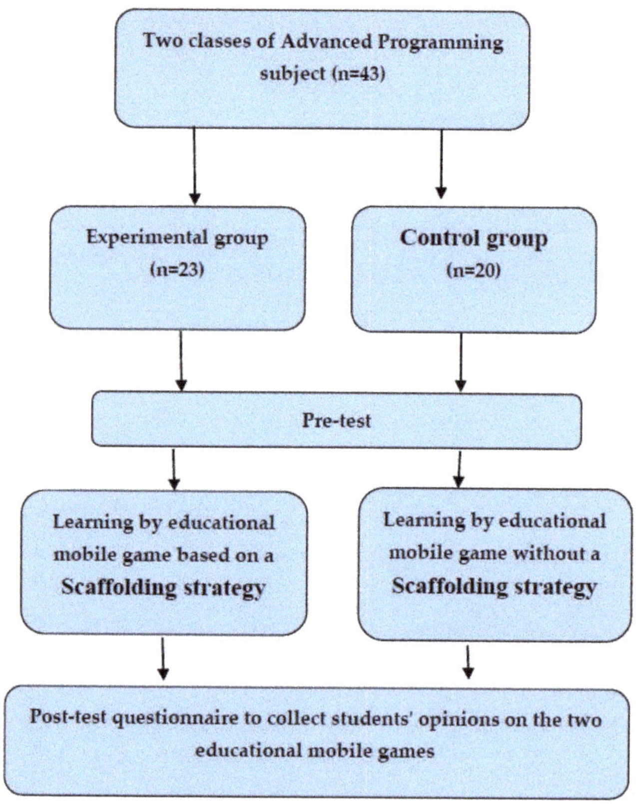

Figure 6. The experimental design procedure.

5. Structural Model Analysis and Results

According to the results of Mann–Whitney U test on the subjects' prior experience of mobile games revealed that the mean and standard deviation were, respectively, 7.32 and 3.25 in the experimental group, and 7.62 and 3.11 in the control group. This indicated that there no significant difference existed between the experimental group and the control group ($p = 0.83 > 0.05$). In addition, the results of Mann–Whitney U analysis on the students' breadth of vocabulary indicated that the mean and standard deviation values were, respectively, 13.21 and 5.21 in the experimental group, and 13.34 and 4.34 in the control group. The results indicated there is no significant difference between the two groups ($p = 0.86 > 0.05$). In general, these findings indicated that the participants in both groups had similar experience of mobile games and breadth of vocabulary.

This study employed structural equation modelling (SEM) to analyse the data collected for both groups, experimental group and control group. Based on that, two structural models were developed to test the hypotheses using path coefficients and R^2 values where path coefficients serve as an indicator of the statistical significance of the hypotheses and R^2 values indicate the models' ability to explain the variations in the dependent variables. Table 1 illustrates the results of path coefficients and R^2 of the structural model for both the experimental group and control group. The results of the structural model by experimental group indicate that eight hypotheses (H1,H2,H3,H4,H5,H7,H8) were accepted, while two (H6 and H9) were rejected. These results also indicate that perceived enjoyment has significant positive effect on perceived usefulness, perceived ease of use, effectiveness and efficiency.

Table 1. Structural Model Analysis Results.

Hypotheses	Experimental Group (Mobile Game with Scaffolding Strategy)			Control Group (Mobile Game without Scaffolding Strategy)		
	(β) Value	T-Value	Results	(β) Value	T-Value	Results
H1:PEU—PEJ	0.66	1.62	Supported	0.42	1.12	Supported
H2:PEU—PU	0.33	1.45	Supported	0.080	0.11	Not Supported
H3:PEU—BI	0.45	1.54	Supported	0.031	0.01	Not Supported
H4:PU—BI	0.78	1.75	Supported	0.021	0.08	Not Supported
H5:PEJ—BI	0.92	1.85	Supported	0.38	1.41	Supported
H6:PEJ—EFF	0.84	1.79	Supported	0.22	1.30	Supported
H7:PEJ—EFC	0.71	1.72	Supported	0.011	0.06	Not Supported
H8:BI—UT	0.62	1.61	Supported	0.42	1.35	Supported
H9:EFF—UT	0.32	1.40	Supported	0.37	1.22	Supported
H10:EFC—UT	0.52	1.51	Supported	0.36	1.39	Supported

The results indicate that students' behavioural intention to use educational mobile games was influenced by perceived enjoyment, perceived usefulness, and perceived ease of use. In addition, the results show that students' utilization of educational mobile games was influenced directly by behavioural intention to use, effectiveness, and efficiency, and utilization was influenced by perceived enjoyment indirectly through effectiveness and efficiency.

6. Discussion

Several studies have indicated that educational mobile games may improve students' learning motivation and effectiveness when they are equipped with appropriate learning strategies. However, investigating the impact of learning strategies in students' utilization of educational mobile games has received little scholarly attention. Hence, this research proposed two learning game scenarios to fill this gap. In the first scenario, students were offered an educational mobile game with a learning strategy called the 'scaffolding strategy'; while in the second scenario, the same game was offered without the strategy. To achieve this objective, an experimental design with a research model was developed by combining TAM and TUT to examine the role of scaffolding learning strategy in students' utilization of educational mobile games. In this experimental study, 43 students from two classes participated in the two learning scenarios. The findings revealed that the introduction of a scaffolding learning strategy into educational mobile games has a significant and positive effect on students' utilization of educational mobile game, and thereby, will improve students' learning effectiveness and their efficiency.

Table 1 presents a comparative analysis between the results of the experimental group (educational mobile game with scaffolding strategy) and control group (without scaffolding strategy). The results showed that the effect of perceived ease of use on perceived enjoyment, perceived usefulness, and behavioural intention to use the mobile game varied in the two groups. Where perceived ease of use had a significant effect on perceived enjoyment, perceived usefulness, and behavioural intention to use in the experimental group, in the control group, it had a significant effect only on perceived usefulness. This implies that the perceived ease of use affected students' behavioural intention to use of educational mobile games significantly in the experimental group more than in the control group. The difference in the results might be explained by the introduction of the scaffolding strategy. In addition, the logical explanation for these results is that the scaffolding strategy makes the mobile game easier to play and learn, and therefore exerts a positive effect on the perceived usefulness of the game, enjoyment in playing the game, and behavioural intention to use the educational mobile game in the experimental group. While, in the control group, students were faced difficult situations due to the lack of a scaffolding strategy (i.e., tips and guidelines to accomplish the mission), thereby making the students tend to give up playing the game to learn. In contrast, the scaffolding strategy in the other group enabled the students to tackle the challenges in the stages of the game more easily, thereby increasing their motivation to learn using the educational mobile game. This can be seen in Figures 4 and 5.

The results also revealed that perceived usefulness had a significant and positive influence on behavioural intention to use in the experimental group, while in the control group; it had a non significant but positive influence on behavioural intention. The explanation for this result is that the mobile game with the scaffolding strategy improved the students' learning effectiveness through providing them timely tips, and thus enhanced the students' willingness to play. On the contrary, the control group did not have such a supporting mechanism, and this could be why the effect of perceived usefulness on the behavioural intention of this group was not as significant as that in the experimental one. This result is inconsistent with [53–57] they found that students' perceived usefulness had no significant and positive influence on their behavioural intention, which might be explained by their lack of interest in learning that led to their ignorance of the usefulness of the games.

The results showed that perceived enjoyment was the only factor that had a significant influence on students' behavioural intentions to use the mobile game, whether the scaffolding strategy was embedded in the mobile games or not. The possible explanation for such a result is that students preferred the educational mobile games as an entertainment tool that brought them fun and pleasure; and this served as the key factor that influenced their intention to use the mobile game. This result is consistent with a [58–61], who investigated the factors that influence individuals' acceptance of mobile learning games. His study showed that perceived enjoyment is the key factor that determines individuals' acceptance of mobile learning games.

The findings also showed that perceived enjoyment had a non significant influence on perceived effectiveness and perceived efficiency of educational mobile games in both groups. This explains that students in the experimental group and control group were interested in the entertainment more than learning, and thus perceived enjoyment is not a significant factor behind enhancing students' learning effectiveness and efficiency. This factor played a significant role in their acceptance of the mobile games. In other words, students did not care about the usefulness of the educational mobile game as a tool for enhancing their learning effectiveness and efficiency; they preferred the educational mobile games as entertainment tool.

Finally, the results revealed that predictive effectiveness and predictive efficiency had significant and positive influence on students' utilization of educational mobile games in the experimental group, while in the control group was insignificant. This implies that the predictive effectiveness and predictive efficiency affected students' utilization of educational mobile games significantly in the experimental group more than in the control group. The difference in the results might be explained by the introduction of the scaffolding strategy. In addition, the logical explanation of these results is that the scaffolding strategy increased students' interest for learning through the mobile game in the experimental group. While students in the control group regarded the mobile game as an entertainment medium rather than a tool for learning due to the lack of a scaffolding strategy (i.e., tips and guidelines to learn), thereby making the students tend to play more than to learn. This result also supports the previous results of this study, which indicates that students' utilization was significantly and positively influenced by perceived enjoyment and perceived usefulness indirectly through the behavioural intention to use for the experimental group. This finding is consistent [62–64] who mentioned that the introduction of learning strategies in educational computer games might not only improve students' learning effectiveness but also reinforce their learning efficiency.

Implications of the Study

This research has several implications for practitioners and researchers. The study findings indicated that students, who were exposed to the mobile game-based learning with the scaffolding strategy, experience a positive effect on learning skills and language aspects. This paper provides evidence that the use of mobile games could support and increase English language-learning outcomes. Hence, mobile game-based learning could achieve the goal of learning effectively. The results show that the learning motivations of students have significant impact on the learning achievement, and the learning achievements of students with mobile game-based learning are better than those who use the traditional face-to-face teaching. This study also found that students who use mobile games improve their motivation to complete learning homework. In this study, mobile games appear to be particularly effective in helping students improve their English knowledge and understanding. In addition, mobile games with the scaffolding strategy offer language practice to help learners to enhance their skills in reading, writing, speaking, and listening. They encourage students to interact and communicate. They create a meaningful context for language use. Mobile games are games that are developed to help students learn about specific subjects, reinforce development, and help them in improving learning skills through play. Mobile game-based learning plays a crucial role in learning and teaching by helping learners to learn, communicate, collaborate, and work in teams. Mobile games establish a dynamic that can allow students to enhance their skills and improve their learning using game strategies. In addition, mobile games improve the functioning of the brain.

7. Conclusions

This study developed a model by combining TAM with TUT to examine the role of scaffolding learning strategy in students' acceptance of educational mobile games. Several findings of this research can be stated as follows: First, in the case of the educational mobile game with the scaffolding learning strategy, the students' perceived ease of use significantly influenced their perceived enjoyment, perceived usefulness, and behavioural intention to use; while in the case without the scaffolding learning strategy, the students' perceived ease of use did not influence their behavioural intention to use. Second, perceived enjoyment was the only important factor that affected the students' acceptance of the mobile games with or without the scaffolding learning strategy. Third, perceived usefulness had a significant and positive influence on behavioural intention to use in the experimental group, while in the control group it had a non significant but positive influence on behavioural intention. Fourth, perceived enjoyment had non significant influence on perceived effectiveness and perceived efficiency of educational mobile games in both groups. Finally, the predictive effectiveness and predictive efficiency had significant and positive influences on students' utilization of educational mobile games in the experimental group, while in the control group the effects were insignificant. In summary, these results show the important role of the scaffolding learning strategy in enhancing students' acceptance of educational mobile games. The results of this study also clarify for designers and developers the important role of the scaffolding learning strategy in designing and developing educational mobile games, which therefore makes this study a practical reference for practitioners and researchers in facilitating the application and development of educational mobile games.

Author Contributions: Conceptualization, M.A.A. and A.A.M.; methodology, M.A.A.; software and validation, M.A.A. and A.A.M.; formal analysis, M.A.A.; investigation, A.A.M.; resources, M.A.A.; data curation, M.A.A.; writing—original draft preparation, M.A.A.; writing—review and editing, A.A.M.; visualization, M.A.A.; supervision, M.A.A.; project administration, M.A.A.; funding acquisition, M.A.A. All authors have read and agreed to the published version of the manuscript.

Funding: This research was funded by the Deanship of Scientific Resarch at King Faisal University under Nasher Track 2020 (Grant No. 206103).

Data Availability Statement: The data presented in this study are available on request from the corresponding author. The data are not publicly available due to privacy.

Conflicts of Interest: The authors declare no conflict of interest.

References

1. Atwood-Blaine, D.; Huffman, D. Mobile Gaming and student interactions in a Science Center: The future of gaming in science education. *Int. J. Sci. Math. Educ.* **2017**, *15*, 45–65. [CrossRef]
2. Baek, Y.; Touati, A. Exploring how individual traits influence enjoyment in a mobile learning game. *Comput. Hum. Behav.* **2017**, *69*, 347–357. [CrossRef]
3. Almaiah, M.A.; Al-Khasawneh, A.; Althunibat, A. Exploring the critical challenges and factors influencing the E-learning system usage during COVID-19 pandemic. *Educ. Inf. Technol.* **2020**, *25*, 5261–5280. [CrossRef]
4. Almaiah, M.A.; Alamri, M.M.; Al-Rahmi, W. Applying the UTAUT model to explain the students' acceptance of mobile learning system in higher education. *IEEE Access* **2019**, *7*, 174673–174686. [CrossRef]
5. Almaiah, M.A.; Jalil, M.A.; Man, M. Extending the TAM to examine the effects of quality features on mobile learning acceptance. *J. Comput. Educ.* **2016**, *3*, 453–485. [CrossRef]
6. Alamri, M.M.; Almaiah, M.A.; Al-Rahmi, W.M. The role of compatibility and Task-Technology Fit (TTF): On Social Networking Applications (SNAs) usage as sustainability in higher education. *IEEE Access* **2020**, *8*, 161668–161681. [CrossRef]
7. Bolliger, D.U.; Mills, D.; White, J.; Kohyama, M. Japanese students' perceptions of digital game use for English-language learning in higher education. *J. Educ. Comput. Res.* **2015**, *53*, 384–408. [CrossRef]
8. Almaiah, M.A.; Alyoussef, I.Y. Analysis of the effect of course design, course content support, course assessment and instructor characteristics on the actual use of E-learning system. *IEEE Access* **2019**, *7*, 171907–171922. [CrossRef]
9. Castillo, R.E.; Cheng, C.J.; Agustin, J.S.; Aragon, M.C.R. Development of an educational mobile game applications for Grade 5 for Knowledge Channel Inc. In Proceedings of the 2019 2nd International Conference on Information Science and Systems, Tokyo, Japan, 16–19 March 2019; pp. 99–104.
10. Almaiah, M.A.; Al-Khasawneh, A. Investigating the main determinants of mobile cloud computing adoption in university campus. *Educ. Inf. Technol.* **2020**, *25*, 3087–3107. [CrossRef]
11. Almaiah, M.A.; Alamri, M.M.; Al-Rahmi, W.M. Analysis the effect of different factors on the development of mobile learning applications at different stages of usage. *IEEE Access* **2019**, *8*, 16139–16154. [CrossRef]
12. Almaiah, M.A.; Almulhem, A. A conceptual framework for determining the success factors of e-learning system implementation using Delphi technique. *J. Theor. Appl. Inf. Technol.* **2018**, *96*, 5962–5976.
13. Chen, H.-R.; Lin, Y.-S. An examination of digital game-based situated learning applied to Chinese language poetry education. *Technol. Pedagog. Educ.* **2015**, *25*, 1–16. [CrossRef]
14. Cheng, Y.-M.; Lou, S.-J.; Kuo, S.-H.; Shih, R.-C. Investigating elementary school students' technology acceptance by applying digital game-based learning to environmental education. *Australas. J. Educ. Technol.* **2013**, *29*, 1. [CrossRef]
15. Almaiah, M.A.; Jalil, M.A. Investigating students' perceptions on mobile learning services. *Int. J. Interact. Mob. Technol. iJIM* **2014**, *8*, 31. [CrossRef]
16. Almaiah, M.A.; Alismaiel, O.A. Examination of factors influencing the use of mobile learning system: An empirical study. *Educ. Inf. Technol.* **2018**, *24*, 885–909. [CrossRef]
17. Crompton, H.; Lin, Y.-C.; Burke, D.; Block, A. Mobile digital games as an educational tool in K-12 schools. In *Mobile and Ubiquitous Learning. Perspectives on Rethinking and Reforming Education*; Yu, S., Ally, M., Tsinakos, A., Eds.; Springer: Singapore, 2017; pp. 3–17. [CrossRef]
18. Almaiah, M.A.; Al Mulhem, A. Analysis of the essential factors affecting of intention to use of mobile learning applications: A comparison between universities adopters and non-adopters. *Educ. Inf. Technol.* **2019**, *24*, 1433–1468. [CrossRef]

19. Almaiah, M.A. Acceptance and usage of a mobile information system services in University of Jordan. *Educ. Inf. Technol.* **2018**, *23*, 1873–1895. [CrossRef]
20. DaCosta, B.; Seok, S.; Kinsell, C. Mobile game-based learning. In *Advanced Methodologies and Technologies in Modern Education Delivery*; IGI Global: Hershey, PA, USA, 2019; pp. 809–824.
21. Davis, F.D.; Bagozzi, R.P.; Warshaw, P.R. User acceptance of computer technology: A comparison of two theoretical models. *Manag. Sci.* **1989**, *35*, 982–1003. [CrossRef]
22. Dele-Ajayi, O.; Strachan, R.; Sanderson, J.; Pickard, A. A modified TAM for predicting acceptance of digital educational games by teachers. In *2017 IEEE Global Engineering Education Conference (EDUCON)*; IEEE: Manhattan, NY, USA, 2017; pp. 961–968.
23. Almaiah, M.A.; Jalil, M.A.; Man, M. Empirical investigation to explore factors that achieve high quality of mobile learning system based on students' perspectives. *Eng. Sci. Technol. Int. J.* **2016**, *19*, 1314–1320. [CrossRef]
24. Alamri, M.M.; Almaiah, M.A.; Al-Rahmi, W.M. Social media applications affecting students' academic performance: A model developed for sustainability in higher education. *Sustainability* **2020**, *12*, 6471. [CrossRef]
25. De Grove, F.; Bourgonjon, J.; Van Looy, J. Digital games in the classroom? A contextual approach to teachers' adoption intention of digital games in formal education. *Comput. Hum. Behav.* **2012**, *28*, 2023–2033. [CrossRef]
26. Almaiah, M.A.; Nasereddin, Y. Factors influencing the adoption of e-government services among Jordanian citizens. *Electron. Gov. Int. J.* **2020**, *16*, 236–259. [CrossRef]
27. Alksasbeh, M.; Abuhelaleh, M.; Almaiah, M.A.; Al-Jaafreh, M.; Abu Karaka, A. Towards a model of quality features for mobile social networks apps in learning environments: An extended information system success model. *Int. J. Interact. Mob. Technol. iJIM* **2019**, *13*, 75–93. [CrossRef]
28. Alghazi, S.; Kamsin, A.; Almaiah, M.; Wong, S.; Shuib, L. For sustainable application of mobile learning: An extended UTAUT model to examine the effect of technical factors on the usage of mobile devices as a learning tool. *Sustainability* **2021**, *13*, 1856. [CrossRef]
29. Shawai, Y.G.; Almaiah, M.A. Malay Language Mobile Learning System (MLMLS) using NFC technology. *Int. J. Educ. Manag. Eng.* **2018**, *8*, 1–7. [CrossRef]
30. Althunibat, A.; Binsawad, M.; Almaiah, M.A.; Almomani, O.; Alsaaidah, A.; Al-Rahmi, W.; Seliaman, M.E. Sustainable applications of smart-government services: A model to understand smart-government adoption. *Sustainability* **2021**, *13*, 3028. [CrossRef]
31. Almaiah, M.A.; Al-Zahrani, A.; Almomani, O.; Alhwaitat, A.K. Classification of cyber security threats on mobile devices and applications. In *Artificial Intelligence and Blockchain for Future Cybersecurity Applications*; Springer: Cham, Switzerland, 2021; pp. 107–123.
32. Al Amri, M.M.; Almaiah, M.A. Sustainability Model for Predicting Smart Education Technology Adoption Based on Student Perspectives. *Int. J. Advance Soft Compu.* **2021**, *13*, 2.
33. Almaiah, M.A.; Al-Khasawneh, A.; Althunibat, A.; Almomani, O. Exploring the main determinants of mobile learning application usage during Covid-19 pandemic in Jordanian universities. *Emerg. Technol. Dur. Era COVID-19* **2021**, *348*, 275–290.
34. Gholizadeh, M.; Taghiyareh, F.; Alvandkoohi, S. Toward a propensity-oriented player typology in educational mobile games. *Int. J. Game Based Learn.* **2018**, *8*, 55–67. [CrossRef]
35. Herodotou, C. Mobile games and science learning: A comparative study of 4 and 5 years old playing the game Angry Birds. *Br. J. Educ. Technol.* **2017**, *49*, 6–16. [CrossRef]
36. Holden, C.L.; Sykes, J.M. Leveraging mobile games for place-based language learning. *Int. J. Game Based Learn.* **2011**, *1*, 1–18. [CrossRef]
37. Bubukayr, M.A.S.; Almaiah, M.A. Cybersecurity concerns in smart-phones and applications: A survey. In *2021 International Conference on Information Technology (ICIT)*; IEEE: Manhattan, NY, USA, 2021; pp. 725–731.
38. Alamer, M.; Almaiah, M.A. Cybersecurity in Smart City: A systematic mapping study. In *2021 International Conference on Information Technology (ICIT)*; IEEE: Manhattan, NY, USA, 2021; pp. 719–724.
39. Al Nafea, R.; Almaiah, M.A. Cyber security threats in cloud: Literature review. In *2021 International Conference on Information Technology (ICIT)*; IEEE: Manhattan, NY, USA, 2020; pp. 779–786.
40. Almaiah, M.A.; Jalil, M.A.; Man, M. preliminary study for exploring the major problems and activities of mobile learning system: A case study of jordan. *J. Theor. Appl. Inf. Technol.* **2016**, *93*, 2.
41. AlMedires, M.; AlMaiah, M. Cybersecurity in Industrial Control System (ICS). In *2021 International Conference on Information Technology (ICIT)*; IEEE: Manhattan, NY, USA, 2021; pp. 640–647.
42. Almudaires, F.; Almaiah, M. Data an overview of cybersecurity threats on credit card companies and credit card risk mitigation. In Proceedings of the 2021 International Conference on Information Technology (ICIT), Amman, Jordan, 14–15 July 2021; pp. 732–738. [CrossRef]
43. Almaiah, M.A.; Alamri, M.M. Proposing a new technical quality requirements for mobile learning applications. *J. Theor. Appl. Inf. Technol.* **2018**, *96*, 19.
44. Huang, Y.-M. Exploring students' acceptance of educational computer games from the perspective of learning strategy. *Australas. J. Educ. Technol.* **2019**, *35*, 3. [CrossRef]
45. Huizenga, J.; Admiraal, W.; Akkerman, S.; Dam, G.T. Mobile game-based learning in secondary education: Engagement, motivation and learning in a mobile city game. *J. Comput. Assist. Learn.* **2009**, *25*, 332–344. [CrossRef]

46. Almaiah, M.A. Thematic analysis for classifying the main challenges and factors influencing the successful implementation of e-learning system using NVivo. *Int. J. Adv. Trends Comput. Sci. Eng.* **2020**, *9*, 142–152. [CrossRef]
47. Hwang, G.-J.; Wu, P.-H.; Chen, C.-C.; Tu, N.-T. Effects of an augmented reality-based educational game on students' learning achievements and attitudes in real-world observations. *Interact. Learn. Environ.* **2015**, *24*, 1895–1906. [CrossRef]
48. Hwang, K.Y.; Kim, H.S. The effects of scaffolding types in wiki-based collaborative learning on creativity. *J. Korea Contents Assoc.* **2019**, *19*, 66–78.
49. Koutromanos, G.; Avraamidou, L. The use of mobile games in formal and informal learning environments: A review of the literature. *Educ. Media Int.* **2014**, *51*, 49–65. [CrossRef]
50. Laine, T.H. Mobile educational augmented reality games: A systematic literature review and two case studies. *Computers* **2018**, *7*, 19. [CrossRef]
51. Laine, T.H.; Suk, H. Designing educational mobile augmented reality games using motivators and disturbance factors. In *Augmented Reality Games II*; Springer: Cham, Switzerland, 2019; pp. 33–56. [CrossRef]
52. Lai, A.-F.; Shih, S.; Hong, C.-R. Developing a mobile-based digital math game for learning number and calculation in elementary school. In Proceedings of the 2018 2nd International Conference on Education and Multimedia Technology, Okinawa, Japan, 2–4 July 2018; pp. 9–13. [CrossRef]
53. Molnar, A.; Kostkova, P. Learning about hygiene and antibiotic resistance through mobile games: Evaluation of learning effectiveness. In Proceedings of the 2018 International Conference on Digital Health, Lyon, UK, 23–26 April 2018; pp. 95–99.
54. Nygren, E.L.; Laine, T.H.; Sutinen, E. Dynamics between disturbances and motivations in educational mobile games. *Int. J. Interact. Mob. Technol.* **2018**, *12*, 120–141. [CrossRef]
55. Nikou, S.; Economides, A. Mobile-based micro-learning and assessment: Impact on learning performance and motivation of high school students. *J. Comput. Assist. Learn.* **2018**, *34*, 269–278. [CrossRef]
56. Prensky, M. Digital natives, digital immigrants. In *On the Horizon*; MCB University Press: Bingley, UK, 2001; Volume 9, pp. 1–6.
57. Rashid, N.A.M.; Salleh, S.M.; Noor, N.M. The role of game elements in improving Jawi Skills through a mobile game 'G-Jawi'. *Int. J. Interact. Mob. Technol. iJIM* **2018**, *12*, 20–30. [CrossRef]
58. Rachels, J.R.; Rockinson-Szapkiw, A.J. The effects of a mobile gamification app on elementary students' Spanish achievement and self-efficacy. *Comput. Assist. Lang. Learn.* **2017**, *31*, 72–89. [CrossRef]
59. Slussareff, M.; Boháčková, P. Students as game designers vs. 'just' players: Comparison of two different approaches to location-based games implementation into school curricula. *Digit. Educ. Rev.* **2016**, *29*, 284–297.
60. Villanueva, K.; Vaidya, J. Power up learning with mobile games: The educational arcade—An innovative approach to learning. In *Advancing Mobile Learning in Contemporary Educational Spaces*; IGI Global: Hershey, PA, USA, 2019; pp. 209–242.
61. Wardaszko, M.; Podgórski, B. Mobile learning game effectiveness in cognitive learning by adults: A comparative study. *Simul. Gaming* **2017**, *48*, 435–454. [CrossRef]
62. Wang, Q.; Sun, X. Investigating gameplay intention of the elderly using an Extended Technology Acceptance Model (ETAM). *Technol. Forecast. Soc. Chang.* **2016**, *107*, 59–68. [CrossRef]
63. Yadav, A.K.; Oyelere, S.S. Contextualized mobile game-based learning application for computing education. *Educ. Inf. Technol.* **2020**, *26*, 2539–2562. [CrossRef]
64. Al Amri, M.M.; Almaiah, M.A. The use of mobile gamification technology for sustainability learning in Saudi higher education. *Int. J. Adv. Trends Comput. Sci. Eng.* **2020**, *9*, 5.

electronics

MDPI

Article

Using Artificial Intelligence to Predict Class Loyalty and Plagiarism in Students in an Online Blended Programming Course during the COVID-19 Pandemic

Jain-Shing Wu [1,*], Ting-Hsuan Chien [1], Li-Ren Chien [2] and Chin-Yi Yang [3]

1 Department of Computer Science and Information Management, Providence University, Taichung City 43301, Taiwan; thchien0616@pu.edu.tw
2 Department of Computer Science and Information Engineering, National Chung Cheng University, ChiaYi 621301, Taiwan; ccumouse@ccu.edu.tw
3 18 DICE Education Technology Co., Ltd., ChiaYi 621006, Taiwan; rebacayang@tkgsh.tn.edu.tw
* Correspondence: jainshingwu1@pu.edu.tw; Tel.: +886-092-873-0417

Abstract: During the COVID-19 epidemic, most programming courses were revised to distance learning. However, many problems occurred, such as students pretending to be actively learning while actually being absent and students engaging in plagiarism. In most existing systems, obtaining status updates on the progress of a student's learning is hard. In this paper, we first define the term "class loyalty", which means that a student studies hard and is willing to learn without using any tricks. Then, we propose a novel method combined with the parsing trees of program codes and the fuzzy membership function to detect plagiarism. Additionally, the fuzzy membership functions combined with a convolution neural network (CNN) are used to predict which students obtain high scores and high class loyalty. Two hundred and twenty-six students were involved in the experiments. The dataset was randomly separated into the training datasets and the test datasets for twenty runs. The average accuracies of the experiment in predicting which students obtain high scores using the fuzzy membership function combined with a CNN and using the duration and number of actions are 93.34% and 92.62%. The average accuracies of the experiment in predicting which students have high class loyalty are 95.00% and 92.74%. Both experiments show that our proposed method not only can detect plagiarism but also can be used to detect which students are diligent.

Keywords: convolution neural network; COVID-19 epidemic; class loyalty

Citation: Wu, J.-S.; Chien, T.-H.; Chien, L.-R.; Yang, C.-Y. Using Artificial Intelligence to Predict Class Loyalty and Plagiarism in Students in an Online Blended Programming Course during the COVID-19 Pandemic. *Electronics* **2021**, *10*, 2203. https://doi.org/10.3390/electronics10182203

Academic Editors: Boni García, Carlos Alario-Hoyos, Mar Pérez-Sanagustín, Miguel Morales and Oscar Jerez

Received: 9 August 2021
Accepted: 8 September 2021
Published: 9 September 2021

1. Introduction

To improve logical thinking in young people, the education policies of many countries have made programming courses compulsory in high school or junior high school for students in grades 1–12 (K12), including in Taiwan, which started in 2019. Therefore, more and more online judgment systems (OJ) have been developed to help students write programs more easily [1–5].

Kafai and Burke [6] introduced a brief history of school programming languages. In the 1980s, some programming languages, such as BASIC, PASCAL, etc., were used as introductory programming languages for students. However, in the mid-1990s, many schools taught students how to surf the Internet and to build web pages instead of teaching programming languages. The reason that "programming languages are hard to teach and hard to understand" is obvious.

Grover, Pea, and Cooper [7] developed the basis for the Foundation for Advancing Computational Thinking (FACT) system, which is an iterative process for an empirical investigation of the computational thinking skills of middle school students. They used Scratch as the starting programming language. They also educated K12 students on what "programming languages" and "algorithms" are. A questionnaire of pre- and post-measures was used to measure students' logical thinking regarding computers and their course experience. Although the feedback was useful, obtaining students' level of enthusiasm from this information was difficult.

Fotaris et al. [8] designed a programming assistance system, similar to the famous TV game show "Who Wants to Be a Millionaire?", to help students understand programming concepts. Similar to the game, the more questions you answer correctly, the higher the reward. They also developed the "Kahoot!" system for students to use their mobile phones to answer questions. Thus, teachers could obtain a distribution of the student answers.

Most OJs help teachers teach outstanding students. However, most students need more help to understand why a program should be written and how to write these programs, which these systems cannot provide. Excellent students have a strong willingness to learn and have strong self-learning abilities; others find difficulty in keeping up with the progress of the class, let alone completing homework. For these students, giving up studying is the easiest solution. Most OJs only know that those students are obtaining fewer and fewer scores, but they do not know why and how these students lose confidence in programming.

1.1. During the Epidemic

In the past two years, educational systems have been forced to change to distance learning; many students have to stay at home and learn remotely due to the impact of the COVID-19 epidemic. In most subjects, "blended learning" is attempted using videos and Internet meetings. Due to distance learning, much chaos has ensued. To overcome this situation, many researchers are studying this topic [9–13].

Tang et al. [14] used the hypothetical model and confirmatory factor analysis (CFA) to analyze several key factors to determine whether students were ready to experience distance learning during the COVID-19 epidemic. In their results, no significant difference was found between the genders. In addition, postgraduate students performed better than undergraduate students. They also recommended increasing the virtual interaction between teachers and students, and the activities among students.

Liu et al. [15] used the Bridge-in, Objective/Outcome, Pre-assessment, Participatory learning, Post-assessment, and Summary (BOPPPS) method as a base to develop their own blended learning approach. They combined BOPPPS to provide interactive, live, and online assignments; online videos for self-study and mind-map summaries; and consolidation exercises via Q&A tests. They also designed a set of criteria for evaluating learning quality. For each stage in the evaluation, a certain percentage of the grades were included in the overall course grades.

AlShamsi et al. [16] showed the effects of online vs. face-to-face teaching on teachers and students during the COVID-19 epidemic. According to their results, students would rather learn via face-to-face teaching than the online course, as do the teachers. They concluded that three reasons influence this preference: the materials offered in each course, fatigue, and the maturity of first-year students.

Peimani and Kamalipour [17] discussed the situation and problems caused because of the fast change from face-to-face teaching to online courses. They also addressed some problems and opportunities between the teacher and students, such as assessments and feedback, and digital platforms for teaching and learning.

These studies all point out the "challenge of rapid transition from face-to-face teaching to online course." Teachers have a hard time monitoring students' behaviors in a classroom, not to mention during distance learning. During distance learning, teachers cannot guide students in summarizing and concluding their knowledge through face-to-face methods but can only help students through video conferences or emails. These only work when students are self-aware and are adept at self-learning.

In addition, distance learning requires high use of the Internet, so unstable Internet connections can cause intermissions in teaching, which is a serious challenge for teachers and students. Therefore, some teachers choose not to have individual students turn on their cameras to reduce some bandwidth use. However, students cannot be guaranteed to be sitting in front of their computer, and some students who do not turn on their camera pretend that they are in front of their monitors but are actually engaged in other activities.

1.2. Programming Education during the Epidemic

Compared with other courses, programming courses have some advantages because of the online programming system [18–23]. However, the problems faced are the same. In a normal computer programming classroom, the teacher can walk up to the students to assess a student's progress in learning. However, when it comes to distance teaching, these typically easy and simple actions become difficult. Currently, teachers can only determine how many questions a student has solved through the online judgment system, but they cannot take the initiative to provide help when the student is unable to understand the course content.

Sefriani et al. [18] used the Edmoo system in a blended learning course for programming. They used the pre-test and post-test to evaluate students' learning performances. In their study, the blended learning course for programming via Edmoo successfully affected Information technology education students during the COVID-19 epidemic.

Dale and Singer [24] used Massive Open Online Course (MOOC) to teach an online programming course. In their course, the MOOC was flexible by providing videos and drop-in sessions and by managing programming assignments and feedback.

Deperlioglu and Kose [25] adopted the Afyon Kocatepe University's Learning Management System (LMS) named @KU-UZEM to fit the requirements of the blended learning system of their "Data Structures and Algorithms" course. In their work, they blended face-to-face teaching and online learning for programming to help students learn. The feedback from the students when the course finished shows that students were satisfied with this blended learning method.

Demaidi, Qamhieh, and Afeefi [26] applied blended learning in a C programming course. They compared students who adopted blended learning and other students who used traditional face-to-face learning. In their study, the students who took the blended learning course had better performances than the students who learned via traditional face-to-face teaching.

The studies mentioned above emphasized the power of online systems used for exercises and of the pre-test and post-test in checking the progress of students. However, students can only ask questions during online meetings or through emails. In addition, teachers cannot confirm whether students really obtained knowledge through these online meetings or emails, and they can only rely on online questions or test results to assess the progress in students' learning.

In addition, another question arises for distance learning in programming. Since the online system only checks whether a student's answers to a question are correct, distinguishing whether the student wrote the program code by themselves or copied the code from other students is difficult. In a classroom, teachers can walk up to the students to prevent plagiarism, but in distance learning, this becomes almost impossible.

Carter et al. [27] developed a method of evaluating the performance of students. They compared the log files of students' behaviors to evaluate the progress in student learning. In addition, they adopted social behaviors to predict student behaviors. Finally, they applied multiple factors to predict students' behaviors. Though in their work they obtained a reliable ability to predict students' behaviors, their work is based on the trust that students do not plagiarize. If students obtained their code from the web or from other students, the evaluation system shows that they perform as well as the students who studied hard.

1.3. Goal of This Paper

Therefore, a system that not only analyzes students' progress in learning and their degree of understanding using an artificial intelligence analysis to monitor students' progress in learning but also prevents plagiarism is useful for improving student learning efficiency in a blended online programming system. Our goal is to provide a novel method combining DICE [28] and an expandable blended precision educational platform to measure the class loyalty of students and to prevent plagiarism. In the following section, some essential definitions and backgrounds are presented. In the Methods section, we introduce the methods used to measure students' enthusiasm towards learning via the logs obtained from DICE. In the Experimental Results section, the results obtained are presented. A conclusion is provided in the last section of this paper.

2. Essential Background and Definitions

In this section, we introduce some essential background and definitions. First, convolution neural networks (CNNs) are introduced in the following subsection. Second, some essential definitions used in the proposed method are given in the next subsection.

2.1. Convolution Neural Network (CNN)

The Convolution Neural Network (CNN) [29] is a kind of deep learning method. CNNs have been applied in many studies, such as car plate identification, person recognition, etc. [30–35].

In 1998, LuChun et al. [29] proposed the first convolution neural network system called LeNet-5, which is based on gradient-based learning. LeNet-5 was applied to recognize text in documents and obtained a good performance. Figure 1 shows an example of the structure of the CNN.

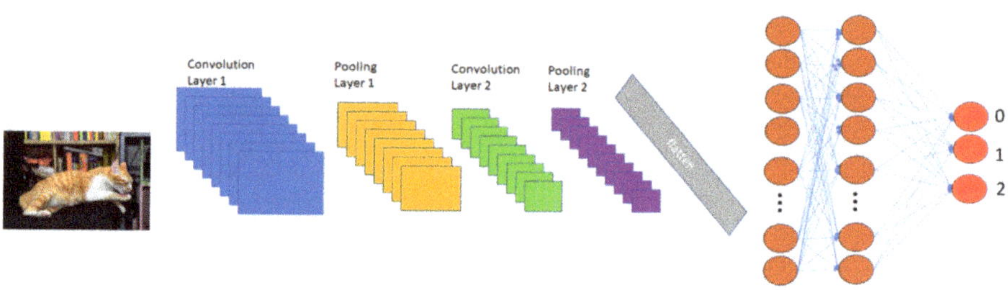

Figure 1. An example of the structure of a convolution neural network.

In their model, the CNN consisted of three different layers: a convolution layer, a sub-sampling (pooling) layer, and a fully connected layer. The picture (cat) is transformed into a matrix and inputted into the CNN. The vector is first convoluted to extract special vectors called feature maps of the picture. Then, the feature maps are sent to the pooling layer to subsample the features. After, the matrix is flattened into a one-dimensional vector and sent to fully connected neural networks. Finally, the fully connected neural network outputs the possibilities of each category. The details of each layer are given as follows.

2.1.1. Convolution Layer

In this layer, the inputted matrix is multiplied with a special matrix called a filter vector to extract a special matrix called a feature map. Figure 2 shows an example of a convolution layer.

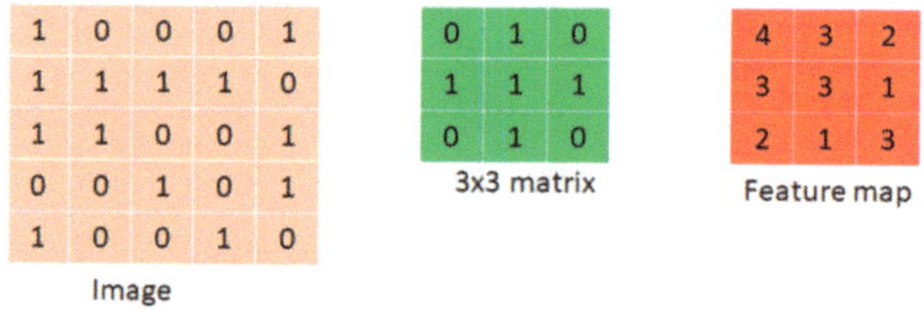

Figure 2. An example of the filter used in the convolution layer of a CNN.

In this example, the submatrix of image ((1, 0, 0), (1, 1, 1), (1, 1, 0)) is multiplied with ((0, 1, 0), (1, 1, 1), (0, 1, 0)) at each position:

$$1 \times 0 + 0 \times 1 + 0 \times 0 + 1 \times 1 + 1 \times 1 + 1 \times 1 + 1 \times 0 + 1 \times 1 + 0 \times 0 = 4.$$

Following this method, the convolution layer finishes the rest of the feature map. Hence, the image can be convoluted into a feature map. After finishing the feature maps, these feature maps are sent to the pooling layer to subsample the feature maps.

2.1.2. Pooling Layer

In the pooling layer, the feature map is subsampled into a small matrix. In this layer, the sampling method can be an average, a maximum, etc. Figure 3 shows an example of the convolution layer.

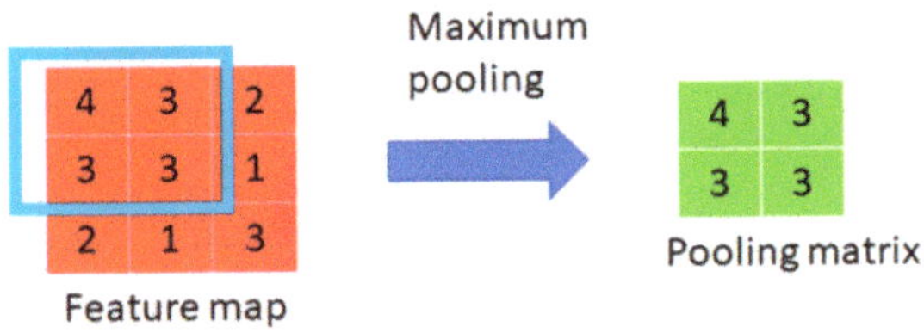

Figure 3. An example of the method used in the pooling layer of a CNN.

In Figure 3, the pooling method adopts the max-pooling method. The feature map is subsampled using a moving window. As Figure 3 shows, the maximum value inside the two-by-two blue rectangle ((4, 3), (3, 3)) is four. Additionally, the maximum value among the next ((3, 2), (3, 1)) matrix is three. The maximum value of the matrix ((3, 3), (2, 1)) is three. Additionally, the maximum value of the matrix ((3, 1), (1, 3)) is three. Hence, the pooling matrix is ((4, 3), (3, 3)).

After the subsampling (pooling) layer, the pooling matrix is flattened into a one-dimension vector and sent to a fully connected neural network.

2.1.3. Fully Connected Neural Network

In a fully connected neural network, the flattened vector is used as input and sent to several hidden layers. In fully connected neural networks, an activation function, such as a sigmoid function, is used. Additionally, a loss function, such as the minimum mean squared error (MSE), is adopted. After the deep hidden layers, the possibilities of each pre-setting category are outputted.

2.2. DICE Expandable Blended Precision Education Platform System

In this paper, we adopted the logs of the DICE expandable blended precision education platform system (DICE) to evaluate students' learning attitudes and behaviors. Hence, we introduce DICE in this section.

DICE [28] is a software system with a default content designed for coding education. It is an online assessment system that automatically and immediately judges students' code throughout the course. This system is a parsing-tree-based automatic programming assessment system with a test-driven development (TDD) environment. TDD is a code development strategy that has been popularized by extreme programming. It writes a test case before writing the code. Additionally, DICE uses the test case to verify the correctness of the code written by the students.

DICE is also an online judgment system for programming. However, unlike other online judgment systems for programming, it was developed to help the teacher teach programming. Teachers can modify the questions quickly, observe how the students solve the questions, discuss the answers with students via the web, help students solve problems via remote control as in the computer classroom, and automatically judge the correctness of the codes written by the students.

Hence, in this paper, we focus on the useful information obtained from the DICE server to help teachers assess the progress of students' learning and to assess for plagiarism.

2.3. Fuzzy Membership Function and Some Essential Definition

A membership function of a fuzzy set is a kind of function that indicates the degree of truth. Sometimes, the boundary between the values of data is unclear. For example, at a temperature of 24°C, one person may feel cool but another may feel comfortable. We often confuse the degrees of truth with probabilities. However, the fuzzy truth only reflects membership in roughly predefined sets but not the likelihood of some situations, such as being cool or comfortable.

In 1965, Zadeh [36] first introduced the fuzzy membership function. In his study, the membership function was used to give grades of membership to each object. Additionally, these grades were set within the range [0~1].

In this paper, we use the concept of fuzzy membership functions to set up the evaluation standards for learning loyalty.

Suppose that a student spends time t_{si} when solving the question i; then, the membership function of time check t_{ci} is represented as in Figure 4.

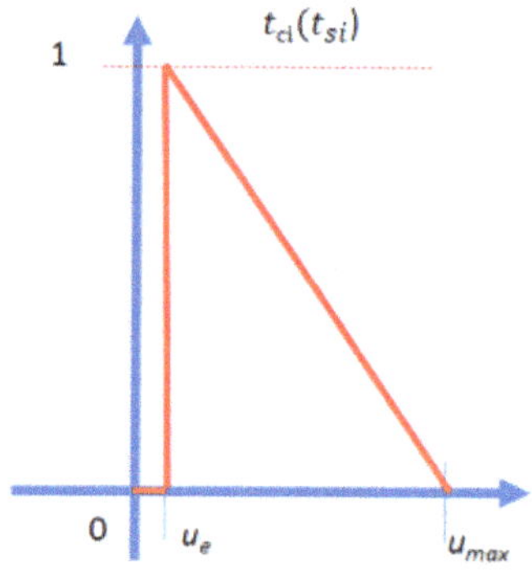

Figure 4. The membership function of time check t_{ci}.

Additionally, it can be represented as follows:

$$t_{ci}(t_{si}) = \begin{cases} 0 & if\ t_{si} \leq u_e \\ 1 - \frac{t_{si} - u_e}{t_{max}} & if\ u_e \leq t_{si} \leq t_{max} \end{cases} \tag{1}$$

where u_e is a constant value and represents the minimum time spent solving this question. This value can be defined by the teacher or by experts. The variable u_e is used so that students cannot copy and paste their answers to the question. In addition, t_{max} is the maximum time spent solving this question. This value can be defined using the maximum time spent solving this question in logs of students in the same class.

Suppose that a student performs some action and the variable a_{si} is the length of the action string when solving the question i. Here, the action string is a record that records how many actions, such as edit, delete, insert, copy, paste, execute, judge, etc., have been performed by student s. Each action represents a character in an action string. For example, if a student performs a set of actions, "insert, insert, delete, edit, insert, insert, copy, insert, edit, execute, judge", then, the action string is recorded as "IIDEIICIEXJ". Note that, if the student keeps performing one action continuously, then we only record actions every thirty seconds or until another action is performed. In this example, the value of a_{si} is eleven.

Additionally, the membership function of action check a_{ci} is represented as follows:

$$a_{ci}(a_{si}) = \begin{cases} 0 & if\ a_i \le a_e \\ 1 - \frac{a_{si} - a_e}{a_{max}} & if\ a_e \le a_i \le a_{max} \end{cases} \tag{2}$$

where a_e is a constant value and it represents the minimum length of the action string for solving this question. Additionally, this value can be defined by the teacher or by experts. The variable a_e is used so that the student cannot copy and paste their answer to the question. In addition, a_{max} is the maximum length of the action string for solving this question. This value can be defined using the maximum length of the action string when solving this question in logs of students in the same class. It can be represented as in Figure 5.

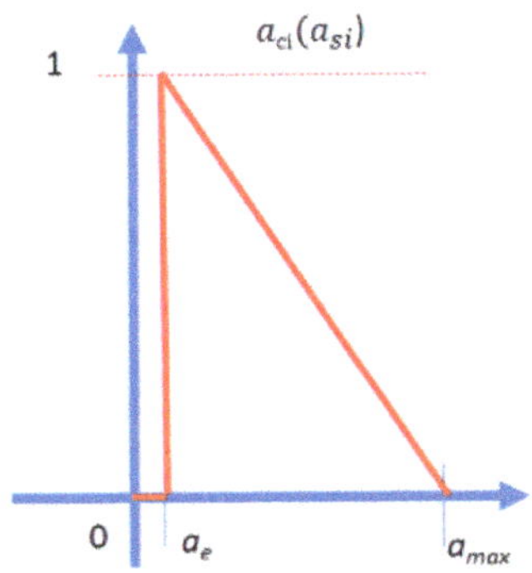

Figure 5. The membership function of action check a_{ci}.

After the values t_{ci} and a_{ci} are obtained, we can further use them to evaluate the class loyalty of students.

2.4. Object Tensor

In this paper, we introduce an important concept called "Object Tensor (OT)." Here, the term "tensor" is not the tensor in physics. The idea of OT is an extended version of the data structure dictionary in python. An Object Tensor could be represented as an array (potentially multidimensional) of Objects. The value in an OT can be null, a value, a vector, a string, or even another OT. $OT_{[d1],[d2],[d3] \ldots ,[dN]}$ is an N-order Object Tensor and N is an integer, where [d1], [d2], ... , [dN] is an ordered set of strings.

For example, an OT could be a set of collections shown as follows:

$$OT_{[user],[question],[items]}$$

where

user = ["Arvin wu", "Rebaca Yang", ...]
question = ["Question1", "Question2", ...]
information = ["Score", "Coding Time", "Actions"]

where the Score records the score that user "Arvin Wu" obtained when answering "Question 1".

In Figure 6, one student's information is stored in different tables. Using the OT, all of the information can be integrated into one OT. We can use the identifier to obtain a particular value in the OT. For example, the value can be obtained using the $OT_{[\text{"Arvin Wu"}][\text{"Question 1"}][\text{"Coding Time"}]}$ to retrieve the coding time of the student "Arvin Wu" in "Question1". Since the OT can add or subtract OT to or from itself easily, integrating the information into OT and applying it to the No-SQL database application is quite easy.

Figure 6 shows the student's record and transforms it into an OT.

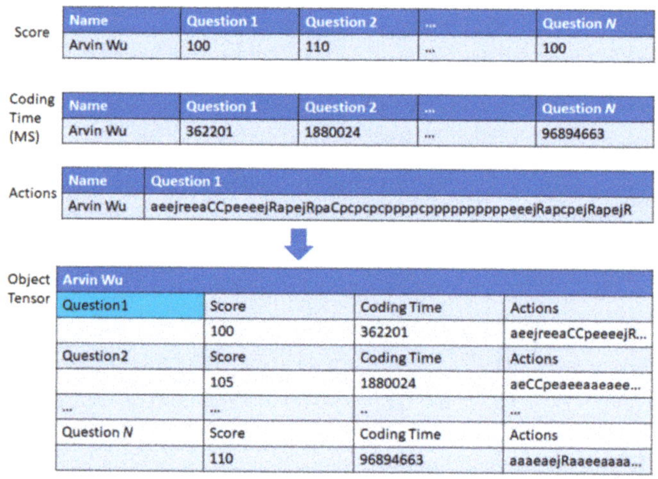

Figure 6. The concept of Object Tensor.

In the concept of OT, if one wants to add some items to the OT, these values can be added to the OT according to a special identifier or even a value. Additionally, if one wants to add a new identifier with values into OT, they can just specify the identifier of the OT and add the new identifier to the OT. In the next section, we provide an introduction on how the OT can be used in the proposed method and the details of our proposed method.

3. Proposed Method

In this section, we provide a brief introduction to the proposed methods. Figure 7 shows a flowchart of the proposed algorithm.

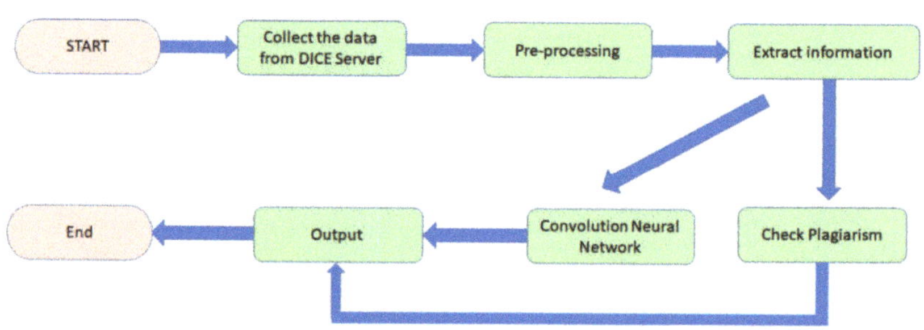

Figure 7. A flowchart of the proposed method.

First, the log files are obtained from the DICE server. Then, all logs are processed to remove empty parts, unrelated data, and outliers in the preprocessing phase. After preprocessing, the data on the durations and action strings of students for each question are extracted. This information is then transformed into OTs. Different vectors are obtained from OTs and sent to the different CNNs to learn the model for learning loyalty and transformed to check the degree of plagiarism. Finally, the classification results are sent to the output. The details of each process are given in the following sections.

3.1. Data Collection from the DICE Server

In this paper, we obtain all of the data from the DICE server. However, for several purposes, the logs are separated into different files. Hence, the first job is to collect the logs and to extract and store the information into Object Tensor (OT). We collect the scores; codes to parsing trees; time taken to answer all questions for all students; and the action logs, which record the actions performed every thirty milliseconds. All of the files are sent to the next process: data preprocessing.

3.2. Data Preprocessing

In the logs, some data should be removed, such as outliers or null data. For example, the data for a student who attends only one class but decides not to take the course should be removed. Additionally, some data should be transformed from their original form into vectors. For example, the time taken by a student to answer each question should be transformed into vectors.

Additionally, the code is another problem. It requires a lot of storage space and time to record the whole code if the code becomes increasingly complicated. To verify plagiarism, some online programming judgment systems adopt a "BLAST"-like technique to check DNA sequences for the same patterns and to check similar patterns within different program codes [37–40]. This technique is powerful for checking codes that just modify the variable names or some structures slightly. However, when students re-order the codes, this technique treats these two codes as different.

A code with slight modifications can be checked easily using the parsing tree technique, as in [41]. However, in this paper, we use functional annotations as the internal nodes instead of using the content as the parsing tree. For example, we compare the differences between the codes of two students. The codes of two students can be obtained from the server. Then, the parsing tree is generated according to the code. Figure 8 shows an example for a code, and Figure 9 is the parsing tree of that code.

```
#include <stdio.h>
#include <stdlib.h>
void show(){
        printf(“There are”);
}
void show1(int cat){
        printf(“ %d cats”,cat);
}
int main(){
        int cat;
        cat=3;
        show();
        show1(cat);
        return 0;
}
```

Figure 8. An example for the code "Show cats".

As Figures 8 and 9 show, the C code is transformed into a parsing tree. Each internal node is the identifier of the variable, function, values, etc. If students copied one others' code and modified the code or even reordered the code, the generated parsing trees of the two codes are similar. Figures 10 and 11 show another example of "copied" code that just reordered the functions show and show1, and its parsing tree.

In Figure 9, we can see that there are 75 internal nodes, 96 branches, 26 different identifiers. Additionally in Figure 11, there are the same 75 internal nodes, 96 branches, 26 different identifiers. Table 1 shows all the different identifiers used in Figures 9 and 11. In Table 1, the identifiers of the internal nodes used in both parsing trees are listed. Comparing Figures 9 and 11, the number of nodes, the number of branches, and the numbers of different identifiers are "the same". If the code was reordered, the generated tree may display a different order, but the number of nodes and the number of branches are still the same. Hence, we use the parsing tree to compare two codes.

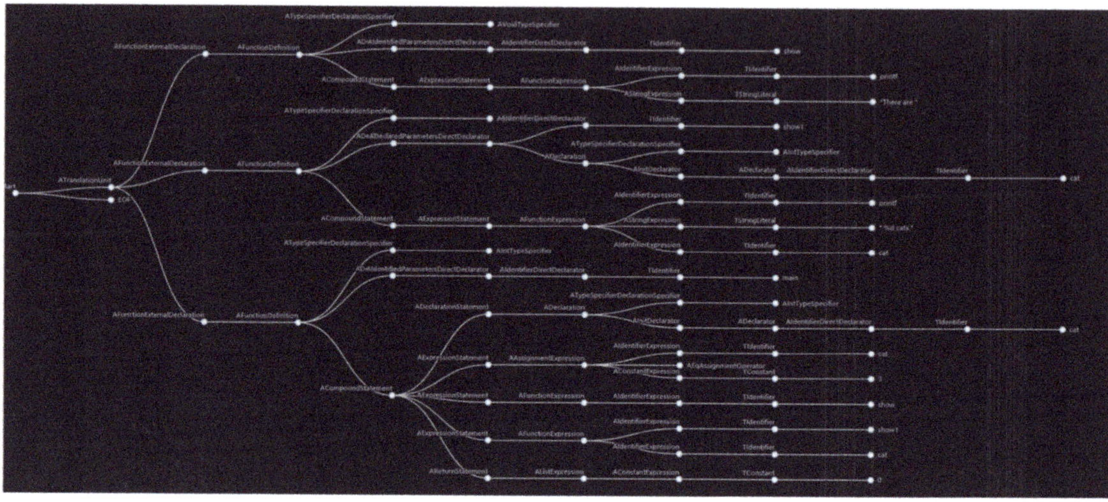

Figure 9. The parsing tree of the code "Show cats".

```
#include <stdio.h>
#include <stdlib.h>
void show1(int cat){
        printf(“ %d cats”,cat);
}
void show(){
        printf(“There are”);
}
int main(){
        int cat;
        cat=3;
        show();
        show1(cat);
        return 0;
}
```

Figure 10. An example of a "copied" code.

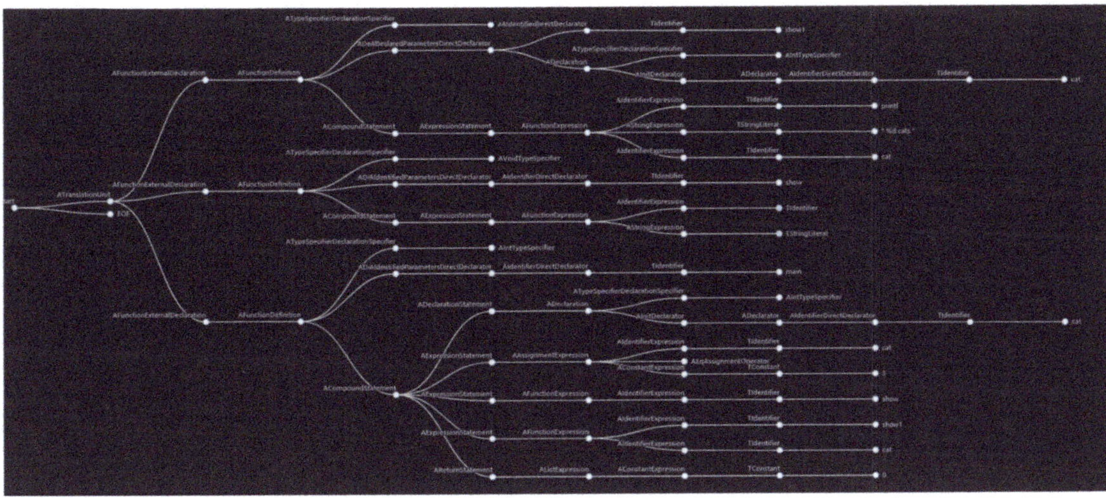

Figure 11. The parsing tree of the copied code "Show cats".

Table 1. The different identifiers used in Figures 9 and 11.

Identifiers	Number of Identifiers Used in Figure 9	Number of Identifiers Used in Figure 11
AAssignmentExpression	1	1
ACompoundStatement	3	3
AConstantExpression	2	2
ADeclaration	2	2
ADeclarationStatement	1	1
ADeclarator	5	5
ADeclaredParametersDirectDeclarator	1	1
AEqAssignmentOperator	1	1
AExpressionStatement	5	5
AFunctionDefinition	3	3
AFunctionExpression	4	4
AFunctionExternalDeclaration	3	3
AIdentifiedParametersDirectDeclarator	2	2
AIdentifierDirectDeclarator	5	5
AIdentifierExpression	7	7
AInitDeclarator	2	2
AListExpression	1	1
AReturnStatement	1	1
AStringExpression	2	2
ATranslationUnit	1	1
ATypeSpecifierDeclarationSpecifier	5	5
EOF	1	1
Start	1	1
TConstant	2	2
TIdentifier	12	12
TStringLiteral	2	2

All of the parsing tree information can be stored into OTs easily. Not all of the information from the parsing tree is needed. If the two trees are the same or similar, the number of nodes and the number of branches should be the same. Hence, in this paper, the number of nodes, the number of branches, and the numbers of the different identifiers are stored in the Question parts of OT.

All of the data for one student in the files are combined into one OT. For a student, the score, time taken, action strings, and parsing tree information of one question are stored into one sub-OT of this student's OT. Finally, all students' OTs are combined into one big OT and sent to the next process: extracting information.

3.3. Extracting Information

In the proposed method, for different purposes, different information is required. For checking the loyalty of the students, the scores, action logs, and the time taken for all questions are required and extracted from the OT according to the question identifiers in the sub-OTs and stored into vectors for processing in the convolution neural network.

Additionally, to check for plagiarism, the information, such as the scores, time taken, action logs, and information from the parsing tree, is extracted and stored into a vector for further verification.

3.4. Plagiarism Checking

In this paper, we check for plagiarism in different processes. However, two situations should be considered when checking for plagiarism.

At the early stages of learning, the codes of all students are similar since the question is simple and students are learning from the teacher's code. At that time, if the only information from the code parsing tree is checked, then we obtain the result that "all codes of students are copied". That is not the goal of education. Hence, at the initial stages, the information from the code parsing tree should not be used.

When the question becomes increasingly complicated, sometimes, teachers use the default code and modify the default code to fit the requirements of the question. At this moment, the teachers obtain a "C" in the action log, which means "copied code". If we check the action logs only, then teachers have also plagiarized.

For these two situations, we cannot only consider the information from the code parsing tree or the action logs. We should combine the information from the code parsing trees and coding action strings to observe the behaviors.

Hence, in this paper, we use the information from the parsing tree as a pre-selection requirement. If the code of one student is similar to the codes of other students, a flag is triggered. Then, the score, action log, and duration are also checked. If the number of actions performed by one student when solving a question is short, the duration is short, and the score is high or perfect, then they may have plagiarized.

Suppose that score s_{si} is the score, t_{si} is the time taken, and a_{si} is the length of the action string of student s when solving question i. The function $Spi(s_{si})$ is used to check the score and is defined as follows:

$$Spi(s_{si}) = (\frac{s_{si}}{s_{maxi}}) \tag{3}$$

where the value s_{maxi} is the highest score of question i. Additionally, the function $Tpi(t_{si})$ is used to check the time taken and is defined as follows:

$$Tpi(t_{si}) = \begin{cases} 1 & if\ t_{si} \le u_e \\ 0 & if\ u_e \le t_{si} \le t_{max} \end{cases} \tag{4}$$

where u_e is a constant value and represents the minimum time spent solving this question. Though this function is similar to Equation (1), the uses of the two functions are different. The function $Api(a_{si})$ is similarly used to check the action string and is defined as follows:

$$Api(a_{si}) = \begin{cases} 1 & if\ a_{si} \le a_e \\ 0 & if\ a_e \le a_{si} \le a_{max} \end{cases} \tag{5}$$

where a_e is a constant value and represents the minimum action string for solving question i. Hence, the plagiarism degree $PD(s_{si}, t_{si}, a_{si})$ is defined as follows:

$$PD(s_{si}, t_{si}, a_{si}) = Spi(s_{si}) \times Tpi(t_{si}) \times Api(a_{si}) \tag{6}$$

Additionally, we defined the function Pcopy($PD(s_{si}, t_{si}, a_{si})$) as in Figure 12.

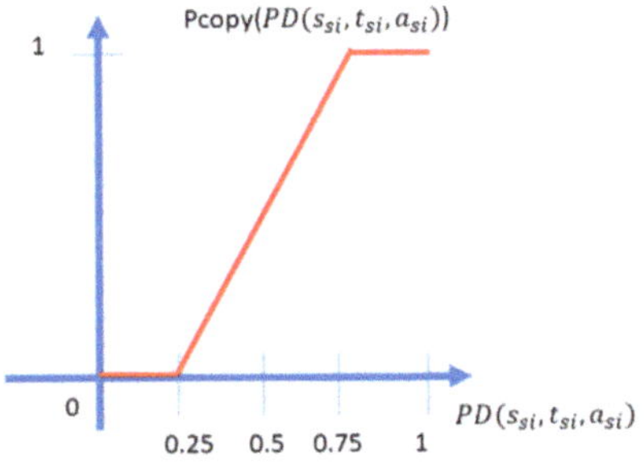

Figure 12. The fuzzy membership function of Pcopy.

The function Pnormal($PD(s_{si}, t_{si}, a_{si})$) is defined as in Figure 13.

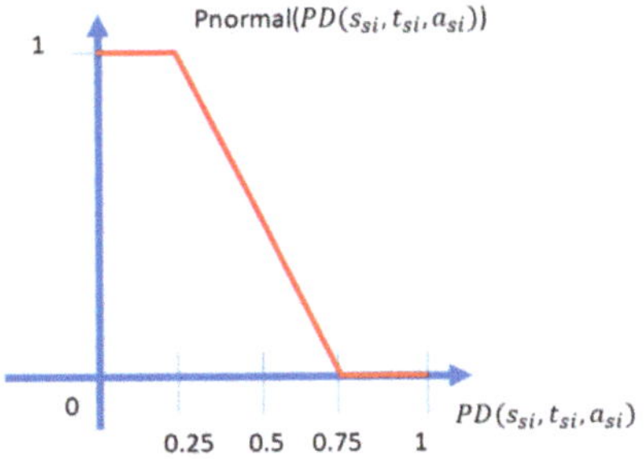

Figure 13. The fuzzy membership function of Pnormal.

Finally, the function PD_f is defined as follows:

$$PD\_f(s_{si}, t_{si}, a_{si}) = \begin{cases} \text{"copy"} & if\ \mathrm{Pnormal}(\mathrm{PD}(s_{si},\ t_{si},\ a_{si})) \le \mathrm{Pcopy}(\mathrm{PD}(s_{si},\ t_{si},\ a_{si})) \\ \text{"normal"} & otherwise \end{cases} \quad (7)$$

Using Equation (7), a judgment on the level of plagiarism can be made. After checking for plagiarism, the results are sent to the output.

Additionally, the class loyalty of the student is checked in the next subsection.

3.5. Convolution Neural Network

In this paper, we use the convolution neural network to predict which students are "good" students. However, defining "good" is hard. Hence, class loyalty and scores are used in this paper to reflect the term "good".

Note that the class loyalty that we defined here is the degree to which a student wants to stay in the class and their will to learn the contents of the course. In order to check the class loyalty of the student, we defined Equations (1) and (2), which were defined in Section 2.3. According to the results of Equations (1) and (2), the measurements of class loyalty of a student *s* is defined in Table 2.

Table 2. The measurements of class loyalty.

t_{ci} (t_{si})\ a_{ci} (a_{si})	a_{ci} (a_{si}) > 0.6	0.6 ≥ a_{ci} (a_{si}) > 0.3	0.3 > a_{ci} (a_{si})
t_{ci} (t_{si}) > 0.6	High	High	Normal
0.6 ≥ t_{ci} (t_{si}) > 0.3	High	Normal	Low
0.3 > t_{ci} (t_{si})	Normal	Low	Low

For example, a student spends three hundred seconds and performs five hundred actions when solving one question. Additionally, the maximum time for solving this question is 1200 s; the shortest time for solving this question is 30 s; the maximum and minimum number of actions for solving this problem are 10,000 and 100. Hence, the t_{ci} (t_{si}) of this student is

$$t_{ci}\ (t_{si}) = t_{ci}\ (300) = 1 - (300 - 30)/1200 = 0.775$$

Additionally, the $a_{ci}(a_{si})$ of this student is

$$a_{ci}\ (a_{si}) = a_{ci}\ (500) = 1 - (500 - 100)/10000 = 0.96$$

When we look up Table 1, the class loyalty of this student for question i is "high" since the t_{ci} is greater than 0.6 and the a_{ci} is greater than 0.6. If the class loyalty of a student is "low", the teacher can give help to the student to increase their class loyalty.

In addition, the convolution neural network is also adopted to evaluate the loyalty of the students. The scores, time taken, and number of actions performed for all questions are used as features. The final scores of the students are used. The scores are sorted and separated into three categories of equal size. Then, the data are labeled according to the scores and the categories. Figure 14 shows the CNN used in the proposed method.

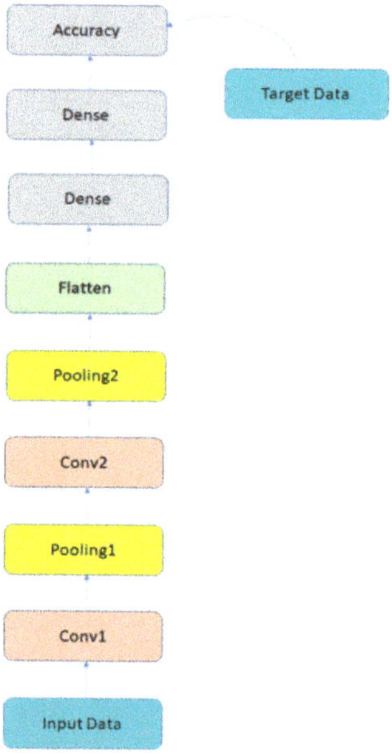

Figure 14. The CNN settings of the proposed method.

In Figure 14, the data are sent to the model by Convolution layer 1 (Conv1), max-pooling 1 (Pooling1), Convolution layer 2 (Conv2), and max pooling 2 (Pooling2). Then, the matrix is sent to the subsampling layer to flatten the matrix into a one-dimensional vector and sent to the fully connected neural network. In the next section, we compare the results obtained using the scores and the results obtained using the fuzzy method to predict the scores.

4. Experimental Results

4.1. Experimental Environment

In this paper, we collected three different classes (a total of two hundred and six samples) of students at National Chung Cheng University in Taiwan for the C language course. The C language course is two hours per week and consists of a total of eighteen weeks. Twenty-eight question sets were used to observe the behaviors of the students. The question sets were designed to teaching programming in C, such as variables, if–else statements, loops, arrays, strings, etc. Each question set consists of five or more questions. At each week, the teacher first demonstrated one or two questions in the question set and the students finished the rest of the questions in the question sets.

The goals of the experiments were to find diligent students who do not use any tricks and students with high scores. However, in order to obtain the natural behaviors of the students, the goals of the two experiments were hidden from the students. The students took the class as usual. Due to roll calls in each course, no absences were found.

After a semester (18 weeks), the data were collected from the DICE servers. All of the actions, scores, and time taken to perform actions for the students were collected. All of the information was used for analysis (plagiarism and class loyalty).

In addition, the CPU used was an Intel i7 9750H 2.6 GHz (Intel, Santa Clara, CA, USA), the RAM was 16 G, the video card was an NVIDIA GeForce RTX 2060 (NVIDIA, Santa Clara, CA, USA), the Python version was 3.8.5, and the "Keras" package of tensorflow was used to perform these experiments.

4.2. Plagiarism Check

In this section, we find out whether students plagiarized. In this experiment, we analyzed all of the scores, action strings, and time taken by the students using Equation (7).

The data consisted of two hundred and six students, and each student finished twenty-eight question sets.

We found that 57 students plagiarized and a total of 575 out of 5768 questions may have been copied. Most students not only have perfect scores but also short times taken and short actions strings. All of these students had at least eight instances of plagiarism. Though almost one in four students plagiarized, many of them were diligent students who did not use any tricks. Hence, we further check the class loyalty of the students in the next section.

4.3. Class Loyalty Experiments

In this section, we perform two experiments to detect which students have high class loyalty and which students obtained high scores. In the first experiment, we examine the relationship between students with high class loyalty and students with high scores. Hence, we transformed the time taken and the number of actions performed for each question to the t_{ci} and a_{ci} of each question and use those values to predict the scores. If the prediction accuracy is higher or equal to the prediction accuracy using the original time taken and number of actions, the fuzzy model is proven to be useful in finding which students obtain high scores and have high class loyalty.

In this experiment, in order to find the values of u_e and a_e, the time taken and number of actions performed for the top 1/3 students in the same class are collected. Then, the times of these students are averaged and divided by twenty, which is the time taken by experts. Similarly, the time taken and the number of actions performed of top 1/3 students are averaged and divided by four, which is the number of actions performed by experts. Once u_e and a_e are set, the fuzzy membership function is applied to evaluate the students. After obtaining the fuzzy values of each student when solving questions, we use these values to predict their scores.

In this experiment, students within the same class were classified into three equal-sized categories according to the final scores of the students. After that, the dataset was randomly separated into 90% training and 10% testing for twenty runs. Figure 15 shows the results of the prediction accuracies.

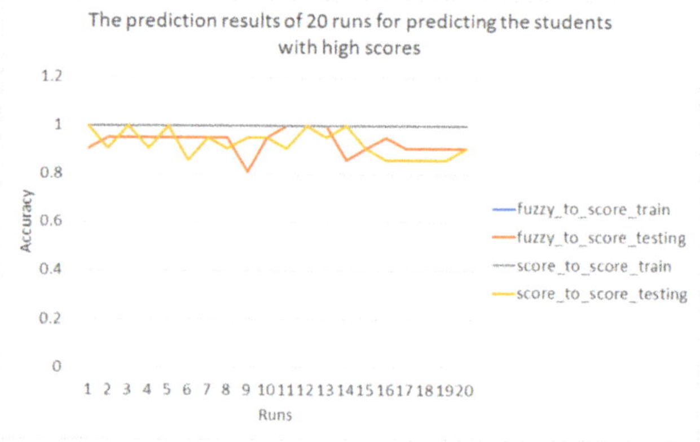

Figure 15. The prediction results of twenty runs for predicting which students obtain high scores. The blue line represents the accuracies of the training dataset using the fuzzy membership function to predict the score; the orange line represents the accuracies of the test dataset using the fuzzy membership function to predict the score; the gray line represents the accuracies of the training dataset using the original time taken and the number of actions to predict the score; and the yellow line represents the accuracies of the dataset using the original time taken and the number of actions to predict the score.

In Figure 15, the line in blue represents the accuracies of the training dataset using the fuzzy membership function to predict the scores; the line in orange represents the accuracies of the test dataset using the fuzzy membership function to predict the scores; the line in gray represents the accuracies of the training dataset using the original time taken and the number of actions to predict the scores; and the line in yellow represents the accuracies of the test dataset using the original time taken and the number of actions to predict the scores. The average accuracies of twenty runs are shown in Table 3.

Table 3. The average scores of twenty runs to predict students with high scores.

	Fuzzy Method	Original Value
Average accuracy (training)	100%	100%
Average accuracy (testing)	93.34%	92.62%

According to the results in Table 3, no significant differences between the fuzzy method and using the original values were found, which means that the fuzzy model can be used to find students with high scores and who have high class loyalty.

In the second experiment, the students with high class loyalty are found using the fuzzy model or using the original time taken and the number of actions. In this experiment, if the student obtained "high" membership in the fuzzy results when solving one question, then two points are awarded for that question; if the student obtained a "normal" membership in the fuzzy results, then one point is awarded for that question; and if the results show "low" membership when solving one question, then 0 points are awarded for that question. Finally, the fuzzy scores were summed as the fuzzy scores of the students. Additionally, we separated the students within one class into three categories according to the fuzzy scores. Similar to experiment 1, the dataset was randomly separated into 90% training and 10% testing for twenty runs. Figure 16 shows the prediction results of the twenty runs for predicting which students have high class loyalty.

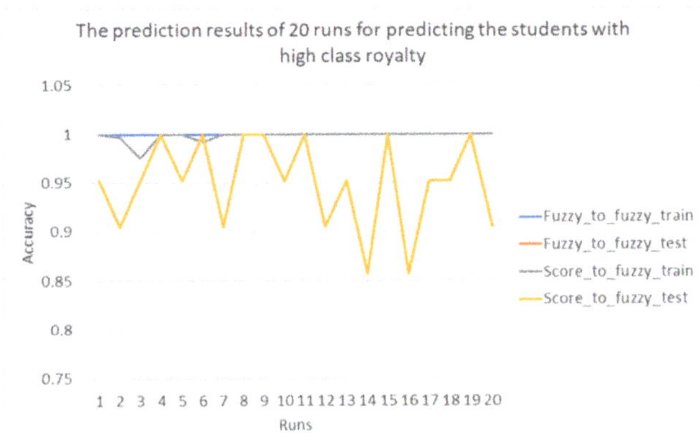

Figure 16. The prediction results of twenty runs for predicting which students have high class loyalty.

In Figure 16, the line in blue represents the accuracies of the training dataset using the fuzzy membership function to predict the higher class loyalty; the line in orange represents the accuracies of the test dataset using the fuzzy membership function; the line in gray represents the accuracies of the training dataset using the original time taken and the number of actions; the line in yellow represents the accuracies of the test dataset using the original time taken and the number of actions. The average accuracies of twenty runs are shown in Table 4.

Table 4. The average score of twenty runs to predict students with high class loyalty.

	Fuzzy Method	Original Value
Average accuracy (training)	100%	99.82%
Average accuracy (testing)	95.00%	92.74%

According to the results in Table 4, a significant difference was found between the fuzzy method and using original values, which means that the time taken and the number of actions performed by students are not the best for predicting which students have high class loyalty.

5. Discussion

In this paper, we did not perform a pre-test and a post-test for the students. Those tests are used to verify the performance of the proposed method. However, it can be achieved using questionnaires but can be misled by the behaviors of students answering the questions. Even if we set some questions to avoid students answering the questionnaires randomly, the questionnaires cannot reflect whether the students plagiarized.

In addition, designing questionnaires to reflect class loyalty is difficult. We directly used the data on the server to verify these two facts. Faking behaviors in each course when a student programs is hard, but faking the results of the questionnaires is easy. If the system does not warn that the students copied code, the students are not aware that their behaviors are monitored by the teacher. This reflects the actual situation in a class during the course. In addition, class loyalty is hardly reflected by the questions in a questionnaire but can be easily observed using the behaviors of a student solving questions. Hence, in this paper, we focus on the data to verify the behaviors and class loyalty of the students.

Though we found a way to observe whether students plagiarized, we are not trying to punish students for their behavior but rather we try to send a warning to the students. In our opinion, students who copy work are trying to pass the question without exerting any effort. If we can stop this behavior when they try to copy code from others, the student can be made aware that their behavior is being monitored and may try to solve questions by themself. Hence, though, in this paper, we proposed a way to assess for plagiarism, we still want to provide opportunities to students who plagiarized by warnings instead of punishing their behavior by decreasing their scores.

During the two experiments, two interesting things are found. First, when observing the class loyalty of a student, we found that some students with lower scores also exhibit a special behavior: they spend over twenty hours and perform over one hundred actions on a beginner question 'Hello World!' To discover the reasons for this behavior, we further observe the behaviors of the students in each course. We found that the students are not really solving the question 'Hello World!' but "pretending" that they are solving the questions while actually waiting for the answers to other questions from other, diligent students.

Since the teachers in the class cannot focus on all students but have to walk around to solve the problems that come from different students, teachers can only assess the progress of a student's learning by observing the situation when solving questions, how many questions the students have solved, and the scores obtained by those students. This kind of student spends a lot of time pretending and waiting on diligent students' answers before copying their answers. This kind of situation worsened during the epidemic since, for distance learning, the teachers and students are not present at the same place. The teachers cannot directly observe whether students are really solving questions. When the students are under surveillance, students tend to solve the problems more seriously. However, when no one is watching, some students become lazy and engage in other activities instead of solving questions, such as playing games.

Based on the class loyalty measured, we can also discover which students have lower-class loyalty and obtain lower final scores, which are students who do not have high class loyalty and just want to pass the course without exerting any effort. Hence, observing the class loyalty of students is another feasible way for a teacher to observe the learning behaviors of their students during the epidemic.

Second, while observing the students' behaviors, we found that some students have high class loyalty for each question but do not obtain high scores. Those students only obtain better than average scores. This situation interests us. The reason why those students have high class loyalty after not obtaining high scores cannot be known via the data stored in the server. Our guess is that those students may simply be happy in that they can overcome hard problems. However, this remains an open issue without any evidence to prove our hypothesis.

6. Conclusions

We proposed a novel method combined with a convolution neural network to predict class loyalty and a novel method to discover whether students plagiarized during distance learning during the COVID-19 epidemic. We first defined some fuzzy membership functions for evaluating class loyalty. Additionally, to detect instances of plagiarism, the parsing trees of programming code were adopted for fast filtering and the fuzzy membership function was used for further evaluation in order to prevent misjudgments. In the experiments, 206 students were assessed for high scores and high class loyalty. Each experiment was randomly separated into training and test datasets for twenty runs. For the high score experiments, the average accuracy of the training datasets using fuzzy membership functions combined with CNN was 100%, which is the same as that using CNN combined with the original time taken and the number of actions. For the test dataset, the accuracies of the two methods were almost the same: 93.34 and 92.62%. For finding which students have high class loyalty, the accuracies of the training datasets of the two methods were 100% and 99.82%. Additionally, the accuracies of the test datasets of the two methods were 95% and 92.74%. Hence, the proposed method can discover which students study hard and do not copy work. In the future, we can apply this system to other educational environments, such as English language learning.

Author Contributions: Conceptualization, J.-S.W. and L.-R.C.; methodology, J.-S.W.; software, J.-S.W. and T.-H.C.; validation, J.-S.W., L.-R.C. and C.-Y.Y.; resources, C.-Y.Y.; data curation, C.-Y.Y.; writing—original draft preparation, J.-S.W.; writing—review and editing, J.-S.W., T.-H.C., L.-R.C. and C.-Y.Y.; project administration, L.-R.C. All authors have read and agreed to the published version of the manuscript.

Funding: This research received no external funding.

Data Availability Statement: In this paper, the dataset used in the experiments is available at https://doi.org/10.6084/m9.figshare.15124056, accessed on 8 September 2021.

Conflicts of Interest: The authors declare no conflict of interest.

References

1. Zawacki-Richter, O.; Latchem, C. Exploring four decades of research in Computers & Education. *Comput. Educ.* **2018**, *122*, 136–152.
2. Lye, S.Y.; Koh, J.H.L. Review on teaching and learning of computational thinking through programming: What is next for K-12? *Comput. Hum. Behav.* **2014**, *41*, 51–61. [CrossRef]
3. Finkelstein, A.C.W.; Finkelstein, L. The convergence of measurement science and computer science: A scientific conversation. *Measurement* **2013**, *46*, 2946–2949. [CrossRef]
4. Azevedo, R.; Gašević, D. Analyzing multimodal multichannel data about self-regulated learning with advanced learning technologies: Issues and challenges. *Comput. Hum. Behav.* **2019**, *96*, 207–210. [CrossRef]
5. Taub, M.; Azevedo, R.; Bradbury, A.E.; Millar, G.C.; Lester, J. Using sequence mining to reveal the efficiency in scientific reasoning during STEM learning with a game-based learning environment. *Learn. Instr.* **2018**, *54*, 93–103. [CrossRef]
6. Kafai, Y.B.; Burke, Q. Computer programming goes back to school: Learning programming introduces students to solving problems, designing applications, and making connections online. *Phi Delta Kappan* **2013**, *95*, 61–65. [CrossRef]
7. Grover, S.; Pea, R.; Cooper, S. Factors influencing computer science learning in middle school. In Proceedings of the 47th ACM Technical Symposium on Computing Science Education, Memphis, TN, USA, 2–5 March 2016; pp. 552–557.
8. Fotaris, P.; Mastoras, T.; Leinfellner, R.; Rosunally, Y. Climbing up the leaderboard: An empirical study of applying gamification techniques to a computer programming class. *Electron. J. e-Learn.* **2016**, *14*, 94–110.
9. Thalluri, J.; Penman, J. Social media for learning and teaching undergraduate sciences: Good practice guidelines from intervention. *Electron. J. e-Learn.* **2015**, *13*, 455–465.
10. Al-Samarraie, H.; Saeed, N. A systematic review of cloud computing tools for collaborative learning: Opportunities and challenges to the blended-learning environment. *Comput. Educ.* **2018**, *124*, 77–91. [CrossRef]
11. Klašnja-Milićević, A.; Vesin, B.; Ivanovića, M. Social tagging strategy for enhancing e-learning experience. *Comput. Educ.* **2018**, *118*, 166–181. [CrossRef]
12. Garcia, R.; Falkner, K.; Vivian, R. Systematic literature review: Self-Regulated Learning strategies using e-learning tools for Computer Science. *Comput. Educ.* **2018**, *123*, 150–163. [CrossRef]
13. Schoor, C.; Bannert, M. Exploring regulatory processes during a computer-supported collaborative learning task using process mining. *Comput. Hum. Behav.* **2012**, *28*, 1321–1331. [CrossRef]
14. Tang, Y.M.; Chen, P.C.; Law, K.M.Y.; Wu, C.H.; Lau, Y.Y.; Guan, J.Q.; He, D.; Ho, G.T.S. Comparative analysis of Student's live online learning readiness during the coronavirus (COVID-19) pandemic in the higher education sector. *Comput. Educ.* **2021**, *168*, 104211. [CrossRef]
15. Liu, S.; Zhang, H.; Ye, Z.X.; Wu, G. Online blending learning model of school-enterprise cooperation and course certificate integration during the COVID-19 epidemic. *Sci. J. Educ.* **2020**, *8*, 66–70. [CrossRef]
16. AlShamsi, A.; Mohaidat, J.; Hinai, N.A.; Samy, A. Instructional and business continuity amid and beyond COVID-19 outbreak: A case study from the higher colleges of technology. *Int. J. High. Educ.* **2020**, *9*, 118–135. [CrossRef]
17. Peimani, N.; Kamalipour, H. Online education and the COVID-19 outbreak: A case study of online teaching during lockdown. *Educ. Sci.* **2021**, *11*, 72. [CrossRef]
18. Sefriani, R.; Sepriana, R.; Wijaya, I.; Radyuli, P.; Menrisal, M. Blended learning with Edmodo: The effectiveness of statistical learning during the COVID-19 pandemic. *Int. J. Eval. Res. Educ.* **2021**, *10*, 293–299.
19. Shi, Y.Q.; Huang, S.; Zheng, C.Y. Research on programming courses teaching based on blended learning. In Proceedings of the 5th International Conference on Frontiers of Educational Technologies, Beijing, China, 1–3 June 2019; pp. 30–34.
20. Lazarinis, F.; Karachristos, C.V.; Stavropoulos, E.C.; Verykios, V.S. A blended learning course for playfully teaching programming concepts to school teachers. *Educ. Inf. Technol.* **2019**, *24*, 1237–1249. [CrossRef]
21. Huda, C.; Hudha, M.N.; Ain, N.; Nandiyanto, A.B.D.; Abdullah, A.G.; Widiaty, I. The implementation of blended learning using android-based tutorial video in computer programming course II. In Proceedings of the IOP Conference Series: Materials Science and Engineering, the 2nd Annual Applied Science and Engineering Conference (AASEC 2017), Bandung, Indonesia, 24 August 2017; Volume 288, p. 012163.
22. Yağcı, M. Impact of the individual innovativeness characteristics on success and contentment at the computer programming course: A web-based blended learning experience. *Malays. Online J. Educ. Technol.* **2018**, *6*, 29–39. [CrossRef]

23. Farag, I. Comparing achievement of intended learning outcomes in online programming classes with blended offerings. In Proceedings of the 13th Annual Conference on Information Technology Education, Calgary, AB, Canada, 11–13 October 2012; pp. 25–30.
24. Dale, V.H.M.; Singer, J. Learner experiences of a blended course incorporating a MOOC on Haskell functional programming. *Res. Learn. Technol.* **2019**, *27*, 2248. [CrossRef]
25. Deperlioglu, O.; Kose, U. The effectiveness and experiences of blended learning approaches to computer programming education. *Comput. Appl. Eng. Educ.* **2013**, *21*, 328–342. [CrossRef]
26. Demaidi, M.N.; Qamhieh, M.; Afeefi, A. Applying blended learning in programming courses. *IEEE Access* **2019**, *7*, 156824–156833. [CrossRef]
27. Carter, A.S.; Hundhausen, C.D.; Adesope, O. Blending measures of programming and social behavior into predictive models of student achievement in early computing courses. *ACM Trans. Comput. Educ.* **2017**, *17*, 1–20. [CrossRef]
28. Yann LeCun, Y.; Bottou, L.; Bengio, Y.; Haffner, P. Gradient-based learning applied to document recognition. *Proc. IEEE* **1998**, *86*, 2278–2324. [CrossRef]
29. Wang, W.W.; Yang, J.; Chen, M.; Wang, P. A light CNN for end-to-end car license plates detection and recognition. *IEEE Access* **2019**, *7*, 173875–173883. [CrossRef]
30. Albashish, D.; Al-Sayyed, R.; Abdullah, A.; Ryalat, M.H.; Almansour, N.A. Deep CNN model based on VGG16 for breast cancer classification. In Proceedings of the 2021 International Conference on Information Technology (ICIT), Bhubaneswar, India, 16–18 December 2021; pp. 805–810.
31. Coşkun, M.; Uçar, A.; Yildirim, O.; Demir, Y. Face recognition based on convolutional neural network. In Proceedings of the 2017 International Conference on Modern Electrical and Energy Systems (MEES), Kremenchuk, Ukraine, 15–17 November 2017; pp. 376–379.
32. Wang, Z.Q.; Li, M.; Wang, H.X.; Jiang, H.Y.; Yao, Y.D.; Zhang, H.; Xin, J.C. Breast cancer detection using extreme learning machine based on feature fusion with CNN deep features. *IEEE Access* **2019**, *7*, 105146–105158. [CrossRef]
33. O'Keefe, W.; Ide, B.; Al-Khassaweneh, M.; Abuomar, O.; Szczurek, P. A CNN approach for skin cancer classification. In Proceedings of the 2021 International Conference on Information Technology (ICIT), Amman, Jordan, 14–15 July 2021; pp. 472–475.
34. Gehlot, S.; Gupta, A.; Gupta, R. A CNN-based unified framework utilizing projection loss in unison with label noise handling for multiple Myeloma cancer diagnosis. *Med. Image Anal.* **2021**, *72*, 102099. [CrossRef]
35. Chien, L.R.; Buehrer, D.J.; Yang, C.Y.; Chen, C.M. An evaluation of TDD training methods in a programming curriculum. In Proceedings of the 2008 IEEE International Symposium on IT in Medicine and Education, Xiamen, China, 12–14 December 2018; pp. 660–665.
36. Zadeh, L.A. Fuzzy sets*. *Inf. Control* **1965**, *8*, 338–353. [CrossRef]
37. Parker, A.; Hamblen, J.O. Computer algorithms for plagiarism detection. *IEEE Trans. Educ.* **1989**, *32*, 94–99. [CrossRef]
38. Ljubovic, V.; Pajic, E. Plagiarism detection in computer programming using feature extraction from ultra-fine-grained repositories. *IEEE Access* **2020**, *8*, 96505–96514. [CrossRef]
39. Santos de Campos, D.; James Ferreira, D. Plagiarism detection based on blinded logical test automation results and detection of textual similarity between source codes. In Proceedings of the 2020 IEEE Frontiers in Education Conference (FIE), Uppsala, Sweden, 21–24 October 2020; pp. 1–9.
40. Iffath, F.; Kayes, A.S.M.; Rahman, M.T.; Ferdows, J.; Arefin, M.S.; Hossain, M.S. Online judging platform utilizing dynamic plagiarism detection facilities. *Computers* **2021**, *10*, 47. [CrossRef]
41. Bartoszuk, M.; Marek Gagolewski, M. SimilaR: R code clone and plagiarism detection. *R. J.* **2021**, *12*, 367–385. [CrossRef]

electronics

Article

The Decision-Making Process and the Construction of Online Sociality through the Digital Storytelling Methodology

Pierpaolo Limone [1], Giusi Antonia Toto [1,*] and Barbara Cafarelli [2]

1 Department of Humanistic Studies, University of Foggia, 71121 Foggia, Italy; pierpaolo.limone@unifg.it
2 Department of Economics, Management and Territory, University of Foggia, 71121 Foggia, Italy; barbara.cafarelli@unifg.it
* Correspondence: giusi.toto@unifg.it

Abstract: Digital storytelling (DST) is a teaching methodology (and tool) that is very widespread in different types of training: formal and informal, professional, and for adults. Presently, education is evolving and moving towards digital storytelling, starting from the models of Lambert and Olher. Today, although DST is usually used in the training that students receive for narrative learning, experimentation on the psychological and social consequences of this online teaching practice is still scarce. The literature acknowledges the widespread use of DST online, from psychology to communication and from marketing to training, providing Lambert's and Olher's models as references. Thus, the purpose of experimentation in this subject has been to try to mix these two models by selecting the phases of the model that focus most on creativity and narrative writing. The purpose of this study is to illustrate the experimentation conducted in the initial training of teachers to monitor the processes of negotiating content, making decisions and building a group atmosphere through the use of a narrative technique in an educational context. The sample was offered comprehension activities on narrative categories, creativity and autobiographical writing. The process in the group choice phase (negotiation) of the story was monitored through a questionnaire that includes three scales (the Melbourne Decision Making Questionnaire, Organisational Attitude, and Negotiations Self-Assessment Inventory). The study concluded that the standardised planning of activities that, to a greater degree of depth, promote participation and emotional involvement allows the creation of strong group thinking and affects the decision-making and negotiation processes of the activities being carried out by the participants.

Keywords: decision-making; digital storytelling; online participation; group; sociality

Citation: Limone, P.; Toto, G.A.; Cafarelli, B. The Decision-Making Process and the Construction of Online Sociality through the Digital Storytelling Methodology. *Electronics* **2021**, *10*, 2465. https://doi.org/10.3390/electronics10202465

Academic Editors: Boni García, Carlos Alario-Hoyos, Mar Pé-rez-Sanagustín, Miguel Morales and Oscar Jerez

Received: 31 August 2021
Accepted: 28 September 2021
Published: 11 October 2021

1. Introduction

Digital storytelling (DST) consists of digital narratives; therefore, it refers to short stories of a personal or academic nature. Digital storytellers transform narrations into videos lasting a few minutes, but there is no standard duration. Over time, in reality, the duration is increasingly compressed. The first cases of digital storytelling, created by Dana Winslow Atchley [1], were real theatrical performances in the 1970s. This format was then gradually modified and condensed, and what emerged was a more contemporary organisation that crystallised, starting from the 1990s, into videos of about two–three minutes long. Today, the duration of these narrations has been reduced further to a maximum of two minutes [2]. These videos include moving images, photographs, titles and effects such as transitions, often accompanied by the author's narrative voice or a background soundtrack [3]. These narratives are not implied to be works of art; in the sense of sophisticated technology, they are not glossy and commercial products but are somewhat distorted and imperfect from a formal and language point of view [4]. These are authentic, personal stories, which depict the true characteristic of DST. Therefore, rather than being fabricated, these stories tell our tales [5].

Digital narratives can potentially merge two apparently different worlds: first, the oral story of the powerful human tool of speech and, second, the orality characterised by new media. The autobiographical narration of these digital narratives blends with innovative and digital tools, such as latest-generation cameras and mobile phones [6]. This amalgamation gives life to another medium, as it is the content that transforms the instrument. Consequently, a device emerges, which is the media that has its own autonomy and specific quality. The structure is a necessary condition to free the creative intentionality and the poetic dimension contained within this narrative [7]. All of this can be transferred to an educational context because when DSTs are effectively applied in schools, we realise that children, who are apparently unmotivated or even marginalised by the formal context of education, become reanimated with these devices. They speak and find the space to express their voice with the tools that are closest to them, such as those of digital media.

The tools available today, namely weblogs, YouTube, Twitter, Instagram, TikTok, etc. allow users to share personal stories and interpretations of reality, comment on stories and expand their social networks. They have the commercial mainstream tools used by the younger generation [8,9]. Students already use these devices, which means that they do not consider engaging in this practice as difficult or didactic. They themselves learnt the elements and media tools that can be reconstructed and reconceptualised critically at school, making them put their creative skills into play on a path that is already programmed [10].

The narrative serves not only to communicate with others or convey emotions but, above all, to understand and interpret the world. This implies that the structure of the narrative is a useful tool in psychological terms to make sense of the complexity of the world [11]. It is a very beautiful and effective formulation in a context in which we are over-exposed to all forms of media stories and are urged to compulsively consume and produce fragments of narration on social networks. Thinking about using these tools in a didactic context is a great challenge that relies on the foreknowledge that students have. The work lies precisely in the fact that teachers have the ability to mediate content and training goals on other dimensions that are merely technical ones of well-made photography. The Center DST in Berkeley has a slogan, which implies that each of us has a story to tell, and technologies enhance the meaning of the story and amplify the narrator's voice [12]. Hence, two main models of digital storytelling emerge. First, the classical model refers to the tradition of digital narratives that provide for the narration of autobiographical stories in digital formats through first-person audio narrations and presents a narrative structure similar to traditional models. On the other hand, the model of digital narratives 2.0 features greater interactivity, as these narratives offer possibilities for modifying the story and co-constructing it [13].

The first model is static and designed for substantially pre-Facebook–era technology. The rationale for this was to produce narratives for distribution through tapes in local communities and project them gradually in churches, parks and schools. Today, however, the workshops, which are carried out on DST, exploit the potential of open narratives, in the sense that collective and collaborative writing methods can be activated. Therefore, stories can incessantly continue to rise on the web with the interaction of other subjects [14]. They are stories that allow the narrative fabric to expand and the grafting of others' perspectives. This is innovation as in the classic narrative by Jason Olher and Joe Lambert; the perspective of the narrator is unique, and then, there are all the dynamics of commenting that take place on Facebook, for example, which characterise the usual practices. There is a continuous inversion and fluidity of roles, as we are authors, producers and actors of narratives, but simultaneously, we are users, commentators, judges and then again producers when our own judgement becomes the object of someone's evaluation [15]. Digital narratives 2.0 are stories in clear discontinuity with the classic model, which may not foresee the author's narration, as the voice of the writer might not be heard. Storytelling mixed with digital tools, such as video resources support teachers in service to teach natural science and technology in a way that is interactive [16].

According to the classical models of motivation, there are at least three conditions that must be respected to generate a truly motivating environment, namely, that students perceive autonomy in the task, they authentically perceive responsibility in realising what they are doing, and this allows them to field relationships between the subjects [17]. In this case, all three dimensions are fulfilled, as they choose the story, are responsible for the result and are the experts in this form of technology. The mechanisms of one's community, groups and relationships facilitate the process as the motivating elements.

This study involves the construction of an online DST laboratory for the construction of individual stories which, through a process of group negotiation and decision-making, become collective history [18]. Negotiation processes are intrinsically linked to one's choice; therefore, sharing the motivation for success with a group and guiding the group's decision-making process makes it possible to achieve a goal through group action [19]. The need to work on digital content rather than tangible content has the advantage of distracting the subject from the judgement mechanism that the activity and the relationship produce, which allows for the group to share a common trajectory in the creation of an intangible product. The construction of an intangible participation relationship (without the participants having ever seen and known each other before sharing the work) in this case, is linked to an intangible cultural product [20]. Not all group decisions and negotiation agreements are the right solutions; to make the right decision, the group must first define the right problem and restructure the problem as the group negotiates points of view and chooses a shared strategy [21,22].

Specifically, DST is used as a tool to build an online collaborative climate and participation in the group through the sharing of stories and the negotiation for the story that will be the final work of the group. The decision-making process, which follows a strong emotional experience that has been seen by the teachers individually and collectively involved in the group activity, will also be monitored. This study aims to add value to the themes already developed, as it builds on the 2.0 evolution of DST and experiments with this practice within the online teaching field and in the initial training of teachers [23]. The examination of DST, which was launched in recent years, can be enriched through the inclusion of cognitive and social psychology components linked to participation and the construction of learning communities.

The innovation of this study lies in examining an innovative teaching methodology in the online context as a tool for building social participation and improving the community climate [24].

This study proposes an innovative intervention that describes and supports the psychological mechanisms of building online participation and online communities and explores the primary effects related to DST and its influence on learning processes that have been investigated in previous studies [25].

The research question focuses on the need to find a DST tool for social participation in online communities (and for building the spirit of belonging).

2. Literature Review

2.1. The Lambert and Olher Model

Joe Lambert, in his book *Digital Storytelling Cookbook*, which is a recipe book on storytelling, theorises seven fundamental steps for storytelling [26]. He found an effective format, which allowed him to obtain the target results. For the DST laboratory and the online experimentation, points 1–3 of the Lambert model and phases 1–2 of the Olher model [27] will be used. In the proposed framework, only points 1–3 of the Lambert model and phases 1–2 of the Olher model were selected because in the writing phase of the storytelling, emotional involvement in the story and the decision-making over the group's goal take place (Tables 1 and 2). In the subsequent phases, skills of an instrumental nature will be developed that will strengthen the process, which is already underway and was built in the previous phases. In fact, as can be seen in Tables 1 and 2, after the writing of the story, the subsequent phases concern the search for the images, the choice of the soundtrack, the editing, etc.

Table 1. Seven-point Lambert model.

Points to Be Developed for DST Description	Description
Point of view Point I	Personal and authentic stories without filter; it is necessary to find narrative nodes that are proper to the author. Centrality of those who tell each part, intentions and opinions must be expressed.
Dramatic questions Point II	Narrative question or narrative instance; it is a question of exposing something that is worth telling. The question must be asked: *Is it really relevant to me?* The narrative instance is the question that must be answered during the two minutes, and the answer must emerge by the end of the story.
Emotional content Point III	A story must have emotional and engaging content, e.g., going to the bar for a coffee in itself is not particularly engaging unless at the bar you met the person you love and that a love story was born from that coffee.
Narrating voice Point IV	The emotional contents are linked to the choice of telling the story with your own perspective and commenting on the salient moments of the narration yourself with your voice, which makes the story authentic.
Soundtrack Point V	The soundtrack resonates on another level than the images and words, and this synaesthesia exercise must be done using soundtracks that are appropriate and consistent with the work.
Economy of the selection Point VI	Elegance, cleanliness and simplicity of the elements that are used; the use of the overabundance of images, words and sounds is typical of neophytes. The effort in DST is to get to the emotional knot and everything else distracts, weakens and annoys because it shifts the register.
Rhythm of the narrative Point VII	For a comic passage, as well as a dramatic passage told without the rhythm, i.e., pauses, acceleration and syncope, the story fails. It is difficult to put into practice as there is no rhythm education. We learn this outside of school and informal contexts. Editing is, therefore, very important, as hooking the music, the timing and the right words are fundamental.

Lambert's point of view on the organisation of the stories is technical but from a perspective of narrative construction [28]. Jason Olher [29] is profoundly technical from the production point of view of multimedia language, as he argues that to build a narrative you must guard five phases that are always valid. In our research project and for our purposes, we will develop only the first two phases (Table 2).

In the educational field, these phases can be explored, and it is useful to dwell on the sharing of personal stories on the internet, privacy, digital identity, personal branding, risks and network opportunities [30,31]. These five sequences are practicable almost always, as even the elderly in nursing homes or primary school children have been able to put them into practice. Often, students are better than teachers in the use of technology. On the other hand, it is difficult to find honesty and authenticity for creating a personal story, as there is resistance from the subject.

Table 2. Olher's DST model.

Stages of Construction of the DST	Description
Story planning Phase I	Plan the story correctly. Have the structure in mind: beginning, body and end as happens in the themes. To learn how to write effectively, it is better for the organisation to be written so as not to risk going off track and losing one's bearings.
Pre-production Phase II	List of media and media materials; retrieval of the media are required to be put into the storyboard, such as videos, soundtracks and photographs. All these raw materials are then put into production, equalised and assembled.
Production Phase III	Closing of the editing work of the various media, followed by the voice, video and messages. Assembly of the product, and analytical and creative revision of the final product.
Post-production Phase IV	Combination and addition of translations and transitions, i.e., fade out fade in, cross fade and all the techniques that allow you to move from one scene to another. Composition with the opening credits, credits and credits effects. Thanks, final review, click and export the file.
Performance Phase V	Presenting oneself in front of an audience; presenting the product, then distributing and sharing. For example, the product is placed on a site.

2.2. Theoretical Framework

For this course, only the classical model will be used, as it is more consistent with the training purposes. Stories, which can be told according to this model, should refer to important people in one's life, regarding people who have significant relevance, and authentic stories that tell life events with an anecdotal, but not necessarily dramatic, colouring [32]. A non-formal and lighter register can be used, as there may be several stories that make one laugh. They may not be comic stories; however, they can narrate happy events. The central point is sharing authentic emotions that the author has experienced and that they want to put into play with other subjects [33]. Stories can also relay what people are doing right now, but it must not be news from other social media posts. It must be something that is being experienced and relevant from the point of view of emotional resonance [34].

The model designed includes eight steps regarding which the subjects will complete activities for enhancing their knowledge of the narrative structure of DST, the creative formulation of the narrative idea and their ability to negotiate personal ideas up to the stage of choosing one or more ideas. A collective story can embody the final work of the group in the laboratory. According to what is reported in Table 3, which is that the purpose of the workshop is to concentrate on the narratological aspects rather than the technical realisation or acquisition of skills, the first three points of Lambert clearly show the desire to focus on the emotional and narrative dimensions of the story and the pre-production and writing of the story presented in Olher's model. These elements in this model have been expanded into eight activities of five exercises and three reflexive activities.

It begins from a dense idea, which expands and maps the story. The maps can be different from each other. A typical example is the *Visual Portrait of a Story* [35], which is a visual representation of a story. It is a model that can be depicted by a little man that functions as the story: each part of the body represents a part of the story, for instance, the feet are the conclusions. The goal of the model (Table 4) is to tell a personal story and then share an autobiographical personal anecdotal story with the audience.

Table 3. Evolution from Lambert and Olher's model to the eight-step UNIFG Model.

Lambert Model	Olher Model	Unifg Model
I. Point view II. Dramatic questions III. Emotional content	I. Story planning	I. Free individual writing II. Exercise I: Writing on incipit III. Group reading IV. Exercise II: Writing on incipit V. Exercise III: Writing on imposed object VI. Exercise IV: Writing on an inherited object VII. Exercise V: Writing the places of memories
	II. Pre-production	VIII. Individual story selection and group story negotiation

Table 4. Eight-step UNIFG Model.

Step		Activity Description
(1)	Free individual writing	Write an autobiographical narrative idea in five lines in ten minutes (time management).
(2)	Exercise I: Writing on incipit	E.g., who taught you to ride a bicycle? Random reading of three shared stories
(3)	Group reading	Sharing of all the stories produced (in the virtual classrooms dedicated to group work a tutor will observe the dynamics of sharing).
(4)	Exercise II: Writing on incipit	As in step 2, writing on incipit. E.g., who taught me to disobey? Five lines in ten minutes. Random sharing.
(5)	Exercise III: Writing on imposed object	As in step 2, write on an object caught by the group leader who will identify the theme of the working groups five lines in ten minutes. Random sharing.
(6)	Exercise IV: Writing on an inherited object	As in step 2, write an individual story on an inherited object five lines in ten minutes. Random sharing.
(7)	Exercise V: writing the places of memories	As in step 2, write an individual story about a place chosen from the memories of the various seasons of life (childhood, youth, today, tomorrow) five lines in ten minutes. Sharing on forums.
(8)	Individual story selection and group story negotiation	Each trainee has five personal stories available and must choose the one they consider best in group sharing. In a group, the individual stories will be read, and a group story will be chosen which will be used for the storytelling useful for the completion of the workshop course.

The effects will presumably be to share this joy with someone and then share the emotions that feed this story. The classical structure begins with the opening, then a problem that can be expressed by our dramatic question, which coincides with the narrative question, and followed by the solution and the end. After having measured themselves in writing five stories, in each working group made up of ten teachers, each teacher will present a chosen story. Among the 10, they will choose the one on which the group will work for the realisation of the multimedia product and the achievement of the training path will be negotiated. The narration, therefore, becomes the pretext for the construction of a group climate in a digital context through emotional tuning. The aim of this study was to monitor the negotiation and decision process in a collaborative learning context.

3. Methodology and Tools

The goal of this research work is to analyse the process of building a group atmosphere (digital learning promotes more significant interaction, collaborative learning, and social relations between students) among the teachers who attended the initial training course for special needs [36,37].

3.1. Measures

The research hypothesis concerns the relationship and mutual influence between three variables, namely, emotional participation, negotiation and the choice of a collective story that will represent the work of the whole group for the general evaluation of the laboratory, particularly with respect to the impact on the professional vision and professionalism of teachers.

The COVID-19 pandemic has caused a profound change in the professional ways of teachers, training paths have been rethought through online teaching. In this context, the University of Foggia has experimented with a working model that facilitates first-person experimentation (by teachers) on innovative teaching methodologies related to media education, gamification and the transfer of knowledge in the field of their daily professional practice [38]. The monitoring of these three dimensions represents a fundamental step for the future of online and hybrid teaching in schools, and the subsequent experimentation will concern the use of this model, both in collaborative teaching in the classroom (differentiated by age groups) and for inclusive teaching, oriented to specific learning needs.

3.2. Participants and Procedure

The research is developed in the Italian context, where the specialisation course of teachers is online and groups teachers on a national scale; therefore, students belong to all areas of Italy (63% from southern Italy; $n = 813$).

In fact, the teachers interviewed received a follow-up to all eight steps of the Foggia model, passing an exercise phase in which they worked individually to a collective phase where they worked in groups. Before the collective phase, the communication and knowledge possibilities between the subjects were very limited. The interviewees were an adult population, belonging to different areas of Italy, who had undergone an extensive training course.

The final version of the questionnaire, in addition to demographic questions on gender, age and grade of school, also presented 74 Likert-type questions relating to the previous three scales with which the user could express various levels of agreement or disagreement.

3.3. Socio-Demographic Characteristics

The data were classified as demographic profile, response processing and education level (childhood, primary school, lower secondary school or upper secondary school). They were provided through Google forms in August 2021. Using the online form made it possible to receive results in real time and to view a summary quickly. The course for preservice teachers at the University of Foggia in July–August 2021 was attended by 813 teachers in four groups per school, and in each grade, the students were divided into work groups of a maximum of ten people. After an introduction on communicative literacy and structure identification, in which the basic elements of communication were addressed, i.e., sender, recipient, objective and message (story), the eight-step guided path for choosing the collective story was described. The maximum age of the respondents was 60 years and the minimum age was 22 years, out of all the respondents, 124 (15.27%) were males and 688 (84.73%) were females.

3.4. Tools

Participants were given a self-report survey, including a socio-demographic scale and the following questionnaires:

I. The Flinders Decision Making Questionnaire consisted of 31 items divided into six scales. Vigilance has six items. Each of the six vigilance items relates to a step into sound decision making, such as defining goals, collecting information, considering alternatives, and checking alternatives. Hyper-vigilance consists of five items. Defensive avoidance consists of five items. The remaining three scales, each comprising of five items, measure different aspects of defensive avoidance (i.e., rationalisation, buck-passing, and procrastination.) Mann et al. [39] reported reliability estimates divided into the following scales: Vigilance alpha 0.80, Buck-passing 0.87, Procrastination 0.81 and Hypervigilance 0.74. The scale was validated on a sample of 2051 university students, comprising of samples from Australia (n = 262), New Zealand (n = 260), the USA (n = 475), Japan (n = 359), Hong Kong (n = 281) and Taiwan (n = 414).

II. Negotiations Self-Assessment Inventory [40] is a questionnaire of principled negotiation. The main variables of the questionnaire of principled negotiation were identified by applying a logical approach. The four dimensions (people, interests, options, and criteria) were obtained from the literature and an in-depth quantitative assessment. This questionnaire of principled negotiation can provide a practical guide for negotiators and researchers who wish to use a scientific measuring tool [41]. If the reliability coefficient is greater than 0.9, the reliability is excellent; if the reliability coefficient is between 0.8 and 0.9, the reliability is good; if the reliability coefficient is between 0.7 and 0.8, the reliability is acceptable; if the reliability coefficient is between 0.6 and 0.7, the reliability is questionable; if the reliability coefficient is between 0.5 and 0.6, the reliability is poor; if the reliability coefficient is below 0.5, it needs to be discarded [42]. The results are for the three scales on which the test is built: People (7 items) alpha 0.643, Interests (11 items) 0.851, Options (8 items) 0.790 and Criteria (6 items) 0.785.

III. The Test team [43] is a questionnaire consisting of 42 items and validated on a group of students trained in managerial action. The questionnaire can be divided into two parts, the first (the one used in this study) of 22 items referring to the level of work structuring (items 1–12), and from 13 and 22, it analyses the strictly relational plan. High scores in this first part (items 1–12) represent that planning, guiding and organising skills are developed, as well as controlling the work of others. High scores in the remaining items indicate attention to comfort and group members. The last 20 questions of the questionnaire (not included in this battery) concern two aspects of leadership: visibility and influence within the group.

In this study, the forward translation of all questionnaires, from English to Italian, was performed by a native English speaker. The existing discrepancies in Italian and in retro translations were then discussed with the authors until a consensus was reached.

3.5. Ethics

The study's procedures were carried out in accordance with the Declaration of Helsinki. The survey was approved by the university ethics committee of the research group.

3.6. Statistical Analysis

Data analysis included descriptive statistics (means and standard deviations), zero-order and partial correlations between the variables of interest for the total sample and correlations for each gender and age group. We used the statistical processing software SPSS.

4. Results

Based on the summary statistics, the maximum age of the respondents was 60 years and the minimum age was 22 years, with the average age being 39.72. The age difference between the youngest and the oldest individual is 38 years. There is a huge discrepancy in the ages of the individuals as shown by a standard deviation of 7.810. The pie chart of gender (Figure 1) shows that out of all the respondents, 124 (15.27%) were males and 688 (84.73%) were females.

In terms of gender, the sample consisted of men and women, and no other genders were considered for this study. Women constituted the largest portion of the respondents, as the mean is 1.85, which deflects towards the women's code of 2.

Based on the ratings of the item of Teams 1 to Team 22 (Table 5), the maximum values and the minimum values for all the cases were 5 and 1, respectively, with a range of 4. Since the codes included 1 = always, 2 = frequently, 3 = sometimes, 4 = rarely, 5 = never, it is evident that Team 6 had the highest mean deflecting towards 2.92, implying that most of the respondents agreed that in group works they sometimes tell people what they expect from them. On the other hand, respondents always seem to respect the opinions and feelings of others, which was the case for Team 15 when they have to negotiate a decision.

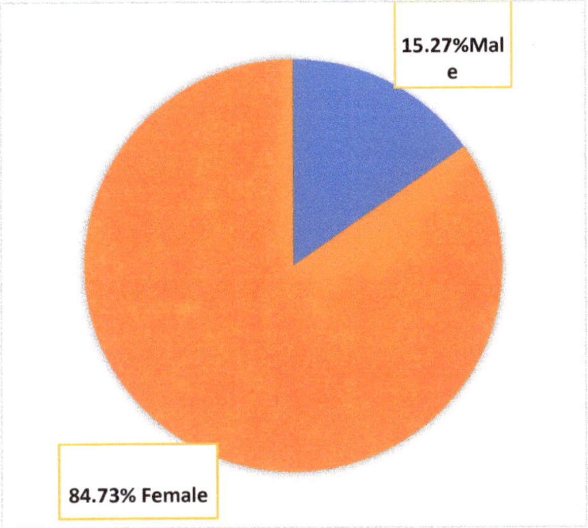

Figure 1. Gender distribution.

The above descriptive statistics table shows the values of summary statistics which include mean, variance, standard deviation skewness, and kurtosis (Table 5).

The first column shows the name of the variable, while the second column shows the number of cases that included all the variables, and a total of 812 observations are included in each variable. The third and fourth columns show the minimum and maximum values for all variables, whereas the fifth column shows the mean values. The second column shows the variance, the third column shows the standard deviation for all variables and the values of skewness and finally, the last column shows the kurtosis values (Kurtosis represents the departure from the normal distribution of values in relation to the scale of values in the data). The most frequent age was 41, with the least frequent being 23 and 60. The scales vary from a minimum to a maximum value. For the TEAM test, the values are from 1 to 5; for the Melbourne test, from 0 to 1; for the negotiation test, from 0 to 5. Only 812 respondents addressed the Melbourne item no. 4. There were 788 respondents. The average of the answers (Column 1) represents the answer as if all the interviewees had positioned themselves on that value; the team scale is 2.5, while the average value (the central answer is in fact 3) for the Melbourne test is 1, and for the negotiation scale it is 3. Taking, for example, the values of the team scale, items 1, 8 and 14 to 22 are below that value. This means that for these items, the distribution ratings of the responses are positioned towards the lower values. The skewness level also measures the level of variation in responses; in the case of the TEAM test, it is low because they differ by 1 point.

Pearson correlations were calculated between the scores of the three scales for the total sample: the gender groups and the fourth-grade school groups. As for the total sample, the test results were significant (Figure 2). Correlation represents the degree of interdependence between two variables. This value varies between −1 and 1; if the value is between −1 and 0, the relationship is negative, i.e., a low evaluation on the first item corresponds to a low evaluation on the second item. The value tends to 1 when the correlation is positive. The table shows that the correlations are all significant for the items except 4 and 13; 6 and 19; 9 and 15; 9 and 16; and finally, 9 and 19.

Table 5. Descriptive statistics.

	Mean	Std. Deviation	Skewness		Kurtosis	
	Statistic	Statistic	Statistic	Std. Error	Statistic	Std. Error
Age	39.7217	7.80963	−0.004	0.086	−0.544	0.171
Degree_of_school	2.8867	1.03310	−0.337	0.086	−1.178	0.171
TEAM1	1.9840	0.96345	0.994	0.086	0.769	0.171
TEAM2	2.5222	1.20960	0.547	0.086	−0.552	0.171
TEAM3	2.0517	1.05334	0.906	0.086	0.261	0.171
TEAM4	2.8559	1.21039	0.241	0.086	−0.847	0.171
TEAM5	2.6552	1.15155	0.366	0.086	−0.606	0.171
TEAM6	2.9224	1.29502	0.172	0.086	−1.064	0.171
TEAM7	2.1490	1.06796	0.784	0.086	0.013	0.171
TEAM8	1.8264	1.02888	1.408	0.086	1.583	0.171
TEAM9	2.8744	1.23511	0.197	0.086	−0.885	0.171
TEAM10	2.3411	1.23077	0.689	0.086	−0.454	0.171
TEAN11	2.3251	1.21040	0.674	0.086	−0.450	0.171
TEAM12	2.7796	1.18225	0.204	0.086	−0.790	0.171
TEAM13	2.0333	1.01293	0.990	0.086	0.610	0.171
TEAM14	1.5640	1.01023	2.091	0.086	3.803	0.171
TEAM15	1.4347	1.05434	2.555	0.086	5.352	0.171
TEAM16	1.5111	1.03746	2.318	0.086	4.581	0.171
TEAM17	1.4988	1.02316	2.389	0.086	4.982	0.171
TEAM18	1.7241	1.01542	1.685	0.086	2.553	0.171
TEAM19	1.4631	1.08225	2.453	0.086	4.817	0.171
TEAM20	1.9828	1.06378	1.076	0.086	0.613	0.171
TEAM21	1.8227	1.03124	1.422	0.086	1.601	0.171
TEAM22	1.5936	1.01158	2.011	0.086	3.568	0.171
melbour1	0.9901	0.57046	−0.001	0.086	0.084	0.171
melbour2	0.8300	0.76655	0.299	0.086	−1.245	0.171
melbour3	1.8695	0.36520	−2.787	0.086	7.419	0.171
melbour4	0.3845	0.59700	1.297	0.087	0.635	0.174
melbour5	0.4667	0.53727	0.515	0.086	−0.957	0.171
melbour6	0.5936	0.62410	0.557	0.086	−0.612	0.171
melbour7	1.6700	0.54800	−1.425	0.086	1.077	0.171
melbour8	0.2611	0.49242	1.693	0.086	1.997	0.171
melbour9	0.3830	0.55730	1.114	0.086	0.246	0.171
melbour10	0.5874	0.57570	0.354	0.086	−0.758	0.171
melbour11	1.8276	0.40930	−2.264	0.086	4.445	0.171
melbour12	0.6268	0.62019	0.458	0.086	−0.655	0.171
melbour13	0.6478	0.65437	0.514	0.086	−0.701	0.171
melbour14	1.8719	0.35927	−2.774	0.086	7.290	0.171
melbour15	0.3079	0.49539	1.225	0.086	0.354	0.171

Table 5. Cont.

	Mean	Std. Deviation	Skewness		Kurtosis	
	Statistic	Statistic	Statistic	Std. Error	Statistic	Std. Error
melbour16	0.5948	0.74465	0.813	0.086	−0.755	0.171
melbour17	0.5456	0.67664	0.852	0.086	−0.450	0.171
melbour18	0.4200	0.57678	1.011	0.086	0.026	0.171
melbour19	1.9015	0.32206	−3.344	0.086	11.239	0.171
melbour20	0.2869	0.50171	1.489	0.086	1.264	0.171
melbour21	0.8153	0.64010	0.185	0.086	−0.643	0.171
melbour22	0.6453	0.58743	0.281	0.086	−0.683	0.171
melbour23	0.6798	0.68338	0.505	0.086	−0.804	0.171
melbour24	1.8805	0.34300	−2.800	0.086	7.308	0.171
melbour25	1.1773	0.63834	−0.173	0.086	−0.627	0.171
melbour26	0.3818	0.59140	1.290	0.086	0.634	0.171
melbour27	0.3695	0.56747	1.255	0.086	0.591	0.171
melbour28	0.7833	0.63581	0.219	0.086	0.646	0.171
melbour29	0.7451	0.70600	0.406	0.086	−0.942	0.171
melbour30	0.5283	0.57737	0.544	0.086	−0.665	0.171
melbour31	1.1133	0.65654	−0.122	0.086	−0.700	0.171
Negoz1	2.6170	1.13670	0.053	0.086	−0.288	0.171
Negoz2	3.2389	1.39386	−0.596	0.086	−0.586	0.171
Negoz3	2.8103	1.41543	−0.444	0.086	−0.652	0.171
Negoz4	3.8534	1.32395	−1.349	0.086	1.190	0.171
Negoz5	3.7796	1.18849	−1.176	0.086	1.154	0.171
Negoz6	2.4754	1.41356	−0.145	0.086	−0.837	0.171
Negoz7	1.2389	1.30807	0.980	0.086	0.244	0.171
Negoz8	1.1170	1.36356	1.084	0.086	0.208	0.171
Negoz9	0.9667	1.24848	1.298	0.086	0.985	0.171
Negoz10	2.1133	1.27364	−0.005	0.086	−0.630	0.171
Negoz11	3.3682	1.11314	−0.478	0.086	−0.176	0.171
Negoz12	3.8116	1.18985	−1.125	0.086	0.963	0.171
Negoz13	3.7081	1.13450	−0.888	0.086	0.577	0.171
Negoz14	3.3264	1.53565	−0.624	0.086	−0.720	0.171
Negoz15	0.6232	1.11745	1.951	0.086	3.254	0.171
Negoz16	1.5603	1.38694	0.500	0.086	−0.743	0.171
Negoz17	2.0998	1.38466	0.153	0.086	−0.852	0.171
Negoz18	2.9532	1.38702	−0.289	0.086	−0.761	0.171
Negoz19	2.7204	1.26878	−0.259	0.086	−0.482	0.171
Negoz20	3.2931	1.22864	−0.772	0.086	0.377	0.171
Negoz21	4.0517	0.99557	−1.314	0.086	2.135	0.171
Negoz22	1.6367	1.43749	0.504	0.086	−0.800	0.171
Negoz23	1.8596	1.26120	0.151	0.086	−0.698	0.171
Negoz24	0.9027	1.34050	1.557	0.086	1.576	0.171
Negoz25	2.7685	1.51461	−0.161	0.086	−0.987	0.171

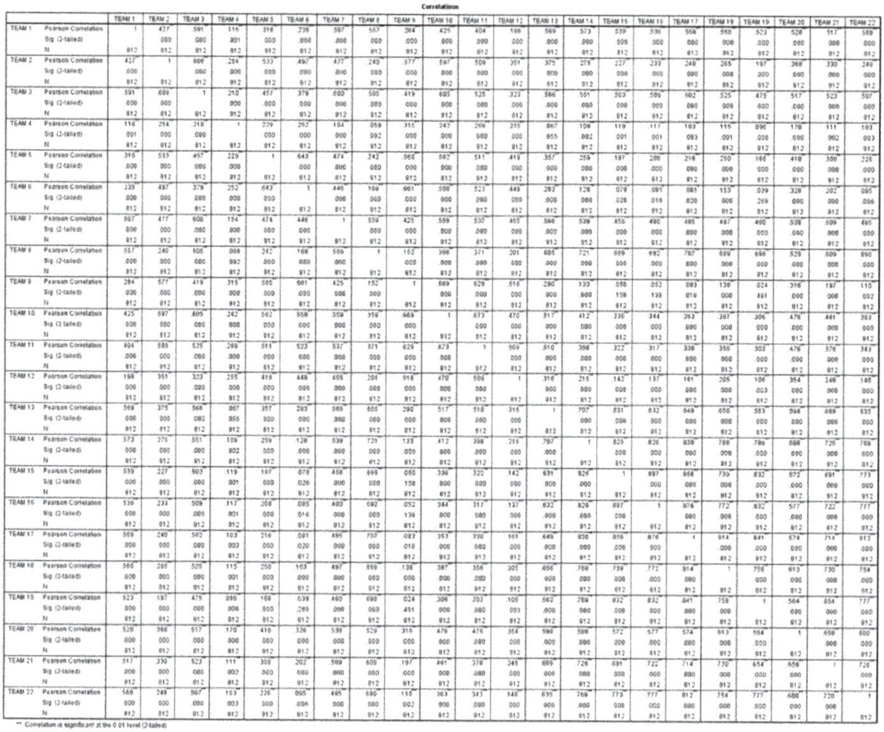

Figure 2. Correlation. ** Correlation is significant at the 00.01 level (2-tailed). * Correlation is significant at the 00.05 level (2-tailed).

Figure 2 shows significant correlations when the values exceed the 0.500 value and are found in items 10–22, relating to the second part of the questionnaire.

Based on the correlation table, the results show a positive linear relationship between the test team scores. Furthermore, there is a strong relationship between, "I am thoughtful and kind to everyone", and "I treat everyone as equals". On the other hand, there is a weak positive linear relationship between Team 15 and 9, i.e., "I assign specific tasks to each one", and "I do what I can to help others". On the basis of these results, the nova was calculated (Table 6). Items 1–22 were considered as predictors and school grade as a dependent variable. However, it was found that the degree factor was not statistically significant in any of those cases.

The Chi-square was also calculated, which produced significant results except for items 4 and 13; 6 and 19; 9 and 15; 9 and 16; 9 and 19. On the basis of this data, three hypotheses have been formulated.

Hypothesis 1 (H1). *Grade of school is not dependent on the team statement (planning, guiding and organising skills are developed, as well as management of other participants).*

Since the F-value is not significant at alpha = 0.05, we accept the null hypothesis and conclude that the school grade is not dependent on the team element of the work (planning, guiding, organising skills and controlling the work of others).

Table 6. Anova [a].

Model		Sum of Squares	df	Mean Square	F	Sig.
1	Regression	28.237	22	1.283	1.209	0.231 [b]
	Residual	837.340	789	1.061		
	Total	865.576	811			

a. Dependent Variable: Grade of school; b. Predictors: (Constant), 22 I am an easily approachable and friendly person in the work team. *, 6 I tell people what I expect from them. *, 4 I organize my work by myself. *. 12 I ask that everyone follow standard rules and procedures. *, 2 I explain the role that everyone plays within the team. *, 1 I always find time to explain how a job needs to be done. *, 11 I plan the work I want done. *, 20 I inform in advance of a change and explain to everyone the impact it will have. *, 5 I inform people about how they are working. *, 7 I encourage the use of uniform procedures so that the whole group can follow them in detail. *, 8 I express clearly what I think. *. 13 I make working on homework enjoyable. *, 3 I make the rules and procedures clear so that everyone can follow them precisely. *, 9 I assign specific tasks to each one. *, 21 I take care of the well-being of the people in my group. *, 10 I make sure everyone understands their role within the team. *, 15 I respect the opinions and feelings of others. *, 18 I am attentive to details that can make being a team member more enjoyable. *, 19 I treat everyone as equals. *, 14 I do what I can to help others. * 16 I am thoughtful and kind to everyone. *, 17 I keep the friendly atmosphere in the team. *

The model summary (Figure 3) shows that the regression is representative of only 3% of the data. The Durbin-Watson statistic is less than two, meaning that a weak correlation exists between the dependent and independent variables.

Model Summary [b]

Model	R	R Square	Adjusted R Square	Std. Error of the Estimate	Durbin-Watson
1	0.181 [a]	0.33	0.006	1.030	0.163

Figure 3. Model summary for regression. a. Predictors: (Constant), 22 I am an easily approachable and friendly person in the work team. *, 6 I tell people what I expect from them. *, 4 I organize my work by myself. *. 12 I ask that everyone follow standard rules and procedures. *, 2 I explain the role that everyone plays within the team. *, 1 I always find time to explain how a job needs to be done. *, 11 I plan the work I want done. *, 20 I inform in ad-vance of a change and explain to everyone the impact it will have. *, 5 I inform people about how they are work-ing. *, 7 I encourage the use of uniform procedures so that the whole group can follow them in detail. *, 8 I ex-press clearly what I think. *. 13 I make working on homework enjoyable. *, 3 I make the rules and procedures clear so that everyone can follow them precisely. *, 9 I assign specific tasks to each one. *, 21 I take care of the well-being of the people in my group. *, 10 I make sure everyone understands their role within the team. *, 15 I respect the opinions and feelings of others. *, 18 I am attentive to details that can make being a team member more enjoyable. *, 19 I treat everyone as equals. *, 14 I do what I can to help others. * 16 I am thoughtful and kind to everyone. *, 17 I keep the friendly atmosphere in the team. b. Dependent Variable: Grade of school.

Hypothesis 2 (H2). *There is no strong relationship between Team 19 and Team 16.*

Tables 7 and 8 shows the Contingency table and Chi-square test for the most positive correlation between items 19 and 16. Since the Chi-square value is 946.321, which is significant at alpha = 0.05, we reject the null hypothesis and conclude that there is a strong relationship between Team 19 and Team 16 (Figure 4).

Hypothesis 3 (H3). *There is no strong relationship between Team 9 and Team 15.*

Tables 9 and 10 shows the Contingency table and Chi-square test for the least positive correlation between items 9 and 15. As the Chi-square value is 46.587, which is significant at alpha = 0.05, we fail to reject the null hypothesis and conclude that there is no strong relationship between Team 9 and Team 15 (Figure 5).

Table 7. Contingency table for the most positively correlated variables.

		16 I Am Thoughtful and Kind to Everyone					Total
		Always	Frequently	Sometimes	Rarely	Never	
19 I treat everyone as equals. *	Always	546	88	5	2	2	643
	Frequently	32	44	3	1	0	80
	Sometimes	5	5	9	0	1	20
	Rarely	2	2	5	7	4	20
	Never	2	0	1	12	34	49
Total		587	139	23	22	41	812

* Correlation is significant at the 00.05 level (2-tailed).

Table 8. Chi-square test for the most positively correlated variables.

	Value	df	Asymp. Sig. (2-Sided)
Pearson Chi-Square	946.321 [a]	16	0.000
Likelihood Ratio	488.963	16	0.000
Linear-by-Linear Association	561.731	1	0.000
N of Valid Cases	812		

a. 14 cells (56.0%) have expected count less than 5. The minimum expected count is 54.

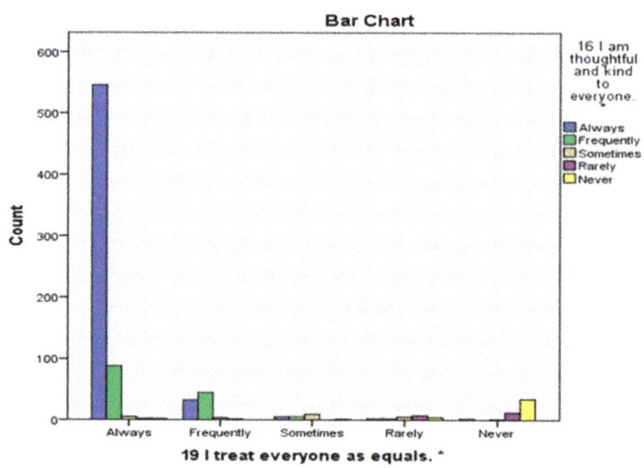

Figure 4. Relationship between items 19 and 6. * Correlation is significant at the 00.05 level (2-tailed).

Table 9. Contingency table for the least positively correlated variables.

		TEAM 15					Total
		Always	Frequently	Sometimes	Rarely	Never	
TEAM 9	Always	106	4	1	0	7	118
	Frequently	170	27	4	3	11	215
	Sometimes	180	27	12	5	13	237
	Rarely	108	12	0	10	5	135
	Never	90	4	1	3	9	107
Total		654	74	18	21	45	812

Table 10. Chi-square test for the least positively correlated variables.

	Value	df	Asymp. Sig. (2-Sided)
Pearson Chi-Square	46.587 [a]	16	0.000
Likelihood Ratio	48.781	16	0.000
Linear-by-Linear	1.992	1	0.000
Association			
N of Valid Cases	812		

[a] 7 cells (28.0%) have expected count less than 5. The minimum expected count is 2.37.

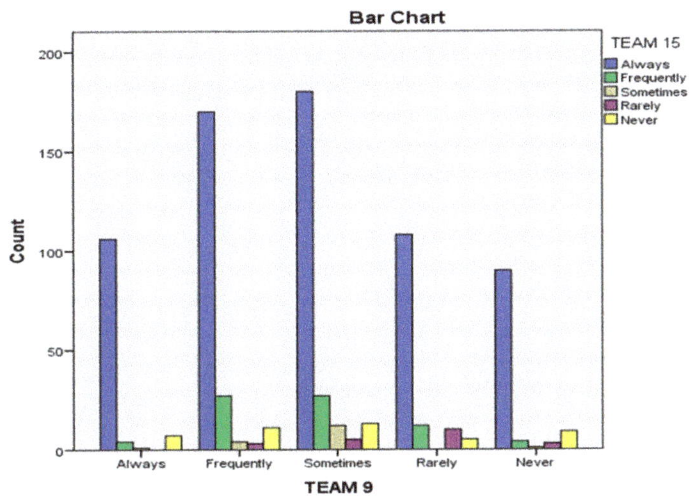

Figure 5. Relationship between items 15 and 9.

5. Discussion and Conclusions

The most frequent age was 41, with the least frequent being 23 and 60. Moreover, the most frequent gender was females, constituting 84.7% of the total respondents, while males constituted 15.3%. Based on the grade of school, 37.9% of the respondents were secondary school, second-degree holders, followed by primary at 28.6%, secondary at 23.1%, and kindergarten constituting 10.3%.

Based on the teams, Teams 1, 2, 3, 4, 5, 6, 7, 10, 11 and 13 have a high percentage of the respondents who *"Frequently"* agreed to the facts mentioned. Nevertheless, Team 9 and 12 had highest percentages of the respondents who *"Sometimes"* agreed to the facts. Lastly, a large number of teams responded as *"Always"*, namely Team 8, 14, 15, 16, 17, 18, 19, 20, 21 and 22.

From an analysis of the correlational data, it is clear that significant results are less evident in the first part of the test items 1–9; the results are much more accentuated when the second part of the test items from 10–22 are considered, i.e., planning, guiding and organising skills are developed, but the management of other participants develops a lot during the creation and sharing phases of DST.

Furthermore, there is a strong relationship between "I am thoughtful and kind to everyone" and "I treat everyone as equals". On the other hand, there is a weak relationship between Team 15 and 9, between "I assign specific tasks to each one" and "I do what I can to help others". In general, the correlations between the three scales bring out high values on the three main elements: encouraging to follow the working procedures in detail and standardising. The focus is still on the division of roles and making the task pleasant. In the negotiation scale, a strong correlation emerged between items 4 and 5, and 7 and 8, which implies that there is a need to actively listen to one's own opinions and of others, and the need to search for a compromising solution. In this scale, an unexpected result also emerged (items 7 and 8), i.e., a very strong spirit of competition in achieving the work objectives. The scores show the reports on negotiation are still strong in the comments on the works proposed by the individuals and the search for a common agreement in the event of disagreement (points 12 and 20) on the final work to be proposed. The Melbourne scale of decision-making offers weaker outcomes [38]; the most significant data are concerned with the clarification of the objectives before choosing (item 18), acceptance of the group decision (21, 27), pessimism about one's personal decision and delegation of responsibility to the group (22, 26, 29). Furthermore, the relationship between the three scales is very strong to highlight that DST has a strong effect on the three constructs investigated.

The intervention, therefore, allows participation and involvement in group activities to emerge, and it is interesting that all the elements for the construction of a group spirit and mentality emerge in this training intervention, like common work objectives, division of the role and pleasure of the task [44,45]. External competition, internal conflict resolution, and finally, the group decision prevails as the final outcome of the triggered process. Creating teaching videos, which the students will produce, has a high educational potential and is a challenge that can motivate students. There is little evidence on the efficacy of the use of this method when applied to creating digital stories. Therefore, the aim of this study was to measure student satisfaction with the creation of audiovisual material through DST, measure its usefulness, and evaluate its impact on students' motivation to study the subject. As a secondary objective, we set out to determine the influence of this learning experience on society's awareness of mental illness by measuring the size of its impact based on the number of views it received on social networking sites.

The goal of participation, group climate, and the resolution of internal conflicts for an external goal and competition was fully achieved in the treatment. Achieving the goal of delivering the task and passing the exam with excellent grades (100% of the test delivery with an average grade of 29.9 out of 30). The effects on group decision-making are still not very visible; it is probably necessary to prolong the activities involving the participants. Future research developments should include testing this treatment and monitoring the model in school and with adolescents [46,47].

The inclusive dimension is certainly another important facet of these tools, and the inclusive dimension can be assessed on at least two levels: the first is the plurality of languages, which, therefore, means cognitive styles and functional abilities of the subjects who can perform image research, such as photographs and oral narration. The other dimension is that of the group where the narratives can be produced by the individual or in small groups, where each member helps the other and amalgamate with the different skills and functional availability to produce a story that is rich and inclusive [48]. There are devices that allow people to work together on a project through a high level of motivation, and if the ingredients are well blended and the process goes well, you see that after some resistance or excessive enthusiasm, the group usually finds a deal [49,50]. A type of harmony is found among the participants and that experience of flow is generated, which is called Psychology of Learning [51]. When the intentions of the subjects are aligned on a task, it can be observed that the components, which manipulate, assemble and search for images move towards a single direction in a collaborative and harmonious way.

Author Contributions: Conceptualization, P.L.; methodology, G.A.T.; software, G.A.T.; validation, and formal analysis G.A.T. and B.C. All authors have read and agreed to the published version of the manuscript.

Funding: This research received no external funding.

Conflicts of Interest: The authors declare no conflict of interest.

References

1. Zheng, J.; Yang, J.; Zhang, H.; Wang, L.; Sun, Z.; Xue, H. Application of digital storytelling in comprehensive practice activity curriculums. In *International Conference on Information Computing and Applications*; Springer: Berlin/Heidelberg, Germany, 2011; pp. 406–414.
2. Kidd, J. Digital Storytelling at the BBC: The reality of innovative audience participation. In Proceedings of the RIPE@ 2006 Conference, Amsterdam, The Netherlands, 2–6 October 2006.
3. Moradi, H.; Chen, H. Digital Storytelling in Language Education. *Behav. Sci.* **2019**, *9*, 147. [CrossRef]
4. Matei, S.A.; Hunter, L. Data storytelling is not storytelling with data: A framework for storytelling in science communication and data journalism. *Inf. Soc.* **2021**, *37*, 312–322. [CrossRef]
5. Weder, F.; Lemke, S.; Tungarat, A. (Re)storying Sustainability: The Use of Story Cubes in Narrative Inquiries to Understand Individual Perceptions of Sustainability. *Sustainability* **2019**, *11*, 5264. [CrossRef]
6. Kallinikou, E.; Nicolaidou, I. Digital Storytelling to Enhance Adults' Speaking Skills in Learning Foreign Languages: A Case Study. Multimodal Technol. *Interact* **2019**, *3*, 59. [CrossRef]
7. Berendsen, M.E.; Hamerlinck, J.D.; Webster, G.R. Digital Story Mapping to Advance Educational Atlas Design and Enable Student Engagement. *ISPRS Int. J. Geo-Inf.* **2018**, *7*, 125. [CrossRef]
8. Tong, C.; Roberts, R.; Borgo, R.; Walton, S.; Laramee, R.S.; Wegba, K.; Lu, A.; Wang, Y.; Qu, H.; Luo, Q.; et al. Storytelling and Visualization: An Extended Survey. *Information* **2018**, *9*, 65. [CrossRef]
9. Dunn, T.M.; Cherup, S. Storytelling and Success: How Storytelling Can be Used to Promote Diversity, Cross-Cultural Understanding, and Confidence in the Classroom. *J. Educ.* **2021**, 00220574211032309. [CrossRef]
10. Rodríguez-Illera, J.L.; Barberà, E.; Martínez, A. Digital storytelling in the ecology of learning (Relatos digitales personales en la ecología del aprendizaje). *Cult. Y Educ.* **2020**, *32*, 390–398. [CrossRef]
11. Rodríguez-Illera, J.L.; Martínez-Olmo, F.; Rubio-Hurtado, M.J.; Galván, C. The content posting practices of young people on social networks. *Pixel-Bit. Rev. Medios Educ.* **2021**, *60*, 135–151.
12. Rodríguez-Illera, J.L.; Barberà, E.; Molas, N. Reasons and mediators in the development and communication of personal digital stories. In *Education and Information Technologies*; Springer Nature: London, UK, 2021; Volume 26, pp. 4093–4109. ISSN 1360-2357.
13. Rahiem, M.D.H. Storytelling in early childhood education: Time to go digital. *Int. J. Child Care Educ. Policy* **2021**, *15*, 4. [CrossRef]
14. de Maurissens, I. Digital Storytelling: Una Narrazione Digitale, una Documentazione Visuale. Digital Storytelling: Creatività e Tecnologia. 2007. Available online: http://eprints.rclis.org/9660/1/Digital_Storytelling-indire_version_3.pdf (accessed on 31 August 2021).
15. Arnedo-Moreno, J.; Garcia-Font, V. A study on the design and application of fictional storytelling in online learning of computer security. *Appl. Sci.* **2021**, *11*, 6185. [CrossRef]
16. Lazar, I.; Panisoara, G.; Panisoara, I.O. Adoption of Digital Storytelling Tool in Natural Sciences and Technology Education by Pre-Service Teachers Using the Technology Acceptance Model. *J. Balt. Sci. Educ.* **2020**, *19*, 429–453. [CrossRef]
17. Spanos, M. Brand storytelling in the age of artificial intelligence. *J. Brand Strategy* **2021**, *10*, 6–13.
18. Cicolini, M.C. Storytelling and Web 2.0: Transforming a Traditional Technique in Order to Achieve 21st Century Learning. Actualización En La Enseñanza Del Inglés. 2013. Available online: http://www.ffha.unsj.edu.ar/wp-content/uploads/2015/11/Actas-de-las-V-Jornadas-de-capacitaci%C3%B3n-en-la-Ense%C3%B1anza-del-Ingl%C3%A9s-2013.pdf#page=23 (accessed on 31 August 2021).
19. Benabdelkader, A. Expert–Novice Negotiation within Learning Opportunities in Online Intercultural Interactions. *Languages* **2019**, *4*, 14. [CrossRef]
20. Ammar, A.; Chtourou, H.; Boukhris, O.; Trabelsi, K.; Masmoudi, L.; Brach, M. ECLB-COVID19 Consortium. COVID-19 home confinement negatively impacts social participation and life satisfaction: A worldwide multicenter study. *Int. J. Environ. Res. Public Health* **2020**, *17*, 6237. [CrossRef] [PubMed]
21. Shakun, M.F.; Martinovski, B. Advances in Defining a Right Problem in Group Decision and Negotiation. In *Handbook of Group Decision and Negotiation*; Springer: Berlin/Heidelberg, Germany, 2021; pp. 137–156.
22. Martinovsky, B. *Emotion in Group Decision and Negotiation*; Series of Advances in Group Decision and Negotiation; Springer: Dordrecht, The Netherlands, 2015.
23. Gaber, S.N.; Nygård, L.; Brorsson, A.; Kottorp, A.; Charlesworth, G.; Wallcook, S.; Malinowsky, C. Social participation in relation to technology use and social deprivation: A mixed methods study among older people with and without dementia. *Int. J. Environ. Res. Public Health* **2020**, *17*, 4022. [CrossRef] [PubMed]
24. Rodríguez-Almagro, J.; Prado-Laguna, M.D.C.; Hernández-Martínez, A.; Monzón-Ferrer, A.; Muñoz-Camargo, J.C.; Martín-Lopez, M. The Impact on Nursing Students of Creating Audiovisual Material through Digital Storytelling as a Teaching Method. *Int. J. Environ. Res. Public Health* **2021**, *18*, 694. [CrossRef] [PubMed]
25. Nair, V.; Yunus, M.M. A Systematic Review of Digital Storytelling in Improving Speaking Skills. *Sustainability* **2021**, *13*, 9829. [CrossRef]
26. Garrety, C.; Schmidt, D. The evolution of digital storytelling: From enhanced oral tradition to genres for education. In *Society for Information Technology & Teacher Education International Conference*; Association for the Advancement of Computing in Education (AACE): Chesapeake, VA, USA, 2008; pp. 916–921.
27. Lambert, J. *Digital Storytelling: Capturing Lives, Creating Community*; Routledge: Abingdon, UK, 2013.

28. Fulwiler, M.; Middleton, K. After digital storytelling: Video composing in the new media age. *Comput. Compos.* **2012**, *29*, 39–50. [CrossRef]
29. Amelia, P.; Rukmini, D.; Mujiyanto, J.; Bharati, D.A.L. Investigating the development of teachers' tpack and the adoption of digital storytelling: A case study of teaching english in elementary school. *J. Asia TEFL* **2021**, *18*, 701–710.
30. Liontas, J.I.; Mannion, P. Voices Heard, Voices Seen: From Classroom Praxis to Digital Stories Worth Sharing. *J. Brand Strategy* **2021**, *10*, 6–13.
31. Bloomfield, E.F.; Manktelow, C. Climate communication and storytelling. *Clim. Chang.* **2021**, *167*, 34. [CrossRef]
32. Alcalde-Peñalver, E.; Santamaría-Urbieta, A. Digital storytelling in ESP: Towards a new literacy in hybrid language learning [El storytelling digital en el aula de inglés para fines específicos: ¿una nueva alfabetización en la formación híbrida en lenguas?]. *Aula Abierta* **2021**, *50*, 567–576. [CrossRef]
33. Devine, K. On country: Identity, place and digital place. *Virtual Creat.* **2021**, *11*, 111–123. [CrossRef]
34. Ohler, J. The world of digital storytelling. *Educ. Leadersh.* **2006**, *63*, 44–47.
35. Mann, L.; Burnett, P.; Radford, M.; Ford, S. The Melbourne Decision Making Questionnaire: An instrument for measuring patterns for coping with decisional conflict. *J. Behav. Decis. Mak.* **1997**, *10*, 1–19. [CrossRef]
36. Cataudella, S.; Carta, S.M.; Mascia, M.L.; Masala, C.; Petretto, D.R.; Agus, M.; Penna, M.P. Teaching in Times of the COVID-19 Pandemic: A Pilot Study on Teachers' Self-Esteem and Self-Efficacy in an Italian Sample. *Int. J. Environ. Res. Public Health* **2021**, *18*, 8211. [CrossRef] [PubMed]
37. Friesen, K.A.; Weiss, J.A.; Howe, S.J.; Kerns, C.M.; McMorris, C.A. Mental Health and Resilient Coping in Caregivers of Autistic Individuals during the COVID-19 Pandemic: Findings from the Families Facing COVID Study. *J. Autism Dev. Disord.* **2021**, 1–11. [CrossRef]
38. Limone, P.; Toto, G.A. Psychological and emotional effects of Digital Technology on Children in Covid-19 Pandemic. *Brain Sci.* **2021**, *11*, 1126. [CrossRef]
39. Fisher, R.; Ury, W. *Getting to Yes: Negotiating Agreement Without Giving in*, 2nd ed.; Houghton Mifflin: New York, NY, USA, 1981.
40. Zhang, S.; Constantinovits, M. Development of a Conceptual Model and Questionnaire of Principled Negotiation. *Bus. Commun. Res. Pract.* **2018**, *1*, 70–77. [CrossRef]
41. Bland, J.M.; Altman, D.G. Statistics notes: Cronbach's alpha. *BMJ* **1997**, *314*, 572. [CrossRef]
42. Slocum, J.W.; Hellriegel, D.; Ballabio, P.; Paradiso, L. Il Comportamento Organizzativo: Fondamenti Psicologici per L'azione Manageriale. Hoepli. 2010. Available online: https://www.libreriauniversitaria.it/comportamento-organizzativo-fondamenti-psicologici-azione/libro/9788820340858 (accessed on 31 August 2021).
43. Frith, J.; Richter, J. Building participatory counternarratives: Pedagogical interventions through digital placemaking. *Convergence* **2021**, *27*, 696–710. [CrossRef]
44. Parsazadeh, N.; Cheng, P.-Y.; Wu, T.-T.; Huang, Y. Integrating Computational Thinking Concept Into Digital Storytelling to Improve Learners' Motivation and Performance. *J. Educ. Comput. Res.* **2021**, *59*, 470–495. [CrossRef]
45. Lopez-Serrano, J.; Díaz-Bóveda, R.; González-Vallespí, L.; Santamarina-Pérez, P.; Bretones-Rodríguez, A.; Calvo, R.; Sara, L. Psychological impact during COVID-19 lockdown in children and adolescents with previous mental health disorders. *Rev. Psiquiatría* **2021**, in press. [CrossRef]
46. Parsons, S.; Ivil, K.; Kovshoff, H.; Karakosta, E. 'Seeing is believing': Exploring the perspectives of young autistic children through Digital Stories. *J. Early Child. Res.* **2021**, *19*, 161–178. [CrossRef]
47. Ofoegbu, T.O.; Asogwa, U.D.; Eseadi, C.; Ogbonna, C.S.; Eskay, M.; Obiyo, N.O.; Nji, G.C.; Ngwoke, O.R.; Agboti, C.I.; Uwakwe, R.C.; et al. Effect of Rational Digital Storytelling Intervention on Depression Among Adolescent-Athletes with Special Educational Needs. *J. Rational-Emot. Cognitive-Behav. Ther.* **2021**, *39*, 217–237. [CrossRef]
48. Okmeydan, S.B. New stories and digital storytelling for new consumers. *Paradig. Shifts Commun. World* **2021**, 205–218.
49. Malmberg, J.; Fincham, O.; Pijeira-Díaz, H.J.; Järvelä, S.; Gašević, D. Revealing the hidden structure of physiological states during metacognitive monitoring in collaborative learning. *J. Comput. Assist. Learn.* **2021**, *37*, 861–874. [CrossRef]
50. Malureanu, A.; Panisoara, G.; Lazar, I. The Relationship between Self-Confidence, Self-Efficacy, Grit, Usefulness, and Ease of Use of eLearning Platforms in Corporate Training during the COVID-19 Pandemic. *Sustainability* **2021**, *13*, 6633. [CrossRef]
51. Chopra, B.; Verma, K.; Singhal, S.; Singla, U. Reality Tales: Facilitating User-Character Interaction with Immersive Storytelling. In Proceedings of the 2021 CHI Conference on Human Factors in Computing Systems, Yokohama, Japan, 8–13 May 2021.

electronics

MDPI

Article

Self-Assessment of Soft Skills of University Teachers from Countries with a Low Level of Digital Competence

Álvaro Antón-Sancho [1], Diego Vergara [2,*] and Pablo Fernández-Arias [2]

1 Department of Mathematics and Experimental Science, Catholic University of Ávila, C/Canteros, s/n, 05005 Avila, Spain; alvaro.anton@ucavila.es
2 Department of Mechanical Engineering, Catholic University of Ávila, C/Canteros, s/n, 05005 Avila, Spain; pablo.fernandezarias@ucavila.es
* Correspondence: diego.vergara@ucavila.es

Abstract: The lockdown of March and April 2020 as a consequence of the COVID-19 pandemic has forced relevant changes in the educational environment in a very short period of time, making it necessary to suspend in-person instruction and generating the need to implement virtual learning mechanisms. In a future post-COVID-19 hybrid educational model, it will be necessary for university teachers to acquire an optimal degree of digital competence, as a combination of different competencies, namely, (i) technical, (ii) digital, and (iii) soft. Soft skills have been shown to have a decisive influence on the development of digital competence. The aim of this study was to analyze the degree of acquisition of soft skills in Latin American university teachers whose countries are less digitally developed. For this purpose, the countries with the lowest Global Innovation Index (GII) were selected: (i) Panama; (ii) Peru; (iii) Argentina; (iv) El Salvador; (v) Ecuador; (vi) Paraguay; (vii) Honduras; and (viii) Bolivia. To achieve this objective, it was necessary to develop a questionnaire on the self-concept of soft skills, based on the soft skills included in the Bochum Inventory of Personality and Competences (BIP). Results obtained from statistical analysis of the data collected from a sample of 219 participants show that university teachers are sufficiently prepared, in terms of their soft skills, for the increase in digital competence required as a result of the COVID-19 crisis, despite the low level of digital development in their respective countries.

Keywords: digital competence; soft skills; university teachers

Citation: Antón-Sancho, Á.; Vergara, D.; Fernández-Arias, P. Self-Assessment of Soft Skills of University Teachers from Countries with a Low Level of Digital Competence. *Electronics* **2021**, *10*, 2532. https://doi.org/10.3390/electronics10202532

Academic Editor: Christos J. Bouras

Received: 26 September 2021
Accepted: 13 October 2021
Published: 17 October 2021

1. Introduction

The global crisis initiated in 2020 due to the COVID-19 pandemic has resulted in structural changes in many dimensions of life in all societies around the world. In particular, it has forced significant adjustments in the work processes of companies and institutions. In the specific field of higher education, in a very short period of time, it has been necessary to suspend in-person instruction, and remote learning has been adopted to reduce the spread of the COVID-19 disease. In the university education system, there is currently a need to implement virtual classroom delivery mechanisms and to introduce technological tools that mediate all teaching and learning processes. The consequence of these factors is that university teachers have to develop skills that are appropriate for the emergence of these new technologies in their teaching work [1].

The specialized literature has portrayed these skills as 21st century skills [2]. These new capabilities comprise specific skills—both technical and digital—and soft skills [3], and this combination of skills has also been referred to as digital competence [4]. Specific skills refer to measurable and job-specific abilities. Soft skills, however, encompass transversal skills, such as those that affect the worker's communicative capacity, critical thinking, collaborative and social skills, and problem-solving abilities. In turn, the concept of soft suggests the opposite of hard [5]. Hard skills are those that are developed at a specific level in a given profession, and are composed of particular and eminently technical or methodological contents, such as knowing the structure of a didactic unit, knowing how to operate a machine, or knowing the side effects of a medication [6]. By comparison, soft skills are not specific to a particular family of jobs, but are transversal to all of them. Industry 4.0 requirements are transferred to the sphere of professional competence profiles while reflecting the need for combining technical and digital skills with soft skills for the 21st century [7]. The professional competences for the 21st century include, but are not limited to, economic competences, project management, and digital skills [8].

For all these reasons, soft skills are essential for university teachers to reach an optimal level of digital competence, which is necessary for the development of any training action in a virtual learning environment. The European Commission supports this idea and links, in its Digital Education Action Plan [9], soft skills with digital skills, stating the need for adequate training at the Higher Level to achieve a solid development level of soft skills and, therefore, an adequate level of digital competence. In fact, soft skills, being transversal competencies, nevertheless have an intimate relationship with digital competencies, because they enable the development and transfer of digital competencies, particularly in the long term. Thus, previous research has highlighted a combined digital–soft competency [10].

Specifically, soft skills are personal and interpersonal competencies linked to the character and personality of individuals [11,12]. As a result, these skills are cross-cutting and transferable between the different areas of learning and the academic and working life of individuals. It is possible to identify soft skills linked to aspects of the subject's personality (such as communication and teamwork skills, motivation, or leadership) and others associated with the intellectual dimension of the person (such as critical thinking or analytical reasoning) [13]. These qualities are required by employers, particularly for the development of technical skills [14], and have a decisive influence not only on professional development, but also on education and training, social relations, and health [15]. Studies show that the most frequent perception of employers is that university graduates have not sufficiently developed these skills to the level required in employment [16,17]. Similarly, university graduates from different areas of knowledge state that they are generally dissatisfied with their training in the aforementioned soft skills, which is a hindrance to their insertion in a technical and digitalized world [18].

Faced with a future hybrid post-COVID-19 educational model and the existing gap between the skills demanded by employers and those of future professionals, this study examined the self-concept of soft skills expressed by a group of university teachers from all areas of knowledge who work in countries with a low level of development, innovation, and digitalization (according to the Global Innovation Index (GII) [19]). University teachers were the focus because they are professionals who are required to have a high level of digital competence (especially since the emergence of COVID-19 and the consequent transition to digital educational environments), and because there is an urgent need to train future professionals who are required to immerse themselves in an increasingly technical and digitized labor market, in which soft skills play a crucial role [20,21]. The main objective was to assess the perception of university teachers in these countries about their soft skills despite the scarce digital development of their environment. This will make it possible to estimate whether, among university teachers in these countries, there is adequate training availability in soft skills that will enable the generation of adequate digital training in the immediate future or whether, on the contrary, it is necessary to work on the training of soft skills prior to the development of digital technologies in the field of higher education. For this purpose, the Bochum Inventory of Personality and Competences (BIP) [22] was used as a reference for the development of the self-assessment questionnaire of soft skills in these university teachers.

The Global Innovation Index (GII) [19] was used to determine the countries with the lowest economic, technological, and digital development. The GII analyzes 130 countries from the perspective of the development of their economies and their innovation, with technical and digital development among its key criteria [23,24]. Latin American countries were chosen because this geographical area, in addition to Africa, comprises the largest concentration and number of countries with low GII (compared to the most developed countries). In Southeast Asia, for example, there are only four countries whose GII is in the range of the countries analyzed in this paper (Indonesia, Cambodia, Laos, and Myanmar) [19]. Furthermore, the case of Africa is very particular because, first, a significant proportion of its countries are not included in the GII. Moreover, it is a region that is heavily dependent on the outside world, with very little technical development. Furthermore, the data measured by the GII do not make it possible to ensure that the innovation index accurately represents the degree of technical development and digital competence of African countries [19]. In Latin America, by comparison, the GII better represents its level of scientific, technical, and digital development.

Examination of the GII scores of the Latin American countries analyzed by the aforementioned index in 2021 [19] (the indexes of the United States and Canada are much higher), shows that the highest index is 33.9 (Chile) and the lowest is 22.4 (Guatemala). The average GII in Latin American countries is 27.91, with a standard error of the mean equal to 1.05. In order to identify the countries that were taken as a reference for this study, i.e., those with a low GII, the mean plus the standard error of the mean, which is 28.96, was taken as the maximum index. Consequently, the self-concept of soft skills was analyzed in a group of Latin American university teachers from countries with a GII less than or equal to this value (Table 1). Panama was also included (with a GII equal to 29) because its index is almost identical to the upper limit that was defined.

Table 1. Number of participants by country, ordered by GII.

Country	GII	Participants	Percentage of Sample
Panama	29.0	2	0.91%
Peru	28.8	96	43.84%
Argentina	28.3	68	31.05%
El Salvador	24.8	10	4.57%
Ecuador	24.1	35	15.98%
Paraguay	24.1	2	0.91%
Honduras	23.0	3	1.37%
Bolivia	22.4	3	1.37%

According to the Gini Index, which is used by the World Bank to evaluate inequality in each country, using a range from 0% (minimum inequality) to 100% (maximum inequality), many of the countries with low GII have high levels of inequality (e.g., 49.8 for Panama, 45.1 for Peru, 42.9 for Argentina, 38.8 for El Salvador, 45.7 for Ecuador and Paraguay). Bearing in mind that the universities are located in favored areas of the cities and their teachers are part of the most affluent sectors of the societies of the respective countries, this fact implies that the results obtained cannot be extrapolated to the entire societies of the geographical area involved, but only to the sector of university teachers. This social sector will, in fact, be more similar in its socio-economic aspects to the social bulk of more technologically advanced societies (e.g., some European countries such as Spain or Italy).

The results of the questionnaire were analyzed at a descriptive level in order to assess the participants' perception of their self-concept regarding their soft skills. An inferential analysis was also carried out to detect whether there are statistically significant differences in these self-concepts when differentiated by certain non-academic characteristics (such as gender or age) or academic characteristics (such as the area of knowledge, the length of teaching experience or the nature—private or public—of the university of the teacher).

2. Materials and Methods

In this work, quantitative research was carried out on the self-concept of soft skills of a group of university teachers from different areas of knowledge located in Latin American countries with a lower Global Innovation Index (GII $\leq$ 29). Consequently, the dependent variables studied were soft skills, which were grouped into the five subscales of skills, as explained. The study was descriptive and involved inferential statistical analysis, and had the following main objectives: (i) to explore the self-concept of university teachers in Latin American countries with lower GII about their soft skills; (ii) to analyze the differences that exist in the self-concepts of the aforementioned university teachers according to their gender, age, teaching experience, area of knowledge and the nature (private or public) of the institution where they carry out their work. These variables were examined in an attempt to describe the influence of certain sociological (gender or age) and academic (area of knowledge, teaching experience, or nature of the educational institution) aspects on the self-concept that the participants express about their soft skills and to identify, if any, the gaps induced in this self-concept by the above variables.

The study was carried out using a questionnaire designed by the Bochum Inventory of Personality and Competences (BIP) [22]. The questionnaire is made up of 19 questions, each of which requests the evaluation of the self-concept on one of the 19 soft skills under consideration. All the questions are Likert-type, from 1 to 5, with 1 corresponding to the lowest rating and 5 to the highest rating. The questions in the questionnaire are grouped into five families or scales, each corresponding to a family of soft skills: (i) Work motivation: results orientation, initiative for change, and leadership; (ii) Work behavior: conscientiousness, flexibility, and action orientation; (iii) Social skills: social intelligence, sociability, relationship development, teamwork, and influence; (iv) Psychic structure: emotional stability, work capacity, and self-confidence; and (v) Additional competences: sense of control, competitiveness, mobility, leisure orientation, and image distortion.

The questionnaire was answered freely, voluntarily, and anonymously by the participants. Table 2 shows the Cronbach's alpha parameters for each of the scales of the questionnaire, which were used to assess the internal consistency of the instrument. All the parameters were between 0.8 and 0.9, which allowed us to assume that the questionnaire was reliable and had good internal cohesion in each of the scales into which it was divided.

Table 2. Cronbach's alpha parameters for the different scales of the questionnaire.

Scale	Cronbach's Alpha
Work motivation	0.8566
Work behavior	0.8553
Social skills	0.8822
Psychic structure	0.8597
Additional competences	0.8017

The participants were selected through a conventional probability sampling process. In total, the sample consisted of 219 teachers from different universities from the eight Latin American countries included in the GII database and whose index is less than or equal to 29 (the mean plus the standard error of the Latin American countries' indexes). Specifically, the distribution of the eight participating countries is shown in Table 1.

The study used independent variables of two different natures: first, variables that affect the sociological profile of the participants. These are gender (dichotomous in nature) and age (grouped in 10 year ranges, from 25 to 74 years; therefore, it is polytomous in nature). The rest of the independent variables concern different dimensions of the academic activity of the participants (Figure 1).

Specifically, the area of knowledge (whose values could be Arts and Humanities, Social and Legal Sciences, Sciences, Health Sciences, or Engineering and Architecture), the years of university teaching experience (grouped in 5 year ranges, from less than or equal to 5 years to more than 25), and the nature of the university institution where the teacher works (private or public). Of these three variables, the first two are polytomous and the latter is dichotomous.

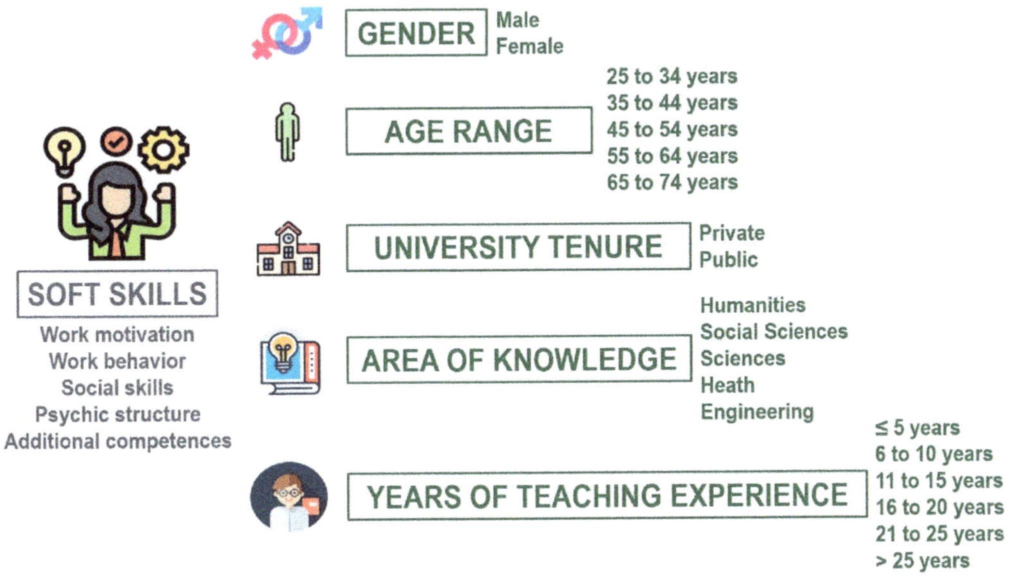

Figure 1. Scheme of the aspects analyzed in the statistical study.

For the inferential study, the Mann–Whitney test (for dichotomous variables) or the Kruskal–Wallis test (for polytomous variables) for comparison of means, and the Levene test for comparison of variances, were applied to compare results when the sample of participants was differentiated by each of the independent variables. The reason for choosing nonparametric tests in inferential analysis for the comparison of means, instead of parametric tests such as the *t*-test or ANOVA, is that the *p*-values of the normality tests applied to the different items of the questionnaire do not allow us to assume that the data follow a normal distribution. This analysis is original and innovative, because previous research was concerned with analyzing the need for the cultivation of soft skills in teachers in their training stage [25], rather than the analysis of these characteristics in active teachers. Figure 1 shows an outline of the methodology followed in this statistical study, and Figure 2 shows the distribution of participants according to each of the independent variables. Sector diagrams are used for nominal variables and histograms for continuous variables grouped by ranges.

In Figure 2a, it can be observed that there are notably more females than males (around 67% more). In terms of age ranges, the largest number of participants is in the middle range, from 45 to 54 years old, with the least number of participants at the extremes (from 25 to 34 or from 65 to 74 years old). It can be seen that the distribution by areas of knowledge tends to be more homogeneous, although with a certain superiority in the Engineering area (Figure 2c). With regard to teaching experience, the sample of participants with more than 25 years of experience is clearly in the majority, followed by those with between 21 and 25 and those with between 16 and 20 years of experience. The lowest frequency is found among participants with less than or equal to 5 years of experience. Finally, although there is a certain superiority in the number of teachers from public universities than from private universities, the difference is very small, in relative terms. In fact, the statistics of the chi-square test of goodness of fit with one degree of freedom (chi-square = 1.3196, *p*-value = 0.2507) allow us to assume, with a significance level of 0.05, that the participants are distributed homogeneously between teachers from private and public universities. For the rest of the independent variables, the chi-square goodness-of-fit test statistics yield *p*-values below 0.05, which does not allow the corresponding distributions of participants to be homogeneous within the sample.

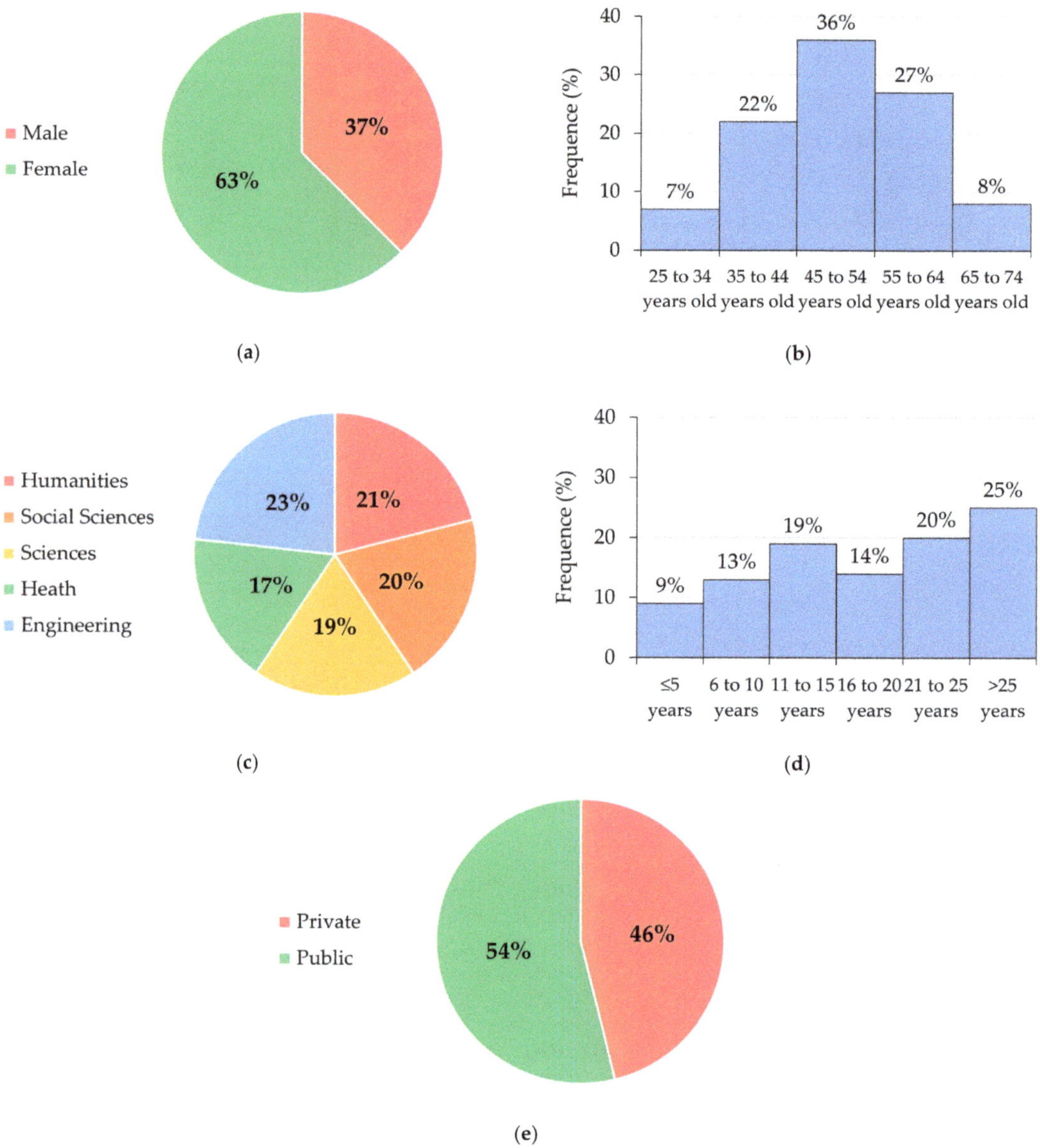

Figure 2. Distributions of the sample participants differentiated according to the different values of the independent variables considered: (**a**) gender; (**b**) age range; (**c**) area of knowledge; (**d**) years of teaching experience; (**e**) university tenure.

3. Results

3.1. Global Results

Table 3 shows the descriptive statistics of the overall answers to the different scales of the questionnaire. The mean answers show that the participants generally have good or very good self-concepts on all the soft skills scales. In fact, all the average answers are above 4, except for the additional skills scale, where the average falls slightly below 4. The scale of additional skills has the greatest dispersion in the answers because it has the greatest standard deviation. It also has the highest coefficient of variation. Nevertheless, it can be assumed that the answers are homogeneously distributed around the mean in all the scales because the coefficients of variation are all less than 30%. It is interesting to note that the work behavior and psychic structure scales present the highest mean answers (4.42 and 4.44, respectively) with the lowest data dispersions (the standard deviations are 0.70 and 0.73, respectively).

Table 3. Overall descriptive statistics of the questionnaire by scales.

	Mean Value	Standard Deviation	Coefficient of Variation	Skewness
Work motivation	4.19	0.85	20.31%	−1.16
Work behavior	4.42	0.70	15.85%	−1.37
Social skills	4.20	0.81	19.30%	−1.11
Psychic structure	4.44	0.73	16.55%	−1.47
Additional skills	3.76	1.04	27.77%	−0.84

Regarding skewness, measured through Pearson's skewness coefficient, all the scales present high negative skewness (or moderate, in the case of the additional skills scale), with moderate skewness being understood as that in which the Pearson coefficient ranges between −0.5 and −1, and high skewness as that in which it is less than −1. This shows a notable polarization of the answers towards values 4 and 5. In fact, the mode is 4 in all the scales. The fact that the mean answer drops to 3.76 in the case of additional skills is due to the fact that the data dispersion is greater and the frequency of answers 1, 2, and 3 is higher, as indicated by the fact that the skewness is closer to 0.

Figure 3 shows the bar charts of the overall relative frequencies of the different scales of the questionnaire. It can be seen that, in the scales of work motivation, social skills, and additional skills, the modal value is 4, and its frequency exceeds 40% in all cases. The second most frequent answer is 5 in the three scales mentioned, and its frequency exceeds 40% in work motivation, is slightly below 40% in social skills, and drops to almost 25% in additional skills. Nonetheless, between the two values corresponding to high self-concepts in these scales, more than 80% of the participants in the work motivation and social skills scales and slightly more than 67% in additional skills are grouped together.

3.2. Gender

When differentiated by gender, the answers show high or very high self-concepts in the scales of motivation and work behavior, social skills, and psychic structure, with moderate or high negative skewness of the distribution (especially in work behavior and psychic structure), and high self-concepts in additional skills, with moderate negative skewness (therefore, with a strong concentration of the answers around values 4 and 5). The modal values are maintained. It should be noted, however, that skewness is higher, in absolute value, in males than in females (Table 4), indicating that in males the concentration of answers around the high and very high self-concepts on all scales is stronger.

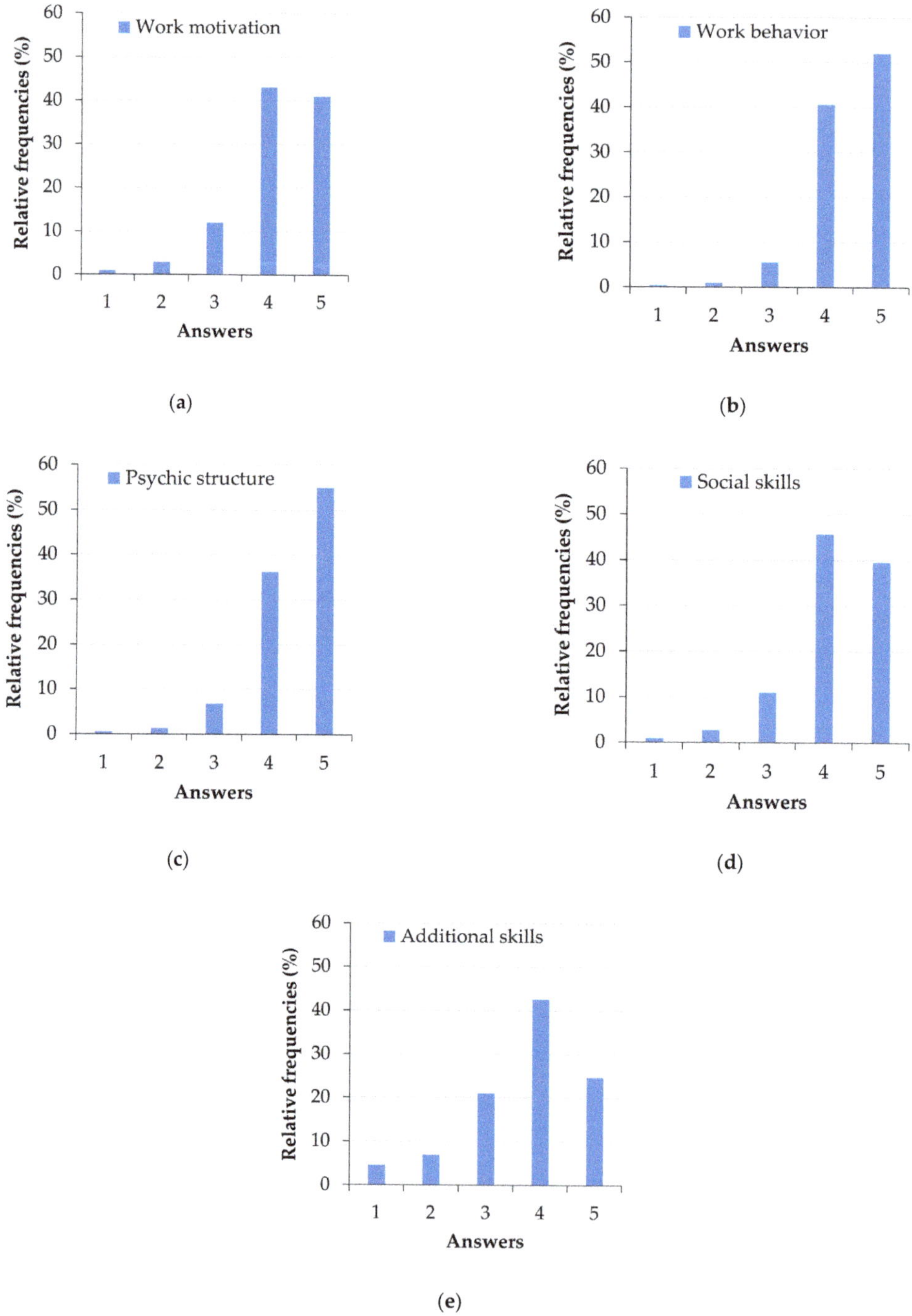

Figure 3. Bar graphs of the relative frequencies of the possible answers for the different scales: (**a**) work motivation; (**b**) work behavior; (**c**) social skills; (**d**) psychic structure; (**e**) additional skills.

Table 4. Means, standard deviations (St.D.), coefficients of variation (C.V.) and skewness (Skew.) of the overall answers of the different scales of the questionnaire when the sample is differentiated by gender.

	Males				Females			
Scale	**Mean**	**St.D.**	**C.V.**	**Skew.**	**Mean**	**St.D.**	**C.V.**	**Skew.**
Work motivation	4.15	0.95	23.00%	−1.27	4.22	0.78	18.59%	−0.99
Work behavior	4.34	0.78	17.97%	−1.61	4.48	0.64	14.41%	−1.01
Social skills	4.16	0.86	20.62%	−1.40	4.23	0.78	18.49%	−0.86
Psychic structure	4.41	0.70	15.84%	−1.76	4.45	0.70	15.84%	−1.22
Additional skills	3.82	1.06	27.88%	−0.93	3.73	1.03	27.68%	−0.80

Nevertheless, there are some significant differences between genders in terms of the distributions of the answers. First, as already noted in the description of the results, males have slightly lower mean self-concepts than females on all scales except for additional skills (Table 4). The Mann–Whitney test statistics in Table 5 show that only a statistically significant difference in the mean values between males and females can be assumed for the work behavior scale because the corresponding p-value is less than 0.05.

Table 5. Mann–Whitney test statistics when the sample is differentiated by gender.

Scale	Mann–Whitney W-Value	p-Value
Work motivation	50,102	0.8358
Work behavior	46,158	0.0359 *
Social skills	87,425	0.4617
Psychic structure	50,052	0.8105
Additional skills	148,893	0.0776

* $p < 0.05$.

In terms of deviations, males have a higher dispersion of answers than females on all scales (Table 4). However, from Levene's test with one degree of freedom, whose statistics are shown in Table 6, it follows that the differences can only be taken as significant in the work motivation scale. From this it can be assumed that the self-concept that females present in terms of work motivation is more solidly formed than that of males.

Table 6. Levene test statistics when the sample is differentiated by gender.

Scale	Levene F-Value	p-Value
Work motivation	4.8796	0.0275 *
Work behavior	1.9908	0.1587
Social skills	0.1531	0.6957
Psychic structure	0.4084	0.5230
Additional skills	0.0001	0.9914

* $p < 0.05$.

3.3. *Age Range*

When differentiated by age range, the answers of the different scales of the questionnaire result in the statistics shown in Tables 7 and 8. From the data in Table 7, it follows that the lowest mean answers and the greatest dispersions are obtained in the additional skills scale for all age ranges. The oldest participants are those who give the lowest mean answers (except for the psychic structure scale, where they tie with the youngest participants), and they are also those with the highest standard deviation. Consequently, it is these participants who have the lowest self-concept of soft skills and, at the same time, the least consolidated self-concept.

Table 7. Means and standard deviations (St.D.) of the overall answers of the different scales of the questionnaire when the sample is differentiated by age range.

	25 to 34		35 to 44		45 to 54		55 to 64		65 to 74	
Scale	**Mean**	**St.D.**	**Mean**	**St.D.**	**Mean**	**St.D.**	**Mean**	**St.D.**	**Mean**	**St.D.**
Work motivation	4.04	0.56	4.18	0.86	4.21	0.89	4.28	0.77	3.96	1.05
Work behavior	4.57	0.55	4.44	0.63	4.41	0.76	4.48	0.67	4.17	0.80
Social skills	4.11	0.82	4.24	0.81	4.20	0.83	4.27	0.70	3.96	1.00
Psychic structure	4.36	0.76	4.54	0.63	4.41	0.78	4.42	0.72	4.37	0.78
Additional skills	3.74	0.97	3.80	1.08	3.84	1.00	3.67	1.06	3.63	1.10

Table 8. Coefficients of variation (C.V., measured in percent) and skewness (Skew.) of the overall answers of the different scales of the questionnaire when the sample is differentiated by age range.

	25 to 34		35 to 44		45 to 54		55 to 64		65 to 74	
Scale	**C.V.**	**Skew.**	**C.V.**	**Skew.**	**C.V.**	**Skew.**	**C.V.**	**Skew.**	**C.V.**	**Skew.**
Work motivation	13.72	0.04	20.53	−1.27	21.27	−1.35	17.94	−0.84	26.38	−0.85
Work behavior	11.97	−0.76	14.23	−0.67	17.17	−1.79	14.88	−1.27	19.08	−0.78
Social skills	20.07	−0.81	19.09	−0.82	19.75	−1.27	16.49	−0.88	25.24	−1.22
Psychic structure	17.43	−1.07	13.93	−1.24	17.80	−1.54	16.27	−1.67	17.94	−1.01
Additional skills	26.01	−0.82	28.55	−0.93	26.08	−0.83	28.99	−0.81	30.16	−0.74

Regarding skewness, as expressed in Table 8, in the work motivation scale, the youngest participants show a slight positive asymmetry (similar to a symmetrical distribution), whereas the remainder show moderate or high negative asymmetry (especially those between 35 and 54 years old). In work behavior, participants younger than 45 years or older than 64 years present moderate negative asymmetries, whereas participants between 45 and 64 years present high negative asymmetry. In social skills, those under 45 years old and those between 55 and 64 years old present moderate negative asymmetry, whereas those between 45 and 54 or older than 64 present high negative asymmetry. In psychic structure, all age ranges present high negative asymmetry, with the highest corresponding to participants between 45 and 64 years old. In additional skills, all age ranges show moderate negative asymmetry, with the smallest asymmetry corresponding to the oldest participants.

Regarding the mean values when differentiating the participants by age, the lowest mean self-concepts are presented by the participants of extreme ages (i.e., the youngest and the oldest), except in work behavior, where the youngest participants show the highest self-concept, and in additional skills, where those older than 54 years show a slightly lower self-concept than the youngest participants. However, the Kruskal–Wallis mean comparison test statistics with four degrees of freedom expressed in Table 9 show that, with significance level 0.05, the differences in the mean answers between the different age ranges cannot be assumed to be statistically significant. The p-value of the work behavior scale is, however, very close to the significance level, so that a certain significant gap by age range can be assumed in this scale.

Table 9. Kruskal–Wallis test statistics when the sample is differentiated by age range.

Scale	Chi-Square	p-Value
Work motivation	7.4424	0.1143
Work behavior	9.4483	0.0508
Social skills	6.2898	0.1785
Psychic structure	3.9054	0.4190
Additional skills	6.0237	0.1970

Table 10 shows the statistics of the Levene's test for comparison of variances with four degrees of freedom when differentiated by age ranges. As can be seen, the test detects significant differences in the deviations of the work motivation scale. Consequently, it follows from the data in Table 7 that the youngest participants, who clearly present the smallest deviation, state that they are more confident than the remainder of their self-concepts in terms of work motivation (although the mean value of their answers is not the highest in the sample on this scale).

Table 10. Levene test statistics when the sample is differentiated by age range.

Scale	Levene F-Value	*p*-Value
Work motivation	4.6029	0.0011 *
Work behavior	0.8797	0.4756
Social skills	1.2987	0.2688
Psychic structure	1.8485	0.1179
Additional skills	1.1739	0.3206

* $p < 0.05$.

3.4. Area of Knowledge

Table 11 shows the means and standard deviations of the overall answers to the different scales when differentiated by area of knowledge. The coefficients of variation and skewness, when differentiating by the same variable, are shown in Table 12. Participants from humanistic-social areas (Humanities and Arts and Social and Legal Sciences) present higher mean answers on the scales of motivation and work behavior, social skills, and additional skills than participants from scientific-technical areas (Sciences, Health Sciences, and Engineering and Architecture). In psychic structure, the means do not show large differences by areas of knowledge. Regarding the deviations, the most dispersed answers on the scale of work motivation, social skills, psychic structure, and additional skills are presented by the participants in Social Sciences and Legal Sciences. In the work behavior scale, the greatest dispersion is presented by the participants from Health Sciences, followed by the Social Sciences area.

Table 11. Means and standard deviations (St.D.) of the overall answers of the different scales of the questionnaire when the sample is differentiated by area of knowledge.

	Humanities		Sciences		Health Sci.		Social Sci.		Engineering	
Scale	Mean	St.D.	Mean	St.D.	Mean	St.D.	Mean	St.D.	Mean	St.D.
Work motivation	4.22	0.86	4.11	0.79	4.19	0.84	4.34	0.92	4.13	0.87
Work behavior	4.48	0.61	4.38	0.67	4.29	0.82	4.53	0.79	4.36	0.66
Social skills	4.24	0.81	4.15	0.78	4.11	0.75	4.33	0.94	4.16	0.81
Psychic structure	4.41	0.68	4.49	0.61	4.37	0.80	4.43	0.88	4.42	0.75
Additional skills	3.84	0.98	3.73	0.89	3.69	1.06	3.88	1.20	3.65	1.09

Table 12. Coefficients of variation (C.V., measured in percent) and skewness (Skew.) of the overall answers of the different scales of the questionnaire when the sample is differentiated by area of knowledge.

	Humanities		Sciences		Health Sci.		Social Sci.		Engineering	
Scale	C.V.	Skew.	C.V.	Skew.	C.V.	Skew.	C.V.	Skew.	C.V.	Skew.
Work motivation	20.31	−1.42	18.72	−0.42	20.16	−0.86	21.22	−1.72	21.08	−1.29
Work behavior	13.70	−0.75	15.31	−0.79	19.06	−1.22	17.38	−2.52	15.25	−0.85
Social skills	19.03	−0.93	18.75	−0.95	18.23	−0.61	21.62	−1.64	19.49	−1.31
Psychic structure	15.42	−1.03	13.65	−0.79	18.29	−1.12	19.78	−1.96	17.04	−1.56
Additional skills	25.58	−1.02	23.82	−0.40	28.70	−0.82	30.93	−1.05	29.74	−0.78

With regard to average self-concepts, significant gaps are observed when participants are differentiated by their area of knowledge. Table 13 shows the statistics of the Kruskal–Wallis test for comparison of mean values with five degrees of freedom. In the p-values column, it can be seen that there are significant differences between the mean answers of the participants from the different areas of knowledge for the scales of motivation and work behavior, social skills, and additional skills. As previously noted in the results section, in these scales the mean answers are perceived to be higher in the humanistic-social areas than in the scientific-technical areas.

Table 13. Kruskal–Wallis test statistics when the sample is differentiated by the area of knowledge.

Scale	Chi-Square	p-Value
Work motivation	11.804	0.0376 *
Work behavior	15.487	0.0085 *
Social skills	15.926	0.0071 *
Psychic structure	2.6767	0.7497
Additional skills	12.290	0.0310 *

* $p < 0.05$.

Consequently, it can be assumed that there is a gap between these two families of knowledge areas in terms of the aforementioned scales, in the sense that participants from humanistic-social areas have higher mean self-concepts than participants from scientific-technical areas. Assuming that the latter are the most sensitive to the level of technical and digital competence and who most realistically assess the possibilities of their immediate environment in this respect, this gap suggests the idea that a low level of competence influences the respondents' perception of their soft skills.

Regarding deviations, the Levene's test statistics with five degrees of freedom (Table 14) show significant differences by area of knowledge in the scales of social skills, psychic structure, and additional skills. Given that, in these scales, it is the participants of Social and Legal Sciences who show more dispersion, it can be concluded that the deviations in the answers of the participants of this area for the aforementioned scales are significantly higher than those of the rest of the areas. Consequently, the self-concepts on social skills, psychic structure, and additional skills are more weakly formed in the Social Sciences participants, even though their mean self-concepts are the highest (or the second highest, in the case of psychic structure) of the different areas for these scales.

Table 14. Levene test statistics when the sample is differentiated by the area of knowledge.

Scale	Levene F-Value	p-Value
Work motivation	0.9372	0.4562
Work behavior	1.7043	0.1314
Social skills	3.0639	0.0095 *
Psychic structure	2.9973	0.0110 *
Additional skills	3.5642	0.0033 *

* $p < 0.05$.

3.5. *Years of Teaching Experience*

The statistics of the distributions of answers when differentiated by the range of years of teaching experience are shown in Tables 15 and 16. In terms of the mean answers, the most notable observation is that the participants with less teaching experience present a lower mean self-concept in all scales. Regarding the remainder of the participants, those with more extensive teaching experience present lower mean self-concepts than the others in terms of motivation and work behavior. In the psychic structure scale, the participants with more experience present a higher mean self-concept, and in additional skills, the highest mean answers are given by participants with between 11 and 20 years of experience.

Table 15. Means (M.) and standard deviations (St.D.) of the overall answers of the different scales of the questionnaire when the sample is differentiated by years of teaching experience.

	≤5		6 to 10		11 to 15		16 to 20		21 to 25		>25	
	M.	St.D.	M.	St.D.	M.	St.D.	M.	St.D.	M.	St.D.	M.	St.D.
Work motivation	3.81	1.04	4.18	0.72	4.09	0.85	4.39	0.59	4.32	0.92	4.19	0.87
Work behavior	4.18	0.89	4.48	0.59	4.41	0.69	4.49	0.54	4.49	0.82	4.40	0.65
Social skills	4.16	0.77	4.18	0.80	4.20	0.75	4.21	0.67	4.23	0.96	4.21	0.83
Psychic structure	4.32	0.71	4.38	0.75	4.47	0.78	4.45	0.66	4.35	0.86	4.54	0.60
Additional skills	3.56	0.96	3.69	1.15	3.80	1.08	3.99	0.86	3.74	1.05	3.73	1.06

Table 16. Coefficients of variation (C.V., measured in percent) and skewness (Skew.) of the overall answers of the different scales of the questionnaire when the sample is differentiated by years of teaching experience.

	≤5		6 to 10		11 to 15		16 to 20		21 to 25		>25	
	C.V.	Skew.	C.V.	Skew.	C.V.	Skew.	C.V.	Skew.	C.V.	Skew.	C.V.	Skew.
Work motivation	27.39	−0.97	17.30	−0.30	20.77	−1.31	13.39	−0.66	21.28	−1.64	20.86	−0.84
Work behavior	21.29	−0.83	13.11	−0.63	15.60	−0.89	12.09	−0.36	18.35	−2.38	14.87	−0.83
Social skills	18.44	−0.83	19.08	−0.65	17.92	−0.79	15.95	−0.43	22.68	−1.49	19.69	−1.30
Psychic structure	16.48	−0.55	17.15	−1.09	17.49	−1.68	14.90	−1.02	19.89	−1.82	13.23	−1.12
Additional skills	27.11	−0.82	31.22	−0.75	28.54	−0.87	21.48	−0.78	28.18	−0.83	28.42	−0.85

The mean data on social skills are very homogeneous when differentiated by areas of knowledge. Moreover, participants with less than or equal to 5 years of experience present moderate negative asymmetry in all scales; that is, a certain concentration of the answers in the values indicating high or very high self-concepts. However, in participants with between 6 and 10 years of experience, this asymmetry is reduced in all scales, except in the psychic structure scale, in which, in fact, the concentration of answers in very high self-concepts increases. The highest deviations in the motivation and work behavior scales were found among participants with teaching experience of 5 years or less, but in social skills, the greatest dispersion was found among participants with more than 20 years of experience.

Tables 17 and 18 show the Kruskal–Wallis test statistics and Levene's test results, respectively, with five degrees of freedom, for the different scales when differentiated by years of teaching experience. In this case, it can be observed that there are significant differences in the mean self-concepts for the scales of work motivation and additional skills. This fact allows us to assume that there is a gap, by time of teaching experience, in the participants' self-concepts of their work motivation and additional skills. This gap can be interpreted in the sense that participants with less teaching experience have significantly lower self-concepts regarding these skills than the rest of the participants. From Table 18 we can deduce that it is not possible to assume homoscedasticity, when differentiating by years of teaching experience, in any scale except that of psychic structure. This reveals that the self-concept on motivation and work behavior of the participants with the least experience is the most weakly formed; in social skills it is the participants with more than 20 years of experience who have the least well-formed self-concept; and in additional skills this position is occupied by participants with between 6 and 15 years of experience.

Table 17. Kruskal–Wallis test statistics when the sample is differentiated by years of teaching experience.

Scale	Chi-Square	*p*-Value
Work motivation	20.600	0.00096 *
Work behavior	9.7579	0.0824
Social skills	3.4043	0.6379
Psychic structure	7.0049	0.2203
Additional skills	13.334	0.0204 *

* $p < 0.05$.

Table 18. Levene test statistics when the sample is differentiated by years of teaching experience.

Scale	Levene F-Value	*p*-Value
Work motivation	3.7392	0.0024 *
Work behavior	2.8129	0.0156 *
Social skills	2.7111	0.0193 *
Psychic structure	2.0113	0.0752
Additional skills	5.2616	0.000088 *

* $p < 0.05$.

3.6. University Tenure

Differentiating by the nature, private or public, of the university where the participants teach, we obtained the statistics described in Tables 19 and 20. In both types of centers, the overall mean answers are above the value of 4 in all scales, except for additional skills, where the mean falls slightly below 4. Regardless of this, the mean self-concepts expressed by teachers from private universities are higher than those of participants from public universities. Although the dispersions are similar, they are slightly greater in private university teachers for the scales of work motivation, psychic structure, and additional skills. The opposite is true for the work behavior and social skills scales.

Table 19. Means and standard deviations (St.D.) of the overall answers of the different scales of the questionnaire when the sample is differentiated by the tenure of the university.

	Private		Public	
Scale	Mean	St.D.	Mean	St.D.
Work motivation	4.24	0.87	4.15	0.83
Work behavior	4.47	0.68	4.39	0.71
Social skills	4.23	0.80	4.18	0.82
Psychic structure	4.49	0.75	4.39	0.71
Additional skills	3.78	1.06	3.75	1.03

Table 20. Coefficients of variation (C.V., measured in percent) and skewness of the overall answers of the different scales of the questionnaire when the sample is differentiated by the tenure of the university.

	Private		Public	
Scale	C.V.	Skewness	C.V.	Skewness
Work motivation	20.56	−1.64	20.06	−0.73
Work behavior	15.31	−1.66	16.28	−1.15
Social skills	18.96	−1.19	19.61	−1.05
Psychic structure	16.81	−2.00	16.28	−0.97
Additional skills	27.98	−0.99	27.61	−0.71

The Mann–Whitney test statistics for comparison of means presented in Table 21 show that there is a significant gap by type of center for the scales of work motivation and psychic structure. It can be assumed, therefore, that, for these soft skills, teachers from private universities manifest a significantly better self-concept than those from public universities. Regarding dispersion, homoscedasticity can be assumed for all scales when differentiating by type of center, as is immediately evident from the Levene's test statistics with one degree of freedom shown in Table 22.

Table 21. Mann–Whitney test statistics when the sample is differentiated by tenure of the university.

Scale	Mann–Whitney W-Value	p-Value
Work motivation	58,067	0.0475 *
Work behavior	56,825	0.1389
Social skills	97,859	0.4628
Psychic structure	58,265	0.0312 *
Additional skills	153,324	0.3789

* $p < 0.05$.

Table 22. Levene test statistics when the sample is differentiated by the tenure of the university.

Scale	Levene F-Value	p-Value
Work motivation	0.0768	0.7818
Work behavior	0.9154	0.3390
Social skills	0.0976	0.7549
Psychic structure	0.2752	0.6001
Additional skills	0.4765	0.4902

4. Discussion

As previously explained in the Results section, the mean is a descriptive statistic suitable for discussing the overall answers given by the participants on the different scales of the questionnaire. Therefore, in light of the data in Table 3 and Figure 3, it is possible to assume that participants' self-concepts about their soft skills are high or very high. In this sense, there are discrepancies with respect to studies that analyze the perception that students have about the skills of their teachers [2,26,27]. This fact highlights that university teachers have a higher confidence in their own soft and digital competencies than they are able to project in the classroom. This suggests that the high mean values detected may be due, at least in part, to social desirability.

In addition, the high dispersion of additional skills with respect to the remainder of the soft skills scales is noteworthy. This fact may be due to the variety of additional skills, in contrast to the rest of the soft skills, which are grouped into families with a common definition. It would be interesting to explore the reasons for this difference by means of a study comparing the teachers' ratings with those of their own students.

When the sample is differentiated by gender, a statistically significant gap is identified on the work behavior scale, in favor of females. Consequently, it can be assumed that females have a higher self-concept of their own abilities in terms of assuming responsibilities, work challenges, and team roles than males. Furthermore, the homogeneity of answers in this sense is greater in females than in males because they have a lower standard deviation, but this difference is not statistically significant. This fact is not accompanied by a significant superiority in the self-concepts on social skills or motivation, which indicates that the perception of females is more optimistic than that of males with regard to taking action. This fact shows that, in the higher education teaching profession, there is no gender gap in favor of males, as some studies indicate with respect to employment in general [28].

The results obtained when the sample is differentiated by age range lead to the following assertions: (i) younger teachers have a significantly higher self-concept on work behavior than the rest of the participants; and (ii) older teachers have the lowest self-concept on this scale. Again, the gap identified refers to the work behavior scale, and in this case the greatest optimism is detected among the youngest teachers. In this case, the youngest teachers demonstrate a higher self-appraisal in the competencies that relate to moving into action, which suggests the idea that their self-concept is conditioned by the outwardness and impetus of youth. However, examination of the data from the variable measuring years of teaching experience shows that the most significant gaps are in the scales of work motivation and additional competencies. Interestingly, it is the most experienced teachers who are more motivated, although they are not the ones who best value their own behavior in action. It is perceived, therefore, that there is a phenomenon derived from the strength and impetus of youth that explains the high self-evaluation of young people's work behavior. However, the experience provides the participants with a more objective awareness of their limitations in terms of their teaching performance and, in addition, induces motivation to continue improving in their teaching activity.

The above analysis represents a novelty with respect to previous studies, which usually have not studied the existence of age or experience gaps in soft skills, but rather analyzed them in students or young professionals [1,2,29]. Alternatively, studies have also focused on digital competence by analyzing gaps by age ranges in university professors. For example, the results obtained here are in line with works such as [30,31], which attribute the highest digital competence rating to the youngest. However, the current results are in contradiction with other studies, such as [32], who found the highest digital competences among teachers under 40 years of age. The explanation for these divergences may lie in the reflection made by [32]: the area of knowledge is probably the variable that most strongly conditions the assessment of this type of competency, because it relates to the use of technologies, and being employed in this area depends strongly on the field in which the teacher is trained.

Undoubtedly, the variable that allows us to identify the greatest differences in the studied population regarding their self-concepts about soft skills is the area of knowledge. Indeed, the data shows that, with the exception of psychic structure skills, all the scales analyzed report significant differences by area of knowledge. In the light of the data, it is possible to corroborate that the observed gap is, in fact, derived from significant differences in their self-concepts between teachers of scientific-technical areas and teachers of humanistic-social areas. The latter are more optimistic in their evaluations on all scales. These observations are in line with works such as [29], which attribute intermediate levels of digital competence to teachers in the area of Health Sciences. However, the current results are in partial contradiction with studies such as that of [30], which attributes the highest levels of digital competence to teachers in the area of Engineering. These differences may be due to the fact that the present study compares all areas of knowledge. In this sense, it can be perceived that teachers from humanistic-social areas are more optimistic in their self-concepts because they usually have less training in digital and communication and information technologies, and are less experienced in them than teachers in the areas of Science or Engineering, who constantly use these technologies. In fact, [29] states that the development of digital competencies is associated with the development, generation, and dissemination of information. Hence, teachers with less experience in these areas show a more optimistic self-concept.

Finally, taking into consideration university tenure, it was found that, both in terms of work motivation and psychic structure, professors at private universities have a significantly more optimistic self-concept than those at public universities. This fact contradicts the results of analogous studies in which the population is made up of students rather than teachers [26,27]. This indicates that a greater effort can be made in public universities to develop students' soft skills. This would explain why professors have a lower self-concept than their own students (because, somehow, students do not miss these skills, which have little presence in the teaching they receive).

This work also has some limitations. These include the lack of an analysis of the answers of the respondents by crossing the different independent variables. This would make it possible to deepen the analysis of the gaps studied and probably identify others. In addition, the article is limited to exploring the self-concepts of university teachers, without taking into account the students' perception of their teachers' skills or of themselves. This latter limitation indicates an interesting future line of research, which involves exploring the self-concept of soft skills of university students in different areas of knowledge and the perception they have of the skills of the faculty, and differentiating the results by the different sociological and academic variables taken into account in this paper. In this same sense, it would also be interesting to explore the opinion of employers, who, in some manner, also evaluate the skills of university graduates.

5. Conclusions

This study found that university teachers from countries with medium or low technical and innovation development (measured through the GII) express high or very high self-concepts on all the soft skills scales. Specialized studies frequently show that students' perceptions of the soft skills (and also the digital competencies) of their teachers are intermediate or low. This discrepancy suggests the need for specific teacher training plans on the use of digital tools in didactic environments and on the development of the different aspects of soft skills. In addition, an interesting line of research would be to study and expose students' suggestions on how teachers could improve their digital skills.

A certain gender gap was identified, in the sense that females expressed a more optimistic self-concept. This gap was especially significant in the work behavior skills: conscientiousness, flexibility, and action orientation. With respect to the area of knowledge, teachers in scientific-technical areas have a worse self-concept of their soft skills than those in humanistic-social areas. This shows that teachers in scientific-technical areas are more insecure about their soft skills. Their insecurity is probably caused by their broader technological knowledge, which makes them more realistic in their perceptions of their abilities.

Age was also identified as a discriminating variable in the self-perception of soft skills. Younger teachers with less teaching experience are more insecure and consequently express the lowest average self-concept in terms of their soft skills. Exceptions to this observation are the skills related to work behavior (conscientiousness, flexibility, and action orientation), in which younger teachers express the highest self-concept (moreover, this is the only scale in which the gap can be assumed to be statistically significant). Older participants also expressed lower self-concepts than those in the middle age ranges. It can be concluded, therefore, that age exerts a certain influence on the assessment of one's own soft skills, in the sense that older teachers have a more pessimistic perception in this regard. The results in the case of long-lived and experienced participants may be due to the dependence caused by the accumulation of years of use of traditional teaching methodologies. However, the insecurity shown by younger and less experienced participants suggests the need for universities to propose training plans on digital competence oriented to novice teachers, and the role for countries with a higher level of digital development to help (for example, through teacher mobility plans or online training) university teachers from countries with less digital development.

A higher self-concept of teachers from private universities was found. This difference was especially significant in the skills of work motivation and psychic structure. This highlights the existence of differences between the two types of centers that result in greater optimism among private university teachers with respect to their soft skills. This is probably because private universities make a greater investment in digital learning resources and faculty training. Consequently, such actions should also be implemented by public universities.

Author Contributions: Conceptualization, D.V. and P.F.-A.; methodology, Á.A.-S. and D.V.; validation, Á.A.-S., D.V. and P.F.-A.; formal analysis, Á.A.-S., D.V. and P.F.-A.; data curation, Á.A.-S.; writing—original draft preparation, Á.A.-S., D.V. and P.F.-A.; writing—review and editing, Á.A.-S., D.V. and P.F.-A. All authors have read and agreed to the published version of the manuscript.

Funding: This research received no external funding.

Data Availability Statement: The data are not publicly available because they are part of a larger project involving more researchers. If you have any questions, please ask the contact author.

Conflicts of Interest: The authors declare no conflict of interest.

References

1. Reis, A.; Leite, C. Digital literacies in higher education: Skills, uses, opportunities and obstacles to digital transformation. *RED* **2021**, *65*, 1–20. [CrossRef]
2. López-Pérez, F.; Bas, M.C.; Escudero-Nahón, A. Self-perception about emerging digital skills in Higher Education students. *Comunicar* **2020**, *62*, 89–98. [CrossRef]
3. Wegerif, R.; Mansour, N. A dialogic approach to technology-enhanced education for the global knowledge society. In *New Science of Learning: Cognition, Computers and Collaboration in Education*; Khine, M.S., Saleh, I.M., Eds.; Springer: Berlin/Heidelberg, Germany, 2010; pp. 325–339. [CrossRef]
4. Gallardo Echenique, E.; Minelli, J.; Marqués Molías, L.; Esteve Mon, F. Digital competence in knowledge society. *MERLOT J. Online Learn. Teach.* **2015**, *11*, 1–15.
5. Fernandes, P.R.d.S.; Jardim, J.; Lopes, M.C.d.S. the soft skills of special education teachers: Evidence from the Literature. *Educ. Sci.* **2021**, *11*, 125. [CrossRef]
6. Rodríguez, A.; Sierra, V.; Falcón, C.; Latorre, C. Key soft skills in the orientation process and level of employability. *Sustainability* **2021**, *13*, 3554. [CrossRef]
7. Maisiri, W.; Van Dyk, L. Industry 4.0 competence maturity model design requirements: A systematic mapping review. In Proceedings of the 2020 IFEES World Engineering Education Forum—Global Engineering Deans Council (WEEF-GEDC), Cape Town, South Africa, 16–19 November 2020; pp. 1–6.
8. Krpálek, P.; Berková, K.; Kubišová, A.; Krelová, K.K.; Frendlovská, D.; Spiesová, D. Formation of professional competences and soft skills of public administration employees for sustainable professional development. *Sustainability* **2021**, *13*, 5533. [CrossRef]
9. European Commission. *Communication from the Commission to the European Parliament, the Council, the European Economic and Social Committee and the Committee of the Regions on the Digital Education Action Plan*; European Commission: Brussels, Belgium, 2018; pp. 1–11. Available online: https://eur-lex.europa.eu/legal-content/EN/TXT/PDF/?uri=CELEX:52018DC0022&from=EN (accessed on 15 September 2021).
10. Freiman, V.; Godin, J.; Larose, F.; Léger, M.; Chiasson, M.; Volkanova, V.; Goulet, M.J. Towards a life-long continuum of digital competencies: Exploring a combination of soft-skills and digital skills development. In Proceedings of the INTED 2017, Valencia, Spain, 6–8 March 2017; pp. 9518–9527. [CrossRef]
11. Claxton, G.; Costa, A.; Kallick, B. Hard thinking about soft skills. *Educ. Leadersh.* **2016**, *73*, 60–64.
12. Sisson, L.G.; Adams, A.R. Essential hospitality management competencies: The importance of soft skills. *J. Hosp. Tour. Educ.* **2013**, *25*, 131–145. [CrossRef]
13. Bartel, J. Teaching soft skills for employability. *TESL Can. J.* **2018**, *35*, 78–92. [CrossRef]
14. Hendarman, A.F.; Cantner, U. Soft skills, hard skills, and individual innovativeness. *Eurasian Bus. Rev.* **2018**, *8*, 139–169. [CrossRef]
15. Kyllonen, P.C. Soft Skills for the Workplace. *Chang. J. High. Educ.* **2013**, *45*, 16–23. [CrossRef]
16. Robles, M.M. Executive perceptions of the top 10 soft skills needed in today's workplace. *Bus. Commun. Q.* **2012**, *75*, 453–465. [CrossRef]
17. Farner, S.M.; Brown, E.E. College students and the work world. *J. Employ. Couns.* **2008**, *45*, 106–114. [CrossRef]
18. Gruzdev, M.V.; Kuznetsova, I.V.; Tarkhanova, I.Y.; Kazakova, E.I. University Graduates' Soft Skills: The Employers' Opinion. *Eur. J. Contemp. Educ.* **2018**, *7*, 690–698. [CrossRef]
19. Cornell University; INSEAD; WIPO. The Global Innovation Index 2020: Who Will Finance Innovation? Ithaca, Fontainebleau, and Geneva. 2020. Available online: https://www.wipo.int/edocs/pubdocs/en/wipo_pub_gii_2020.pdf (accessed on 29 September 2021).
20. Warrner, J. Integrating Soft Skills into an Academic Curriculum. *Am. Assoc. Adult Contin. Educ.* **2021**. Available online: http://www.aaace.org (accessed on 30 September 2021).
21. Green-Weir, R.R.; Anderson, D.; Carpenter, R. Impact of Instructional Practices on Soft-Skill Competencies. *Res. High. Educ.* **2021**, *40*, 1–20.
22. Batista-Foguet, J.; Sipahi-Dantas, A.; Guillén, L.; Martínez Arias, R.; Serlavós, R. Design and evaluation process of a personal and motive-based competencies questionnaire in Spanish-speaking contexts. *Span. J. Psychol.* **2016**, *19*, 1–12. [CrossRef]
23. Wang, X.; Wang, Z.; Jiang, Z. ConFiguretional Differences of National Innovation Capability: A Fuzzy Set Qualitative Comparative Analysis Approach. *Technol. Anal. Strateg. Manag.* **2021**, *33*, 599–611. [CrossRef]
24. Ionescu, G.H.; Firoiu, D.; Pîrvu, R.; Enescu, M.; Rădoi, M.-I.; Cojocaru, T.M. The Potential for Innovation and Entrepreneurship in EU Countries in the Context of Sustainable Development. *Sustainability* **2020**, *12*, 7250. [CrossRef]
25. Ngang, T.K.; Yunus, H.M.; Hashim, N.H. Soft Skills Integration in Teaching Professional Training: Novice Teachers' Perspectives. *Procedia Soc. Behav. Sci.* **2015**, *186*, 835–840. [CrossRef]
26. Gallardo-Echenique, E.; Poma, A.; Esteve, F. La competencia digital: Análisis de una experiencia en el contexto universitario. *Academicus* **2018**, *12*, 6–15.

27. Garzón, E.; Sola, T.; Ortega, J.; Marín, J.; Gómez, G. Teacher training in lifelong learning. the importance of digital competence in the encouragement of teaching innovation. *Sustainability* **2020**, *12*, 2852. [CrossRef]
28. Hong, R. Soft skills and hard numbers: Gender discourse in human resources. *Big Data Soc.* **2016**, *3*, 1–13. [CrossRef]
29. Tang, K.N. The importance of soft skills acquisition by teachers in higher education institutions. *Kasetsart J. Soc. Sci.* **2020**, *41*, 22–27. [CrossRef]
30. Cabero-Almenara, J.; Guillén-Gámez, F.D.; Ruiz-Palmero, J.; Palacios-Rodríguez, A. Digital competence of higher education professor according to DigCompEdu. Statistical research methods with ANOVA between fields of knowledge in different age ranges. *Educ. Inf. Technol.* **2021**, *26*, 4691–4708. [CrossRef] [PubMed]
31. Kingsley, K.V.; Kingsley, K. A case study for teaching information literacy skills. *BMC Med Educ.* **2009**, *9*, 1–6. [CrossRef] [PubMed]
32. Eloy, S.; Dias, M.S.; Lopes, P.F.; Vilar, E. Digital technologies in architecture and engineering: Exploring an engaged interaction within curricula. In *Architecture and Design: Breakthroughs in Research and Practice*; IGI Global: Hershey, PA, USA, 2019; pp. 390–426. [CrossRef]

electronics

MDPI

Article

Applicability of Collaborative Work in the COVID-19 Era: Use of Breakout Groups in Teaching L2 Translation

Iman Oraif [1,*] and Tariq Elyas [2]

1 Department of English Language and Literature, College of Languages and Translation, Imam Mohammad Ibn Saud Islamic University (IMSIU), Riyadh 3204, Saudi Arabia
2 Department of Foreign Languages and Literature, Faculty of Arts and Humanities, King Abdulaziz University, Jeddah 21589, Saudi Arabia; telyas@kau.edu.sa
* Correspondence: IMOraif@imamu.edu.sa

Abstract: Social distancing became a must during the pandemic, which not only had implications for people's social lives, but also for their learning. Collaborative work was almost impossible, especially in the classroom, despite a great need for this approach. For example, in their translation classes, the learners needed to collaborate with their peers, assisting each other in translating texts. Thus, the use of breakout groups is proposed in this study, although there is no guarantee that learners will accept this online approach. Consequently, the current research looks at learners' acceptance of breakout groups on Blackboard in a translation class. To examine their acceptance, an existing scale was used, developed by Davis (1989) to measure two factors of technology acceptance: perceived usefulness and ease of use. A sample of 54 students on a Translation course at Al-Imam Mohammed Ibn Saud Islamic University in Riyadh, Saudi Arabia, participated in this study. The results show that the learners found breakout groups on Blackboard to be useful and easy to use.

Keywords: TAM; collaboration; breakout groups; COVID-19; translation; L2 learners; KSA

Citation: Oraif, I.; Elyas, T. Applicability of Collaborative Work in the COVID-19 Era: Use of Breakout Groups in Teaching L2 Translation. *Electronics* **2021**, *10*, 2846. https://doi.org/10.3390/electronics10222846

Academic Editors: Carlos Alario-Hoyos, Mar Pérez-Sanagustín, Miguel Morales, Oscar Jerez and Boni García

Received: 26 October 2021
Accepted: 12 November 2021
Published: 19 November 2021

1. Introduction

Collaboration between learners may be noted throughout the literature as having a positive impact on learning. This is especially the case on courses that involve a high level of thinking and problem-solving skills [1] such as translation courses. In these instances, learners may prefer to collaborate traditionally by meeting face to face, but due to the pandemic, this was not possible. Instead, all learning was transferred online, with classes continuing in various online environments, known as virtual learning environments (VLEs), wherein learners could only collaborate via the virtual world. The most commonly used VLE in Saudi universities is Blackboard. This is a learning management system (LMS) with tools for developing a collaborative environment, such as breakout groups. Breakout groups enable learners to be allocated to different groups, whereupon they continue to work alone, while collaborating virtually with fellow group members. They cannot see the activities of other groups, but the teacher can enter each virtual room and check the learners' performances. On this basis, the present study explored the extent to which learners in a translation class find breakout groups on Blackboard useful and easy to use for working collaboratively in an online environment. An attempt will, therefore, be made to answer the following questions:

1. To what extent do learners in a translation class find breakout groups on Blackboard useful for working collaboratively?
2. To what extent do learners in a translation class find breakout groups on Blackboard easy to use for working collaboratively?

2. Literature Review

2.1. Online Learning in KSA throughout the Pandemic

COVID-19 lockdowns caused school closures and the cancelling of commercial passenger flights [2], with tourism industries receiving a harsh hit, contributing to the crippling of the world economy. According to the World Health Organization [3], more than 37 million COVID-19 cases and 1 million deaths were reported globally from December 2019 to October 2020. Meanwhile, according to [4], a range of needs have demanded an imperative reform to education in Saudi Arabia and worldwide. The shift towards a new medium of teaching, without the physical presence of the teacher or student, has proven to be a major challenge to this change in educational settings. The physical medium that once gave presence, used by traditional methods to enact authority and control, is becoming obsolete [5]. According to few researchers [6], on the concept of online teaching in Saudi Arabia five years before the pandemic, they claim that the classroom environment has shifted to something different, something distant, taking away some control from the teachers and putting it entirely in the hands of the students. It also revealed how Saudi education had been slowly advancing conceptually. This manifested as incompatibility with the medium. It has been proven that teachers were struggling to cope with the changes to online teaching [6] many years before the impact of the COVID-19 pandemic. This mere transition of Saudi classrooms and the teacher's role remained supplemented with suitable changes that reflected present day societal advancements. The existence of COVID-19 forced a change to unfamiliar grounds [4]. However, it is not COVID-19 that made these new grounds unfamiliar, but the role itself had been stuck in a loop for generations, neglecting the potential for appropriate changes [7].

The teacher's role in online teaching both 'during COVID-19' and 'before the pandemic' is not merely to present the material, but rather to introduce students to creative methods [8], where the physical interaction is missing. Hence, the role would need to change from being authoritative to a cooperative, engaging, playful, and fun learning environment. In a cooperative classroom, teachers are part of an interactive environment that focuses on aural and visual stimulation [8]. Students become creators, shaping their own experiences independently and actively contributing to the shape of the classroom's environment through visual and aural manipulation, while the teacher would act as a creative guide who enhances their experiences through feedback [8].

COVID-19 has in fact created a new opportunity for teachers and Saudi education; in particular, they are presented with the opportunity look at education through an array of creative lenses. For instance, ref. [9] claims that 'without the outbreak of the pandemic, our schools and universities would have not practiced distance learning in such a fluent way'. The Saudi government has made sure that all its sectors (public and private) need to work together to prevent the spread of COVID-19. According to [10], the Ministry of Education (MoE) was in a dilemma. However, this type of 'online' education has its own culture, ideologies, and mechanisms that might be absent among some people [11], which may prove to a challenge for parents who are facilitating the learning of their stay-home children. A few researchers claim that online education services began years ago [4], but the systems were upgraded recently to deal with the COVID-19 pandemic [12]. Therefore, the Ministry of Education (MoE) has introduced several learning strategies to facilitate the distanced learning process for the students. They applied mechanisms and solutions of distance learning—establishing *Ien* TV channels, *Ien* on YouTube, the *Ien* virtual gate, and other electronic platforms, to accomplish eight million teaching hours, three million items of digital content, and three and half million virtual classrooms [13].

Many Saudi researchers have praised the work of MoE [13,14] for their upscaling of educational strategies during the pandemic. With the beginning of the current academic year, the Ministry of Education was ready to launch the new *Madrasati* platform and 23 TV channels, with a specific channel for each level, complementing what was set to be available and beneficial for each student, such as Future Gate, Saudi Virtual School, and the Virtual Kindergarten. The MoE has applied many educational policies to effectively enable students to make the most of these electronic and educational possibilities, assigning morning hours to the intermediate and secondary school students, and the afternoon hours to the elementary school students, in order for parents to follow up with their kids and help them through the process of distanced learning in the evening [14]. In the same vein, ref. [13] points out the important fact that the MoE is working closely with other ministries to ensure that each student has the necessary tools and resources to complete the school year via distanced learning. The aim is to broadcast during school time and accommodate students who lack internet access or who have no computers to access the platform [14]. In fact, the MoE has been doing the best job in dealing with education during the COVID-19 pandemic. Commenting on the ministry's pledge that students who could not afford computers would not be left behind, Alfaries stated that 'there are a number of initiatives and collaborations between the ministry and other organizations such as Takaful to help students who cannot afford the basic equipment to access the platform' [14]. The MoE was successful in providing 6 outstanding educational platforms—from which 3 million students benefitted—with TV channels achieving 61 million views, the Ien national educational gate having 45 thousand courses and two thousand school digital books, and the *Ien* channel on YouTube having 750,000 subscribers, with 5400 recorded lessons and 4000 live hours [15]. In fact, Barakat [16] stated that 'distance learning provided many educational courses as well as recorded lessons which students can accomplish at any time to increase their educational capabilities; and it is not restricted by the time of the class' [16]. Moreover, Barakat stated that 'distance learning does not support the idea of direct communication between the teacher and the student which may result in making the student lazy in his/her learning process because he/she is not under the domination of the teacher and the school administration' [16].

Regarding the same point, 'we may need to explore methods incorporate with online teaching to enhance students' engagement' [17]. The authors of this paper believe that there is still a huge gap between the students and the prospects of remote learning processes. It is also observed that some parents no longer follow the education of their kids these days because they underestimate the importance of remote learning.

During the period of the pandemic, universities and educational institutions proved their effectiveness in the educational process, especially in university education. A study carried out at King Khalid University aimed to assess academic staff satisfaction with suspending traditional teaching and shifting to online education, revealing that 55.9% out of the participants agreed or strongly agreed that the sudden shift to web-based education was done smoothly, and 57.6% agreed or strongly agreed that giving lectures remotely was more flexible than delivering face-to-face lectures [18]. Moreover, ref. [15] points out that the MoE is working, in cooperation with expertise international organizations, to conduct an extensive validating assessing study for distanced learning in universities and schools during the period of the pandemic and afterwards. In addition, ref. [4] conducted a study to explore learners' levels of engagement in online courses using a designated school platform within the context of Saudi Arabia. A reliable measure was implemented in their study, based on the Student Course Engagement Questionnaire (SCEQ), with a sample of 379 female English as a foreign language (EFL) learners, studying a general English language course. The results revealed a high level of engagement among EFL Saudi learners during the COVID-19 pandemic. Ref. [19], in their recent study, explored the challenges and support methods for d/Dhh students during their distanced education in Saudi Arabia. A qualitative research study, using semi-structured interviews, was conducted with 37 parents of d/Dhh students to answer research questions. Three themes emerged from the parents' responses, as follows: (1) the challenges faced by d/Dhh students in distanced education; (2) the specific needs of d/Dhh students in distanced education; and (3) the supports provided to d/Dhh students in distanced education. They concluded that d/Dhh students require various forms of ongoing support from both their families and schools to ensure that they succeed and benefit from their experiences. In her study, ref. [20] investigated the experiences of English major students at the College of

Language and Translation (COLT) in King Saud University, Saudi Arabia, with Emergency Remote Learning and Teaching (ERLT) during the COVID-19 crisis. A total of 150 students participated in the study. An online survey was carried out in the form of questionnaires to elicit their responses. The results showed that laptop computers were the dominant devices students used and preferred. They also revealed that the Zoom platform came first in students' preferences, followed by Blackboard. Additionally, the findings showed that, although participants used smartphones in their ERL, they did not recommend it, and it came last compared with laptops, tablets, and desktop computers in their order of preference.

2.2. Breakout Groups and Language Learning

The COVID-19 pandemic is forcing many institutions to consider remote, virtual instruction for the safety of employees and students [21]. Many online language teaching users found the plethora of learning technologies available to be overwhelming [22]. Helping English language teachers select appropriate technologies is imperative now, more than ever. Many educators are divided about the benefits of online teaching and learning, especially in the current COVID-19 pandemic when such instruction is necessary and often accompanied by insufficient preparation [23].

Many instructors struggling to get their content online, for the first time ever tasks are carried out under great pressure, and crucial success factors for online teaching, which are being overlooked. Teaching online involves much more than creating a library of learning materials, streaming a lecture, or overloading learners with assignments [24]. Active learning is a broad term that includes a variety of different methods, which teachers can apply in the classroom [25]. A general breakdown of active learning will reveal three interrelated subsets, as follows: (a) collaborative learning; (b) cooperative learning; (c) problem-based learning (PBL) [25]. All of which are methods of instruction that aim to engage students in the learning process [26].

Meanwhile, breakout groups are employed across a wide range of settings, with most published reports describing their use in conferences [27], workshops [28]. However, it was claimed that breakout groups have been widely used on undergraduate programs [26]. Breakout sessions have therefore been a staple of face-to-face class sessions, and more recently have been employed in both asynchronous and synchronous online courses [28–31]. While the use of breakout rooms is widely studied, the intersection between breakout rooms and e-learning is left almost unexplored [32].

At this point in time, very little has been published regarding the use of breakout groups in a range of educational settings during the COVID-19 pandemic response to the use of breakout groups [26]. Because of the scarcity of research on breakout rooms (virtual or face-to-face), as described by [26], it is almost difficult to know how often students are provided with little or no structure to guide collaboration in real breakout room environments at this point. Therefore, this study seeks to fill the gap in the research on the use of breakout groups in Saudi Arabia during the COVID-19 pandemic.

In their study, ref. [33] used an exploratory case study design to gain insight into instructors' experiences with web-based synchronous communication using two-way audio and direct messaging. They conducted semi-structured interviews with eight instructors. They concluded that the use of text-based messaging and two-way audio raised issues in relation to the need to divide one's attention and in promoting student–student interaction during the breakout groups. Breakout groups have become very well known in Zoom software, used by many educators around the world. According to a couple of researchers [34], 'educators have utilized a range of online synchronous meeting tools (SMTs) to facilitate student learning'. One of the popular, immersive, and easy-to-use SMTs these days is Zoom, as it includes several features, such as annotation tools, polls, breakout rooms, and video and screen sharing. It has proved that using Zoom in an educational setting can facilitate communicative language learning through the use of authentic language instruction in interactive synchronous classes. However, other researchers have called upon future goals for blended learning, post the COVID-19 pandemic. Conversely, ref. [35] advocate for a more blended synchronous learning, claiming it has the potential to increase students' co-presence and improve upon the flexibility and accessibility of course offerings, if it is designed well.

In breakout research, a study has established an 'interactive, no-prep' approach, using breakout rooms within a videoconferencing system and requiring prior learner preparation. The authors found that, during the breakout groups, the learners produced perfectly effective study designs to answer the research question(s). Their results revealed that, as learners, they were well engaged in facilitated open discussion, and their learning outcomes were even [36].

3. Methodology

To gather data to address the research questions in this study, a survey was conducted with an existing case. The data were collected during March 2021 from a translation class in the English Department at Al-Imam Mohammed bin Saud University in Riyadh, KSA. The translation course was at the second level and focused on technical translation, before which, the learners were obliged to complete a prerequisite course in 'Cultural Translation'. The second language (L2) English learners involved used breakout groups for the final two classes of term, in order to collaborate in translating a text specified by their instructor. Two different Departmental sections were sampled. Ultimately, 54 students agreed to participate, out of the total population of 463 students who were studying translation at the time. The participants comprised a homogeneous sample of female students at university level six, studying Translation in the English Department, aged between 20 and 22 years. These study participants represented 11% of the whole population, which is considered to be a representative sample. The participants were exclusively female, due to the restrictions that are inherent to KSA.

Comprehensive sampling was applied in this current study, whereby all students in the translation class were included. A consent form was also signed by all the participants before completing the questionnaire. At the end of the second class, the instructor herself sent a link to the participants via the Blackboard chat window, so that they could access and complete the questionnaire.

The participants were all taught by the same teacher and covered the same material during both classes. The students met their teacher twice a week for three hours. On the Translation course, different types of texts were studied and translated throughout the 12 weeks of teaching, for example, political, medical, financial, and law texts. Breakout groups were created for the classes. Here, the learners practiced translating texts such as tenancy agreements. In the first class, these law texts were translated from Arabic (the source text) to English (the target text). Meanwhile, in the second class, the students translated from English (the source text) to Arabic (the target text).

The researcher observed the classroom procedure with the instructor's permission and without interrupting or distracting her class. During the breakout groups, the instructor randomly allocated the students to different groups, with each group consisting of 5–7 students. These groups worked on translating the text for half an hour at the end of the lesson, after which they sent their work to the instructor. The groups then reviewed their translations with the instructor at the beginning of the next translation lesson, as a whole class discussion. Examples of collaboration between the students were recorded by the researcher without interrupting or disturbing the participants, and only after gaining permission from the instructor (Figures 1 and 2).

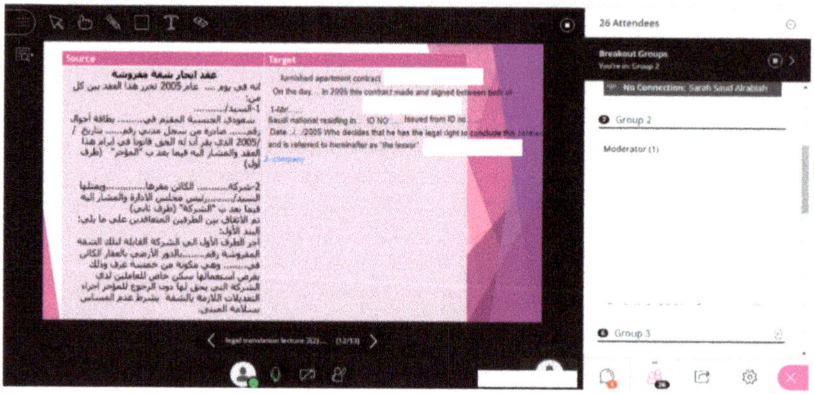

Figure 1. Sample of students' interaction in one of the groups.

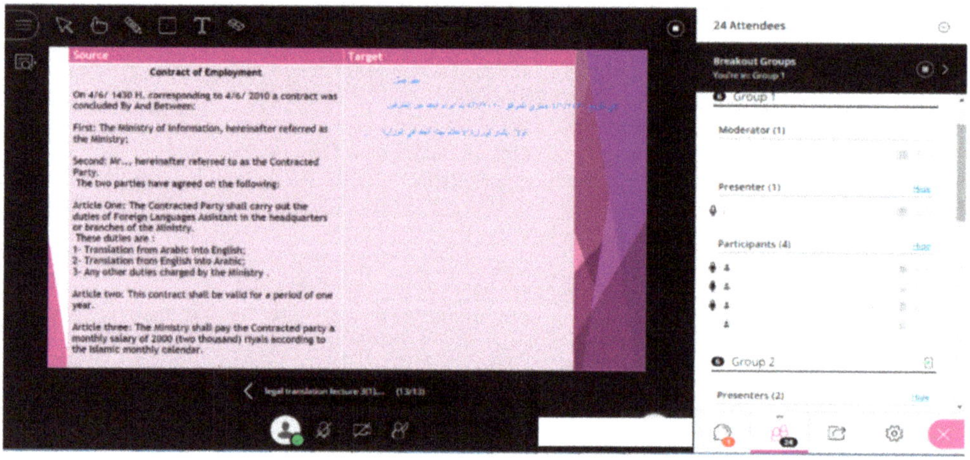

Figure 2. Sample of students' interaction in one of the groups.

The Survey

The survey used in this current study was adapted from [37], who stated that there are many variables that can affect people's acceptance of technology. However, previous research suggests that the most common variables include the extent to which people find an application helpful for performing their job more effectively (perceived usefulness) [37] and the extent to which people find an application easy to use (perceived ease of use) [37]. In this regard, Davis defined these two variables as follows: '[Perceived usefulness is] the degree to which a person believes that using a particular system would enhance his or her job performance'; while perceived ease of use refers to 'the degree to which a person believes that using a particular system would be free of effort' [37]. The original scale developed by Davis consisted of a 7-point Likert scale based on the results from his pilot study conducted with 120 participants, and it was adopted in the current research as it was found to be easily understood by the participants (Extremely Unlikely, Quite Unlikely, Slightly Unlikely, Neither Likely nor Unlikely, Slightly Likely, Quite Likely, and Extremely Likely) [37].

The theoretical foundation of perceived usefulness was driven by [38–42]. Conversely, the importance of ease of use is supported by Bandura's research on self-efficacy [40]. In Davis' study, the perceived usefulness scale attained a Cronbach's alpha reliability of 0.97

for both the electronic mail and XEDIT systems, while the perceived ease of use achieved a reliability of 0.86 [37]. This scale has been adopted in many studies to examine the applicability of information technology (IT) in relation to ease of use and usefulness [41,42]. However, few studies have been conducted on the adoption of this model to test L2 learners' acceptance of implementing IT in the classroom. More specifically, the use of breakout groups to help teach translation has been under researched.

4. Data Analysis

In this chapter, statistical data analysis is conducted, as follows: the main characteristics of the sample are presented in Section 4.1 to describe the participants' personal information (i.e., gender, age, and university level). Reliability and validity analysis is then conducted in Section 4.2 to evaluate the reliability and validity of the study dimensions: usefulness and ease of use. Meanwhile, Section 4.3 is devoted to confirmatory factor analysis, with the aim of testing how well the measured variables represent their dimensions. Finally, the results obtained to answer the research questions are presented.

4.1. Sample Characteristics

This section presents the main sample characteristics. The following tables offer a picture of the participants' personal information (i.e., gender, age, university level).

The following, Table 1, illustrates that the whole sample is female.

Table 1. Frequency table for gender.

	Frequency	Percentage
Female	54	100.0
Male	0	0.0
Total	54	100.0

The following, Table 2, illustrates that the percentage of sampled participants aged between 18 and 20 years was 1.9%; 79.6% of the sampled participants were aged between 20 and 22 years; 16.7% were aged between 22 and 24 years; 1.9% were aged over 24 years. Figure 3 below illustrates percentages for age for the participants.

Table 2. Frequency table for age.

	Frequency	Percentage
18–20 years	1	1.9
20–22 years	43	79.6
22–24 years	9	16.7
Over 24 years	1	1.9
Total	54	100.0

The following, Table 3, illustrates that the percentage of participants in the sample at university level 6 was 98.1%, and the percentage of sampled participants at university level 7 was 1.9%. This level is the final level before graduation. Figure 4 below illustrates Percentages for university level.

Table 3. Percentages for university level.

	Frequency	Percentage
Level 6	53	98.1
Level 7	1	1.9
Total	54	100.0

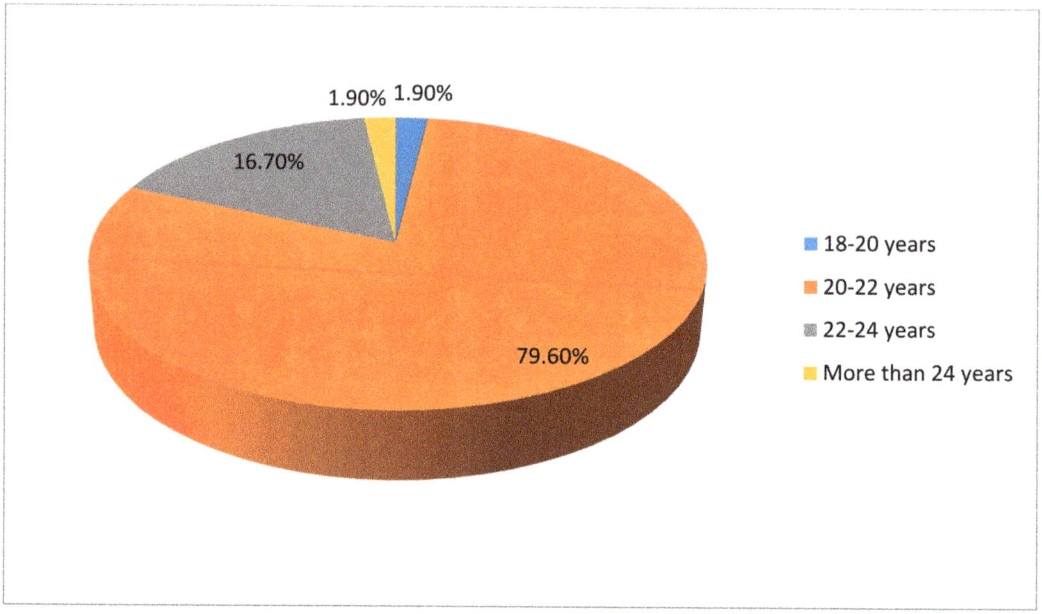

Figure 3. Percentages for age.

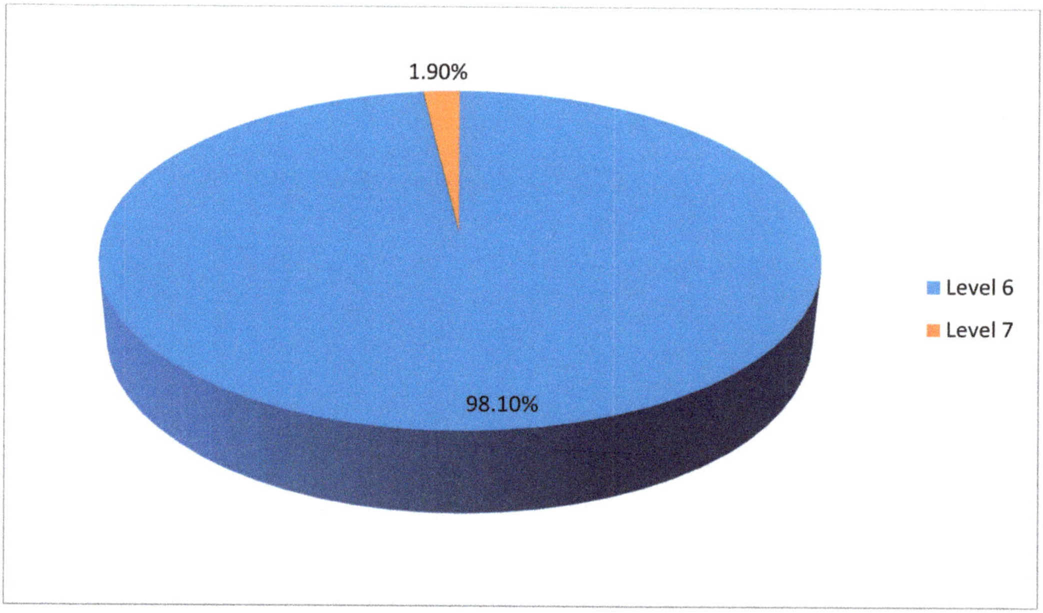

Figure 4. Percentages for university level.

The following, Table 4, illustrates that all the participants in the sample were studying translation, more specifically, at the College of Language and Translation at Al-Imam University.

Table 4. Frequency table for courses.

	Frequency	Percentage
Translation	54	100.0
Other courses	0	0.0
Total	54	100.0

4.2. Reliability and Validity

In the original study conducted by Davis [37], Cronbach's alpha was calculated to measure reliability, the results indicated 0.98 for the perceived usefulness scale and 0.94 for the perceived ease of use scale. A pilot study was also conducted in the current study, and only 9 out of the 20 participants involved in the pilot study responded. These participants shared the same characteristics as those sampled for the main study, in terms of gender, age, and year of university study, but they did not participate in the main study. As they had not previously participated in the breakout groups, a video was sent to them to explain how breakout groups operate, and the researcher also demonstrated how breakout groups would be used in the translation class. Cronbach's alpha (α) was calculated for each dimension, with results determined in values ranging between 0 and 1, and range of between 0.7 and 1 was determined to be acceptable. Thus, the original reliability measure from Davis was used in the current study [37].

The following, Table 5, illustrates that the data were evaluated as reliable, having satisfied the reliability test, as all Cronbach's alpha values exceeded the acceptable value. The Cronbach's alpha value for perceived usefulness was 0.783 (more than 0.7) and the Cronbach's alpha value for perceived ease of use was 0.875 (greater than 0.7).

Table 5. Reliability analysis.

	No. of Items	Cronbach's Alpha
Usefulness	5	0.783
Ease of use	6	0.875

Conversely, in order to evaluate validity, Pearson's correlation coefficients were calculated for each item with its dimension (see Table 6, below).

The above, Table 6, illustrates that the validity of the study was supported. The correlation coefficients of the following items, within the dimension of 'perceived usefulness', were greater than 0.5: work more quickly, increases productivity, effectiveness, makes job easier, and useful. The correlation coefficients of the following items, within the dimension, 'perceived ease of use', were greater than 0.5: easy to learn, controllable, clear and easy to understand, flexible, easy to become skillful, and easy to use.

Table 6. Validity analysis.

	Correlation	Significance
Usefulness		
Work more quickly	0.806 **	0.000
Increases productivity	0.680 **	0.000
Effectiveness	0.593 **	0.000
Makes job easier	0.765 **	0.000
Useful	0.803 **	0.000

Table 6. *Cont.*

	Correlation	Significance
Ease of use		
Easy to learn	0.816 **	0.000
Controllable	0.768 **	0.000
Clear and easy to understand	0.682 **	0.000
Flexible	0.788 **	0.000
Easy to become skillful	0.873 **	0.000
Easy to use	0.817 **	0.000

** Correlation is significant at the 0.01 level (two-tailed).

4.3. Answering the Research Questions

The following, Table 7, presents the participants' opinions in relation to perceived usefulness and ease of use.

Table 7. Descriptive statistics for perceived usefulness.

Items	Mean	Median	Min	Max	SD	%	Response	Rank
1. Using breakout groups on Blackboard in my translation class would enable me to accomplish tasks more quickly.	6.13	6	3	7	0.97	87.6%	Quite likely	2
2. Using breakout groups on Blackboard in my translation class would increase my productivity.	6.11	6	3	7	0.95	87.3%	Quite likely	3
3. Using breakout groups on Blackboard would enhance my effectiveness in performing tasks.	6.11	6	4	7	0.90	87.3%	Quite likely	4
4. Using breakout groups on Blackboard would make it easier to perform my tasks.	6.11	6	3	7	1.02	87.3%	Quite likely	5
5. I would find breakout groups useful in my translation class.	6.30	7	3	7	0.98	89.9%	Extremely likely	1
Usefulness	6.15	6	3	7	0.71	87.9%	Quite likely	

4.3.1. Research Question One

To what extent do learners in a translation class find breakout groups on Blackboard useful for working collaboratively?

The above, Table 7, illustrates that the fifth item in usefulness had the highest mean value (6.3 out of 7).

The following, Table 8, presents the participants' opinions with regard to the items relating to perceived usefulness—work more quickly, increases productivity, effectiveness, makes job easier, and useful.

The following table presents the main descriptive statistics (means, standard deviations) for the following items, relating to perceived usefulness: work more quickly, increases productivity, effectiveness, makes job easier, and useful. In order to answer the first research question, a one-sample *t*-test or a Wilcoxon signed rank test were used. The one-sample *t*-test contained four main assumptions, as follows: the dependent variable must be normally distributed, the dependent variable must be continuous, the observations must be independent, and the dependent variable must not contain outliers. If any of these assumptions were violated, the Wilcoxon signed rank test was used as an alternative to the one-sample *t*-test.

Table 8. Frequency table for perceived usefulness.

		Extremely Unlikely	Quite Unlikely	Slightly Unlikely	Neither Likely nor Unlikely	Slightly Likely	Quite Likely	Extremely Likely
Usefulness								
1. Using breakout groups on Blackboard in my translation class would enable me to accomplish tasks more quickly.	F	0	0	2	1	7	22	22
	P	0.0	0.0	3.7	1.9	13.0	40.7	40.7
2. Using breakout groups on Blackboard in my translation class would increase my productivity.	F	0	0	1	3	6	23	21
	P	0.0	0.0	1.9	5.6	11.1	42.6	38.9
3. Using breakout groups on Blackboard would enhance my effectiveness in my tasks.	F	0	0	0	4	7	22	21
	P	0.0	0.0	0.0	7.4	13	40.7	38.9
4. Using breakout groups on Blackboard would make it easier to perform my tasks.	F	0	0	2	2	7	20	23
	P	0.0	0.0	3.7	3.7	13.0	37.0	42.6
5. I would find breakout groups useful in my translation class.	F	0	0	2	1	5	17	29
	P	0.0	0.0	3.7	1.9	9.3	31.5	53.7

The following, Table 9, presents the results of the first assumption (the normality assumption). In this table, the normality assumption for perceived usefulness was evaluated using Kolmogorov–Smirnov tests. From these results, it may be noted that the normality assumption for usefulness was violated, as the value of significance was less than 0.05. Thus, the Wilcoxon signed rank test was used as a non-parametric alternative to the one-sample *t*-test.

Table 9. Normality tests.

	Statistic	Significance
Usefulness	0.147	0.000

The following Table 10 demonstrates that perceived usefulness was a significant factor, as the value of significance was less than 0.05. The learners in the translation class found that breakout groups on Blackboard were useful for working collaboratively.

Table 10. Wilcoxon signed rank tests.

	Mean	Std Dev	Statistic	Significance
Usefulness	6.15	0.71	1483.0	0.000

4.3.2. Research Question Two

To what extent do learners in a translation class find breakout groups on Blackboard easy to use for working collaboratively?

The following, Tables 11 and 12, presents the main descriptive statistics (means, standard deviations) for the following items relating to perceived ease of use: easy to learn, controllable, clear and easy to understand, flexible, easy to become skillful, and easy to use.

Table 11. Descriptive statistics for perceived ease of use.

Items	Mean	Median	Min	Max	SD	%	Response	Rank
1. Learning how to perform the task using breakout groups would be easy for me.	6.02	6	2	7	1.17	86.0%	Quite likely	3
2. I would find it easy to achieve what I want through breakout groups.	5.85	6	1	7	1.31	83.6%	Quite likely	6
3. My interaction using breakout groups would be clear and easy to understand.	5.93	6	1	7	1.20	84.7%	Quite likely	5

Table 11. *Cont.*

Items	Mean	Median	Min	Max	SD	%	Response	Rank
4. I would find breakout groups flexible in the way that they enabled me to interact with my classmates.	6.26	6.30	3	7	0.99	89.4%	Extremely likely	1
5. It would be easy for me to become skillful at using breakout groups.	6.02	6	2	7	1.17	86.0%	Quite likely	4
6. I would find breakout groups easy to use.	6.20	6	3	7	0.96	88.6%	Extremely likely	2
Ease of use.	6.05	6	3	7	0.90	86.4%	Quite likely	

Table 12. Table of frequency for perceived ease of use.

		Extremely Unlikely	Quite Unlikely	Slightly Unlikely	Neither Likely nor Unlikely	Slightly Likely	Quite Likely	Extremely Likely
1. Learning how to perform tasks using breakout groups would be easy for me.	F	0	1	2	2	8	18	23
	P	0.0	1.9	3.7	3.7	14.8	33.3	42.6
2. I would find it easy to achieve what I want through breakout groups.	F	1	0	2	5	8	17	21
	P	1.9	0.0	3.7	9.3	14.8	31.5	38.9
3. My interaction using breakout groups would be clear and easy to understand.	F	1	0	1	4	7	22	19
	P	1.9	0.0	1.9	7.4	13.0	40.7	35.2
4. I would find the breakout groups flexible in the way that they enabled me to interact with my classmates.	F	0	0	2	2	3	20	27
	B	0.0	0.0	3.7	3.7	5.6	37.0	50.0
5. It would be easy for me to become skillful at using breakout groups.	F	0	1	2	2	8	18	23
	B	0.0	1.9	3.7	3.7	14.8	33.3	42.6
6. I would find breakout groups easy to use.	F	0	0	1	2	8	17	26
	P	0.0	0.0	1.9	3.7	14.8	31.5	48.1

The above table illustrates that the fourth item (flexible) had the highest mean value (6.26 out of 7).

The following table presents the participants' opinions with regard to the following items, relating to perceived ease of use: easy to learn, controllable, clear and easy to understand, flexible, easy to become skillful, and easy to use.

In order to answer the second research question, one-sample *t*-test or Wilcoxon signed rank test was used. The one-sample *t*-test contains four main assumptions, as follows: the dependent variable must be normally distributed, the dependent variable must be continuous, the observations must be independent, and the dependent variable should not contain outliers. If these assumptions were violated, a Wilcoxon signed rank test would be used as an alternative to the one-sample *t*-test.

The following Table 13 presents the results for the first assumption (the normality assumption). In this table, the normality assumption for the 'ease of use' dimension was evaluated using Kolmogorov–Smirnov tests. From these results, it may be noted that the normality assumption was violated for ease of use, as the value of significance was less than 0.05. Thus, the Wilcoxon signed rank test was used as a non-parametric alternative to the one-sample *t*-test.

Table 13. Normality tests.

	Statistic	Significance
Ease of use	0.220	0.000

The following Table 14 illustrates that perceived ease of use was a significant factor since the value of significance was less than 0.05. The learners in the translation class found that breakout groups on Blackboard were useful and easy to use for working collaboratively.

Table 14. Wilcoxon signed rank tests.

	Mean	Std Dev	Statistic	Significance
Ease of use	6.05	0.90	1475.0	0.000

5. Discussion

The results revealed the learners' opinions of using breakout groups on Blackboard. Two factors in the survey were considered: usefulness and ease of use. The learners subsequently expressed positive views of breakout groups.

Regarding the first factor, concerning usefulness, several items received positive responses, such as 'Using breakout groups on Blackboard in my translation class would enable me to accomplish tasks more quickly'. This advantage could be very helpful, specifically because the learners would be able to enjoy privacy in their groups without feeling embarrassed about making mistakes. They would also negotiate solutions to any problems that they faced in translation. This aspect of active learning has been covered by [25].

Based on that was agreed on by the participants, for example, 'I would find breakout groups useful in my translation class', it was gathered that they could finish their work more quickly, as they would be assisting and making suggestions to each other. Moreover, the participants 'Strongly agreed' on the following item: 'Using breakout groups on Blackboard in my translation class would increase my productivity'. In particular, the learners could share their audio content and written text on the screen, whereupon they would all comment on each other's work, thereby leading to better learning and better work being produced within the limited time spent in class. In some ways, this is similar to the study conducted by [33], where it was found that text and audio messaging helped the learners in their learning.

In relation to the second factor mentioned above, namely ease of use, the participants were also positive in their responses, agreeing on most of the items. The most highly ranked item, with which most of the participants agreed, was 'I would find breakout groups flexible in the way that they enabled me to interact with my classmates'. This indicated that interaction was the most important outcome of using breakout groups. This is because it highlights scaffolding and collaborative learning as being important for active learning [25].

Another important factor of using breakout groups in teaching is their simplicity in implementation. The participants 'Strongly agreed' on this survey item, stating, 'I would find breakout groups easy to use'. For example, all the learners had to do was interact with each other and save their interactions to send to their teacher later, as described by the course instructor. The item ranked third in the survey, which also received highly positive responses, was the following: 'Learning how to perform the task using breakout groups would be easy for me'. If the technology itself is easy to use, then it will promote concentration on the task and the ability to perform it properly within the required time.

All other items were positively agreed with by the participants and the highest-ranking items have been discussed here. It was found that maximizing the potential of any technology that is used will lead to a better teaching and learning experience, as it will add variety to the process, more specifically in an EFL environment.

6. Conclusions

Based on the data presented and the ensuing discussion in this study, it may be concluded that educators need to invest more thought and self-development to advance their teaching, which would also mean that some expectations might need to be readjusted. Despite the many obstacles identified in e-learning, however, the educators sampled for this study highlighted its opportunities as well as its challenges. One especially pragmatic solution proposed was increased flexibility, which would enable students to learn independently. Additionally, e-learning has triggered reflection among educators with regard to their teaching, which could lead to improved practice [43,44].

7. Limitations and Implications for Practice

In the current study, the following limitations were noted:

1. The participants were exclusively female, due to the context of the study and social restrictions of the country.
2. The data were collected over a period of two weeks only, as the teacher used the breakout groups.
3. The data were only collected from translation classes.
4. Furthermore, according to the results, the following implications may be drawn for actual practice.
5. Breakout groups could be included in online translation classes to help learners to collaborate in groups.
6. Attempts could be made to use this feature on other online courses that might require collaboration, like L2 writing classes, where the learners need each other's help under the teacher's guidance.
7. Breakout groups should be included in a male section to examine their suitability, using the same scale.

Author Contributions: Conceptualization, I.O. and T.E.; methodology I.O. and T.E.; software, I.O. and T.E.; validation, I.O. and T.E.; formal analysis, I.O. and T.E.; investigation, I.O. and T.E.; resources, I.O. and T.E.; data curation, I.O. and T.E.; writing—original draft preparation, I.O. and T.E.; writing—review and editing, I.O. and T.E.; visualization, I.O. and T.E.; supervision, I.O. and T.E.; project administration, I.O. and T.E.; funding acquisition, I.O. and T.E. All authors have read and agreed to the published version of the manuscript.

Funding: This research was supported by the Deanship of Scientific Research, Imam Mohammad Ibn Saud Islamic University (IMSIU), Saudi Arabia, Grant No. 21-13-18-019.

Informed Consent Statement: Informed consent was obtained from all subjects involved in the study.

Conflicts of Interest: The authors declare no conflict of interest.

References

1. Laal, M.; Ghodsi, S.M. Benefits of collaborative learning. *Procedia-Soc. Behav. Sci.* **2012**, *31*, 486–490. [CrossRef]
2. Nicola, M.; Alsafi, Z.; Sohrabi, C.; Kerwan, A.; Al-Jabir, A.; Iosifidis, C.; Agha, M.; Agha, R. The socio-economic implications of the coronavirus pandemic (COVID-19): A review. *Int. J. Surgery.* **2020**, *78*, 185–193. [CrossRef]
3. World Health Organization. COVID-19 Weekly Epidemiological Update, 3 November 2020. Available online: https://www.who.int/emergencies/diseases/novel-coronavirus-2019?adgroupsurvey (accessed on 3 October 2021).
4. Oraif, I.; Elyas, T. The impact of COVID-19 on learning: Investigating EFL learners' engagement in online courses in Saudi Arabia. *Educ. Sci.* **2021**, *11*, 99. [CrossRef]
5. Basalamah, O.; Elyas, T. The Emergent Effects of a Wired World to an Educational Paradigm Shift in Saudi Arabia: A Case Study. *Procedia-Soc. Behav. Sci.* **2012**, *47*, 1534–1538. [CrossRef]
6. Basalamah, O.; Elyas, T. Conceptualizing the virtual learning space(s) in Saudi Arabia: A Foucauldian panoptic approach. *Int. J. Engl. Lang. Teach.* **2014**, *2*, 1–15.
7. Aldwaihi, A. Available online: https://www.okaz.com.sa/articles/authors/2034871 (accessed on 26 July 2020).
8. Al-Bogami, B.; Elyas, T. Promoting Middle School Students' Engagement through Incorporating iPad Apps in EFL/ESL Classes. *SAGE Open* **2020**, *10*, 2158244020926570. [CrossRef]
9. Saati, A. Distance Learning and Coronavirus Crisis. Aleqt. Available online: https://www.aleqt.com/2020/04/05/article_1797576.html (accessed on 5 April 2020).
10. MOE. Ministry of Education-My School Virtual Classrooms. Available online: https://www.moe.gov.sa/ar/news/pages/mn-2020-876.aspx (accessed on 26 July 2020).
11. Alamer, O. The Culture of Distance Learning. *Al-Jazirah Newspaper*, 24 March 2020. Available online: https://www.al-jazirah.com/2020/20200324/ar3.htm (accessed on 20 August 2020).
12. Khalid, T. Distance Learning Amid Coronavirus to Help School Dropouts in Saudi Arabia: Minister. Al Arabiya (English). Available online: https://english.alarabiya.net/en/coronavirus/2020/05/15/Distance-learning-amid-coronavirus-to-help-school-dropouts-in-Saudi-Arabia-Minister (accessed on 14 May 2020).
13. Alfifi, A. After Coronavirus: Our Education towards the Best. *Alriyadh Newspaper*. Available online: http://www.alriyadh.com/1816856 (accessed on 20 April 2020).

14. Al-Mayman, H. Saudi E-Learning Portal Ushers in New Normal with Virtual Classrooms. 512 Arab News. Available online: https://www.arabnews.com/node/1730541/saudi-arabia (accessed on 6 September 2020).
15. AlSalih, K. Saudi Study for Assessment of Remote Learning Process. *Alwatan Newspaper*. Available online: https://www.alwatan.com.sa/article/1048932 (accessed on 20 June 2020).
16. Barakat, R. Extension of Distance Learning in Saudi Arabia. Mhtwyat. Available online: https://bit.ly/35gxtPG (accessed on 11 August 2020).
17. Alshehri, Y.A.; Mordhah, N.; Alsibiani, S.; Alsobhi, S.; Alnazzawi, N. How the Regular Teaching Converted to Fully Online Teaching in Saudi Arabia during the Coronavirus COVID-19. *Creat. Educ.* **2020**, *11*, 985–996. [CrossRef]
18. Almaghaslah, D.; Alsayari, A. The Effects of the 2019 Novel Coronavirus Disease (COVID-19) Outbreak on Academic Staff Members: A Case Study of a Pharmacy School in Saudi Arabia. *Risk Manag. Healthc. Policy* **2020**, *13*, 795–802. [CrossRef] [PubMed]
19. Alqraini, F.M.; Alasim, K.N. Distance Education for d/Deaf and Hard of Hearing Students during the COVID-19 Pandemic in Saudi Arabia: Challenges and Support. *Res. Dev. Disabil.* **2021**, *117*, 104059. [CrossRef] [PubMed]
20. Al-Samiri, R.A. English language teaching in Saudi Arabia in response to the COVID-19 pandemic: Challenges and positive outcomes. *Arab. World Engl. J. Spec. Issue Covid* **2021**, *19*, 147–159. [CrossRef]
21. Vollbrecht, P.J.; Porter-Stransky, K.A.; Lackey-Cornelison, W.L. Lessons learned while creating an effective emergency remote learning environment for students during the COVID-19 pandemic. *Adv. Physiol. Educ.* **2020**, *44*, 722–725. [CrossRef]
22. Gruber, A.; Bauer, E. Fostering interaction in synchronous online class sessions with foreign language learners. *Teach. Technol. Teach. Educ. Dur. COVID-19 Pandemic Stories Field* **2020**, 175–178.
23. Marshall, H.W.; Kostka, I. Fostering teaching presence through the synchronous online flipped learning approach. *Tesl-Ej* **2020**, *24*, n2.
24. De Pryck, D.; De Pryck, K. Building a community and facilitating social presence in an online learning environment: Why? and how? In *Society for Information Technology & Teacher Education International Conference;* Association for the Advancement of Computing in Education (AACE): Waynesville, NC, USA, 2021; pp. 1188–1190.
25. Prince, M. Does active learning work? A review of the research. *J. Eng. Educ.* **2004**, *93*, 223–231. [CrossRef]
26. Lougheed, J.; Kirkland, J.; Newton, G. Using Breakout Groups as an Active Learning Technique in a Large Undergraduate Nutrition Classroom at the University of Guelph. *Can. J. Sch. Teach. Learn* **2012**, *3*, 6. [CrossRef]
27. Daniel, E.A. Exploring the process of assessment: Report on the ATS workshop on assessing theological learning. *Theol. Educ.* **2003**, *3*, 3–100.
28. Pritchard, G.E. *Improving Learning with Information Technology: Report of a Workshop (Washington, DC, January 24–25, 2001)*; National Academy Press: Washington, DC, USA, 2002.
29. Chandler, K. Using Breakout Rooms in Synchronous Online Tutorials. *J. Perspect. Appl. Acad. Pr.* **2016**, *4*, 16–23. [CrossRef]
30. Martin, F.; Parker, M.A. Use of synchronous virtual classrooms: Why, who, and how? *J. Online Learn. Teach.* **2014**, *10*, 192–210.
31. Saltz, J.; Heckman, R. Using Structured Pair Activities in a Distributed Online Breakout Room. *Online Learn.* **2020**, *24*, 227–244. [CrossRef]
32. Rahayu, D. Students' E-Learning Experience through a Synchronous Zoom Web Conference System. *J. ELT Res. Acad. J. Stud. Engl. Lang. Teach. Learning.* **2020**, *9*, 68–79.
33. Murphy, E.; Ciszewska-Carr, J. Instructors' experiences of web based synchronous communication using two-way audio and direct messaging. *Australas. J. Educ. Technol.* **2007**, *23*. [CrossRef]
34. Kohnke, L.; Moorhouse, B.L. Facilitating Synchronous Online Language Learning through Zoom. *RELC J.* **2020**, 0033688220937235. [CrossRef]
35. Angelone, L.; Warner, Z.; Zydney, J.M. Optimizing the Technological Design of a Blended Synchronous Learning Environment. *Online Learn.* **2020**, *24*, 222–240. [CrossRef]
36. Cadieux, M.; Campos-Zamora, M.; Zagury-Orly, I.; Dzara, K. Journal Club Using Virtual Breakout Rooms: Interactive Continuing Education with No Learner Preparation during COVID-19. *J. Contin. Educ. Health Prof.* **2020**, *40*, 217–219. [CrossRef] [PubMed]
37. Davis, F.D. Perceived usefulness, perceived ease of use, and user acceptance of information technology. *MIS Q.* **1989**, *13*, 319–340. [CrossRef]
38. Robey, D. User attitudes and management information system use. *Acad. Manag. J.* **1979**, *22*, 527–538.
39. Bandura, A. Self-efficacy mechanism in human agency. *Am. Psychol.* **1982**, *37*, 122. [CrossRef]
40. Davis, F.D.; Bagozzi, R.P.; Warshaw, P.R. Extrinsic and intrinsic motivation to use computers in the workplace 1. *J. Appl. Soc. Psychol.* **1992**, *22*, 1111–1132. [CrossRef]
41. Venkatesh, V. Determinants of Perceived Ease of Use: Integrating Control, Intrinsic Motivation, and Emotion into the Technology Acceptance Model. *Inf. Syst. Res.* **2000**, *11*, 342–365. [CrossRef]
42. Venkatesh, V.; Morris, M.G.; Davis, G.B.; Davis, F.D. User acceptance of information technology: Toward a unified view. *MIS Q.* **2003**, *27*, 425–478. [CrossRef]
43. Müller, A.M.; Goh, C.; Lim, L.Z.; Gao, X. COVID-19 Emergency eLearning and Beyond: Experiences and Perspectives of University Educators. *Educ. Sci.* **2021**, *11*, 19. [CrossRef]
44. Khalifé, K. Pragmatic Tips for Optimizing Remote Classroom Instruction in High School and College. *Fr. Rev.* **2020**, *94*, 17–21. [CrossRef]

electronics

MDPI

Article

A Competency Framework for Teaching and Learning Innovation Centers for the 21st Century: Anticipating the Post-COVID-19 Age

Mar Pérez-Sanagustín [1,2,*], Iouri Kotorov [1,3], António Teixeira [4], Fernanda Mansilla [2], Julien Broisin [1], Carlos Alario-Hoyos [5], Óscar Jerez [6,7], Maria do Carmo Teixeira Pinto [4], Boni García [5], Carlos Delgado Kloos [5], Miguel Morales [8], Mario Solarte [9], Luis Magdiel Oliva-Córdova [10] and Astrid Helena Gonzalez Lopez [11]

1 Institut de Recherche en Informatique de Toulouse (IRIT), Université de Toulouse, 31062 Toulouse, France; iouri.kotorov@karelia.fi (I.K.); julien.broisin@irit.fr (J.B.)
2 Department of Computer Science, Pontificia Universidad Católica de Chile, Santiago 7810000, Chile; mvmansilla@uc.cl
3 Department of International Business, Wärtsilä Campus, Karelia University of Applied Sciences, 80200 Joensuu, Finland
4 Centre of Philosophy, School of Arts and Humanities, Universidade Aberta, 1600-214 Lisboa, Portugal; antonio.teixeira@uab.pt (A.T.); maria.pinto@uab.pt (M.d.C.T.P.)
5 Telematics Department, Campus Leganés, Universidad Carlos III de Madrid, 28911 Leganés, Spain; calario@it.uc3m.es (C.A.-H.); bogarcia@it.uc3m.es (B.G.); cdk@it.uc3m.es (C.D.K.)
6 Economics and Business School, Universidad de Chile, Santiago 8320000, Chile; ojerez@uchile.cl or oscar_jerez@harvard.edu
7 Laspau Affiliated with Harvard University, Cambridge, MA 02138-6095, USA
8 Institute Von Neumann, Universidad de Galileo, Ciudad de Guatemala 01010, Guatemala; amorales@galileo.edu
9 Telematics Department, Universidad del Cauca, Popayán 190002, Colombia; msolarte@unicauca.edu.co
10 Information and Communication Unit, Medical School, Universidad San Carlos de Guatemala, Ciudad de Guatemala 01012, Guatemala; moliva@profesor.usac.edu.gt
11 Education and Pegagogy Department, Universidad San Buenaventura de Cali, Cali 764501, Colombia; ahgonzalezl@usbcali.edu.co
* Correspondence: mar.perez-sanagustin@irit.fr

Citation: Pérez-Sanagustín, M.; Kotorov, I.; Teixeira, A.; Mansilla, F.; Broisin, J.; Alario-Hoyos, C.; Jerez, Ó.; Teixeira Pinto, M.d.C.; García, B.; Delgado Kloos, C.; et al. A Competency Framework for Teaching and Learning Innovation Centers for the 21st Century: Anticipating the Post-COVID-19 Age. *Electronics* **2022**, *11*, 413. https://doi.org/10.3390/electronics11030413

Academic Editor: Flavio Canavero

Received: 22 December 2021
Accepted: 19 January 2022
Published: 29 January 2022

Abstract: During the COVID-19 pandemic, most Higher Education Institutions (HEIs) across the globe moved towards "emergency online education", experiencing a metamorphosis that advanced their capacities and competencies as never before. Teaching and Learning Centers (TLCs), the internal units that promote sustainable transformations, can play a key role in making this metamorphosis last. Existing models for TLCs have defined the competencies that they could help develop, focusing on teachers', students', and managers' development, but have mislead aspects such as leadership, organizational processes, and infrastructures. This paper evaluates the PROF-XXI framework, which offers a holistic perspective on the competencies that TLCs should develop for supporting deep and sustainable transformations of HEIs. The framework was evaluated with 83 participants from four Latin American institutions and used for analyzing the transformation of their teaching and learning practices during the pandemic lockdown. The result of the analysis shows that the PROF-XXI framework was useful for identifying the teaching and learning competencies addressed by the institutions, their deficiencies, and their strategic changes. Specifically, this study shows that most institutions counted with training plans for teachers before this period, mainly in the competencies of digital technologies and pedagogical quality, but that other initiatives were created to reinforce them, including students' support actions.

Keywords: teaching learning center (TLC); competence; pandemic

1. Introduction

"Transformation" and "metamorphosis" are in essence the same word. "Transformation" is of Latin and "metamorphosis" (μεταμόρφωσις) is of Greek origin. They both have a prefix "trans" or "meta", which means "on the other side" or "beyond". The second part refers to form, "forma" in Latin and "morphe" (μορφή) in Greek. Although etymologically closely related, "metamorphosis" is often used with more abrupt changes, such as the evolution in butterflies from larvae to chrysalis and adult butterflies. "The Metamorphosis" is also the chosen translation for Kafka's book "Die Verwandlung" in languages such as English or Spanish. In this book, the main character, Gregor Samsa, wakes up one morning converted into a huge insect. He struggles with his new condition, where even simple gestures, such as opening a door, become nearly impossible for him.

In mid-March 2020, with the COVID-19 pandemic, professors all over the world must have experienced feelings such as Kafka's character when universities across the globe moved abruptly to the "emergency online education" [1]. For better or worse, this rapid transition forced teachers to transform their teaching practices, students to adapt to new forms of learning, and Higher Education Institutions (HEIs) to advance their organization and infrastructures [2]. For a while, HEIs suffered a metamorphosis that advanced their capacities and competencies as never before.

In Kafka's book, Gregor Samsa dies. However, some authors such as Drüke (2013) [1] believe that the title of the book refers to the transformation of Grete, Gregor's sister, who experiences along the book a transformation to adapt to the new family circumstances. As Drüke, we believe that the metamorphosis experienced by teachers during the pandemic might serve as a catalyst for developing capabilities that last and promote a deep and sustainable transformation of Higher Education Institutions into organizations of the 21st century.

From the 1960s, institutions have invested in the creation of Teacher and Learning Centers (TLCs) as the solution for supporting this transformation. Although the concept of TLC has evolved, the spirit remains in the institutions as internal units with capabilities and positions to promote sustainable transformation of teaching and learning practices or, as Ringer defines, the hubs of educational reforms [3]. In the traditional paradigm, TLCs focus on the improvement of teaching skills and transfer of knowledge about student learning [4]. However, a new paradigm of TLCs emerged in the past years that stresses the importance of involving other stakeholders, such as students or managers, in the center activities [4], and emphasizes the idea of leadership role that they might play for having more pervasive effects [5]. Authors such as Holt et al. (2013) [5] stress the idea that TLCs "need to see their strategic leadership contribution as the designers and sustainers of open teaching and learning networks encompassing powerful forms of learning both across, and up and down the organization".

The role of the Higher Education institutions is to develop practical knowledge to be transferred to their students to prepare skilled workforce adapted to the continuously changing market needs [6]. According to a recent study by Dondi et al. (2021) [7], today, digital skills are one of the four categories that will help students to thrive in the future of work. So, HEIs need to expose students to a variety of learning methods (online, face to face blended, theoretical, and practical) to assure job-readiness [1]. They need to assure a sustainable digital transformation, since the transformation taking place in HEIS around the globe may become the blueprint for innovation and creativity in the next decade [8]. TLCs are the units that should guide institutional transformations from the core to this end.

To support the continuous development of TLCs, prior work proposes different models defining the competencies that these centers should develop, mainly focusing on teachers' and students' development of digital and pedagogical practices. Although these models have shown to be very effective for defining the plans for teachers' professional development, they are still anchored in the traditional paradigm of TLCs and mislead key aspects needed for making these centers evolve. Within these aspects, current literature highlights the need for supporting leadership and cultural change at the institution [5], the technological infrastructure for education [9], or the concept of supporting evidence-based practices to promote scholarly teaching practices [10]. There is a need for new models able to provide a more holistic perspective on the competencies of the new TLCs, more focused on providing guidelines and support for defining strategic plans for facing the challenges to come.

To advance on this challenge, this paper presents the PROF-XXI framework. This framework is one of the results of the large-scale European project PROF-XXI (http://profxxi.org/ (accessed on 21 December 2021)), which aims to build capacity for the development of TLCs adapted to the 21st century in Latin American Higher Education Institutions. The PROF-XXI framework describes the competencies that TLCs should consider for defining the strategies and actions allowing for support of teaching and learning innovation. The goal of this paper is to present and evaluate the framework with four Latin American institutions. Specifically, this paper shows how this framework can be used for analyzing the current level of competencies of an institution in terms of innovation and education from the perspective of different stakeholders. Using mixed methods, we cross-analyze quantitative and qualitative data of a workshop with 83 participants and analyzed the initiatives conducted by four TLC leaders to answer two research questions: (RQ1) How does the PROF-XXI framework helps with analyzing the competence level of the teaching and learning centers at an institution from the managers, teaching and administrative staff perspectives? (RQ2) How does the PROF-XXI framework help with identifying the competencies of the TLC developed before and during the COVID-19 pandemic lockdown? The result of this analysis aims at serving as a validation of the framework as well as a showcase for other institutions to apply it.

2. Teaching and Learning Centers: History and Models

At the end of the 1950s, international concern for teaching and learning at the higher education level was raised [11–15], which manifested itself in many ways: generation of taxonomies for assessment, support units for teaching and learning, vocational guidance and curriculum design, among other initiatives. However, the Anglo-Saxon universities began to configure in the 1960s [4] the Teaching and Learning Centers (TLCs) as we know them today. From the 1970s onwards, open universities have contributed to the further consolidation of the concept, by developing advanced models of teacher support and especially learner support, embedding those activities in their core organizational service structure [15,16].

In recent years, the globalization of higher education combined with the dissemination of digital technologies generated a strong political and social pressure for universities to continuously innovate their teaching and learning practices. A growing need to identify, build and develop strategic actions and contributions of support units for teachers and students has become a means for higher education institutions to guarantee quality and competitiveness [16–18].

Consequently, the need for frameworks to guide these practices and types of experiences has been emerging. Some of them focused on benchmarking performance [17,18], others on good practices and institutional policies [17,19]; some centered on the development and accompaniment of teachers [11,19–21], the systematization and maturation of learning about practice as significant referential elements for teacher development [11,20,22–24], or from a vision of co-creation of added value and relevance for institutions and internal collaboration [4,25]. In Table A1 in Appendix A, we present several alternative models that represent these different approaches indicating the stakeholders they focus on (teachers, students, or managers).

Despite all these experiences, the gap of not having an articulated frame of reference to guide leadership and decision-making has become evident and necessary in recent times [4,18,22,26,27]. In fact, as literature shows, teaching and learning innovation can only be successfully embedded in higher education organizational culture and practices if supported by strategically driven systemic change [28]. The need to use a holistic approach for such purpose was particularly evident in the context of the pandemic as higher education institutions across the world realized they should move from randomly selected emergency remote teaching practices to more sustainable, evidence-based digital transformation processes involving strategically their entire operations, infrastructure, and staff. Recent studies analyzing the changes conducted by HE institutions during the COVID-19 pandemic also align with this idea. For example, the work by Alan and Parvin (2021) [28] proposes a policy framework for managing higher education during emergency periods based on the idea that "only a substantial policy framework will enable online technology to play a constructive role". That is, the metamorphosis of an organization can only be complete if it involves the entire body in a process of change.

This was the background which led to the development of the PROF-XXI framework, described in detail in the following section. At the core of its design, two basic references were used, which represent the holistic and organic nature of the PROF-XXI TLC model: the European Framework for Digitally Competent Educational Organization (DigCompOrg: https://ec.europa.eu/jrc/en/digcomporg (accessed on 21 December 2021)) and the European Digital Competence Framework for Educators (DigCompEdu: https://publications.jrc.ec.europa.eu/repository/handle/JRC107466 (accessed on 21 December 2021)). In fact, the PROF-XXI framework was designed having in mind the critical interdependence that should be established between the institutional vision, strategy, and policies; infrastructure, processes, and organization; faculty, educational practices, teacher, and learner support; and, finally, quality and evaluation. These four basic pillars of higher education institution operation informed the five dimensions of the PROF-XXI TLC competencies framework (first published in [29]).

3. The PROF-XXI Framework

The change of teaching and learning practices as the metamorphosis metaphor suggests a dynamic process with different stages that could have different paces. A TLC should not only be able to address each of the different changes and paces but also be organized as a complex, multilayered, and multipurpose unit. Therefore, the PROF-XXI competencies have been structured according to five levels, each representing a TLC type of activity and stage of development. Specifically, the PROF-XXI framework proposes a set of competencies that institutions can take as a reference to develop the actions and strategies of their TLCs into these five levels and five dimensions that interrelate to define a total of 50 competencies (see Figure 1). Levels are from 1 to 5, where 1 means the lowest level of competence and 5 means the highest. See a complete definition of each level and dimensions in Appendix B.

While Level 1 "Development" defines the basic competencies that any TLC should have to start its innovation in teaching and learning, Level 5 "Public Accountability of Impacts for Continuous Improvement" refers to those competencies needed to monitor the actions carried by the TLCs to assess their impact and assure transparency. In the middle, there are Level 2 "Innovation", Level 3 "Value Generation", and Level 4 "New Challenges and Opportunities". The "Innovation" level refers to those competencies able to generate and promote educational innovation at the institution. The "Value Generation" level makes explicit those competencies that will enable the institution to add value to its teaching and learning practices, generating changes that affect its culture. The "New Challenges and Opportunities" level refers to those competencies that institutions need to identify new horizons on teaching practice and quality learning scenarios, enabling the identification of indicators and metrics for evaluating these innovations.

LEVEL OF COMPETENCE	OBJECTIVE	DIMENSIONS (A–E) AND COMPETENCES (1–5)				
		A. Teacher support	**B. Student support**	**C. Leadership, Culture and Transformation**	**D. Technology for Learning**	**E. Evidence-based practices**
1	**Development**	A1. Identify innovative good practices with high impact considering the institutional educational project and local, national, and international trends and good practices. A2. Design a model and/or actions for teacher training and counselling based on good practices and the characteristics of the disciplines. A3. Initially implement actions designed considering different scenarios and contexts.	B1. Diagnose student characteristics and needs to enhance teaching and learning processes. B2. Develop actions to support students considering the diagnosis, the educational project, and the characteristics of the disciplines. B3. Implement actions or initiatives, considering emerging situations, and adjustments for continuous improvement.	C1. Identify stakeholders, considering characteristics, socio-cultural scenarios, leadership, and cultural particularities within the institution. C2. Analyse previous experiences within the institution, linked to organizational cultural changes, identifying strengths, weaknesses, and challenges. C3. Conceive a plan for the TLC adapted to the University's strategy, considering international reference models and good practices.	D1. Systematise prior learning in digital education and TEL, implemented within the institution. D2. Design a pedagogical model for quality digital education and TEL, D3. Implement digital education and TEL actions, in an initial way, considering the contexts and available resources.	E1. Identify available evidence and good practice in relation to teaching and learning at higher education level and disseminate these findings within the education community. E2. Collect initial results and effects on the actions taken, considering feedback from stakeholders and external non-participating peers.
2	**Innovation**	A4. Set up benchmarking and innovative experiences. A5. Promote innovation processes among teachers that have an impact on learning. A6. Encourage the coverage of teachers who implement innovations.	B4. Promote meaningful practices and tools for learning among students in the various disciplines. B5. Foster the articulation of teaching innovation with the effective learning experience of students.	C4. Promote groups of innovative teachers and support their transformational dynamics. C5. Generate collaboration and work networks among the various members of the university community.	D4. To train the various actors of the university community for the development of innovative practices with the use of digital technology. D5. To observe new TEL trends at international level.	E3. Use available evidence on teaching and learning. E4. Promote peer exchange related to innovative practices in teaching and learning.
3	**Value Generation**	A7. Disseminate innovative internal models and experiences of excellence for teaching practices. A8. Generate changes in teaching practices and their effects on the student experience.	B6. Contribute to increasing the quality of learning. B7. To bring about positive effects on the student experience	C6. Actively participate in the generation of an institutional culture of sustainable transformation and quality. C7. Position the unit as relevant within the educational process.	D6. Encourage the increased implementation of TEL in everyday learning. D7. Model TEL best practices among the educational community.	E5. Generate evidence on results, outcomes, and impacts. E6. Share the evidence collected among the different levels and actors of the organization.
4	**New Challenges and Opportunities**	A9. To visualize new horizons on teaching practice, capable of fostering transformative pedagogical practices.	B8. Visualize new challenges and scenarios in student learning as lifelong practice. B9. Ensure the overall increase of student learning outcomes.	C8. Define metrics and indicators to evaluate the impact of pedagogical innovation. C9. Contribute to the process of transforming the University into a learning and innovative organization,	D8. Sustainable involvement of the teaching staff in TEL pedagogical practices	E7. Systematise the new challenges arising from the work, the available evidence, and the good practices in actions for the institutional strategy.
5	**Public accountability of impacts for continuous improvement**	A10. Ensure tools to monitor and report on the quality of innovative teaching practices. A11. Support institutional decision-making based on challenges and good practices.	B10. Implement evaluation surveys on the university and student learning experience. B11. Support institutional decision-making based on the student experience.	C10. Evaluate the transformative impact of innovative teaching practices. C11. Generate spaces for dialogue and meetings that foster co-responsibility for results, effects and impacts among the different actors in the educational community.	D9. Implement public reports that account for the effects of TEL on educational practice.	E8. Communicate scientifically in different formats and external academic communities, the processes implemented within the unit.

Figure 1. PROF-XXI Framework organizing the competencies of a TLC into five levels of competency and five dimensions. Extracted from [30].

The PROF-XXI framework also organizes the competencies into five dimensions that refer to the institutional aspects that are affected by the different competencies developed by the TLC. The "Teacher Support" (Dimension A) refers to the competencies related to the support of teachers, while the "Student Support" (Dimension B) refers to those related to students' support. The dimension "Leadership, Culture and Transformation" (Dimension C) is one of the most innovative added by this framework. It refers to the competencies needed for leading and promoting cultural transformations at the institution through the definition of new policies and actions that affect its current processes. The dimension "Technology for Learning" (Dimension D) refers to those competencies that an institution should have to manage educational initiatives supported by technology, including the definition of technological processes and infrastructures. Finally, the dimension "Evidence-based Practice" (Dimension E) refers to the competencies needed to be able to collect data and information for understanding the effect of the transformation conducted by the institution.

The PROF-XXI framework was defined to be used in three different ways, depending on the objective of the institution: (1) as a self-assessment tool to help institutions understand their overall competence level, (2) as a reference for strategic planning definition to identify the strategic aspects to develop, or (3) as an accreditation framework to certify the innovation competence level of their TLC. This paper will focus on evaluating how the PROF-XXI framework can be used as a self-assessment method.

4. Methods

4.1. Research Objective and Design

The main aim of this paper is to evaluate the PROF-XXI framework as a self-evaluation tool to help HEIs understand the competencies of their teaching and learning centers and identify the institutional changes regarding their teaching and learning innovation policies conducted during the COVID-19 pandemic. Two research questions guided the data collection process and analysis: (RQ1) How does the PROF-XXI framework help with analyzing the competence level of the teaching and learning centers at an institution from the managers, teaching and administrative staff perspectives? (RQ2) How does the PROF-XXI framework help with identifying the competencies of the TLC developed before and during the COVID-19 pandemic lockdown?

To address these research questions, we adopted a mixed method research approach. Mixed methods are used in research that involves collecting, analyzing, and interpreting both qualitative and quantitative data from a single study to understand a phenomenon in its context. This research approach has become popular since the 1960s in disciplines such as education [30], in which a unique paradigm of research (qualitative or quantitative) is not enough to understand a complex phenomenon. For our study, we chose a fully mixed sequential method approach with a dominance of the qualitative dimension [31] to complement quantitative and qualitative data collected from a large sample of participants with qualitative information from a small sample.

We organized the research design into two phases (see Figure 2). The first phase consisted of a two-hour workshop with participants from four Latin American universities. During the workshop, the organizers introduced the PROF-XXI framework and conducted two practical activities with the participants. In the first activity, the participants were asked to answer a questionnaire to evaluate the competencies of their institution regarding the different dimensions of the PROF-XXI framework. In the second activity, they were grouped in teams of five to eight people of the same institution to compare their perceptions of the different competencies, identify activities and/or initiatives existing at their institutions and classify them into the different dimensions of the framework.

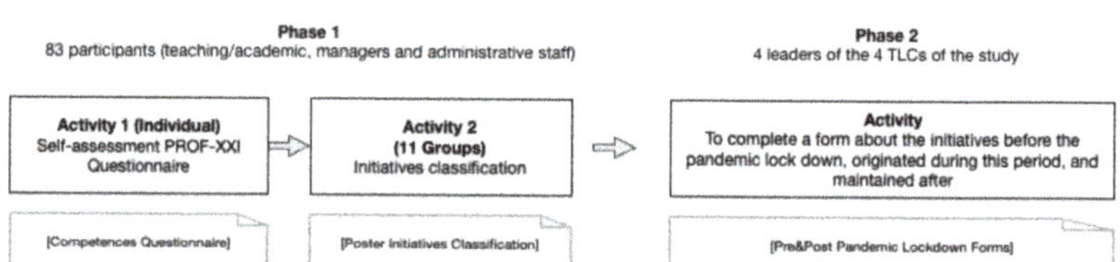

Figure 2. Phases of the different activities conducted for gathering all the data for the study.

The second phase occurred 2 months after the workshop and after three researchers analyzed the data from Phase 1 to extract general conclusions of the collected data so far. The researchers prepared a document to be completed by the main leaders of the Teaching and Learning Centers of each institution. In this document they had to indicate what kind of initiatives related to education and innovation existed at their institution before the pandemic and which of them were created specifically during the pandemic and maintained nowadays.

Further details about the data gathering techniques, instruments and original data can be found in the following sections and are accessible in the Supplementary Material.

4.2. Participants and Sample

Four universities participated in this study: two universities in Guatemala (U1 and U2) and two in Colombia (U3 and U4). These universities differ in size, type of administration (public or private) and year of foundation, which represent contrasting higher education systems (see Table A7 in Appendix F for more information about the universities). They all have TLCs in charge of supporting teachers' professional development. Eighty-three people from these four universities participated in the first phase of the evaluation. Table 1 shows the role of the different participants in each university. Only the leader of the TLC in each institution participated in the second phase of the evaluation (four people in total).

Table 1. Participants from the different universities in the different phases.

University	1st Phase of the Analysis				2nd Phase of the Analysis
	Administrative	Manager	Teaching/Academics	Total	Teaching and Learning Center Leader
U1 (Universidad de San Carlos de Guatemala)	11	6	9	**26**	**1**
U2 (Universidad de Galileo)	8	2	-	**10**	**1**
U3 (Universidad de San Buenaventura Cali)	5	5	13	**23**	**1**
U4 (Universidad del Cauca)	1	1	22	**24**	**1**
Total	**25**	**14**	**44**	**83**	**4**

4.3. Data Analysis

Different data were collected during the two different phases of the evaluation. Table 2 shows the different data sources and codes used to refer them, its nature (qualitative or quantitative), the links to the original files used for the data collection and the collected data. All the data are accessible publicly via the following link: https://osf.io/e742r/ (accessed on 21 December 2021). Three researchers participated in the different phases of the analysis for answering the different research questions addressed.

Table 2. Data gathering techniques and nature.

Code	Description	Nature of the Data Collected	Original Instrument
[Competencies Questionnaire]	Questionnaire including 50 questions in which the participants have to value from 1 to 4 each of the competencies in the PROF-XXI framework in their institution	Quantitative	[Phase1-Activity1-CompetenciesQuestionnaire-EN]: https://osf.io/zdw3e/ (accessed on 21 December 2021). [Phase1-Activity1-CompetenciesQuestionnaire-ES]: https://osf.io/ehr2t/ (accessed on 21 December 2021).

Table 2. *Cont.*

Code	Description	Nature of the Data Collected	Original Instrument
[Poster Initiatives Classification]	Collaborative digital poster created with Lucid.app for the participants to classify the different activities and initiatives conducted by their institution within the PROF-XXI framework competencies (See Appendix C). Participants had 20 min to add and discuss about the initiatives existing in their institution and associate them to a particular competence of the PROF-XXI framework.	Qualitative	[Phase1-Activity2-PosterInitiativeClassification]: https://osf.io/mfjtg/ (accessed on 21 December 2021). [ANNEX 1] For accessing the original poster used during the sessions and the main contributions.
[Pre & Post Pandemic Lockdown Forms]	For to be completed by the TLC leaders. It includes two sections: (1) a table for listing the initiatives carried out for the institution to encourage the transformation and innovation for the teaching and learning processes, indicating whether they existed before the pandemic lockdown, whether they were maintained during the this period, whether they were originated with the pandemic lockdown, whether they are currently maintained in the institution; (2) a table for indicating, for each of the initiatives in the first table, to which dimension and competencies of the PROF-XXI framework they are associated. Only those responsible of the TLC of each institution completed this form.	Qualitative	[Phase2-PosCovidForm-EN]: https://osf.io/jxhc5/ (accessed on 21 December 2021). [Phase2-PosCovidForm-ES]: https://osf.io/2trk5/ (accessed on 21 December 2021). [Phase2-U1-PosCovidForm-ES]: https://osf.io/p6rk9/ (accessed on 21 December 2021). [Phase2-U2-PosCovidForm-ES]: https://osf.io/5rac3/ (accessed on 21 December 2021). [Phase2-U3-PosCovidForm-ES]: https://osf.io/g45ty/ (accessed on 21 December 2021). [Phase2-U4-PosCovidForm-ES]: https://osf.io/s684y/ (accessed on 21 December 2021).

To address RQ1 about how the PROF-XXI framework helps managers, teaching and administrative staff with analyzing the competence level of the teaching and learning centers at their institution, data collected during the first phase from both the [Competencies Questionnaire] and [Poster Initiatives Classification] were analyzed. Firstly, one researcher analyzed the data from the questionnaire using Excel for calculating the average value given to each competence in the model per university (Administrative staff, Manager Staff and Teacher Staff) (Table A3 in Appendix D). Then, using these processed data, two researchers worked independently to extract a list of partial results about how institutions perceive their competence level (Table A4 in Appendix D). Secondly, two researchers analyzed the list of activities indicated by the participants in the [Poster Initiatives Classification] to understand what type of activities were associated with each of the competencies in the framework. Table A5 in Appendix D summarizes the partial results of this analysis, indicating some of the supporting data for each result. Finally, the quantitative results of the [Competencies Questionnaire] and the qualitative analysis of the [Poster Initiatives Classification] were triangulated to deepen our understanding of how the framework can help with analyzing the current competencies of a TLC. This process consisted of contrasting evidence obtained from the different stakeholders and from the different data sources. Three researchers participated in this process. Table 3 shows the main findings obtained from this process.

Table 3. Findings from RQ1. Cross-analysis of Tables A2 and A3 for extracting findings about RQ1.

Finding Code	Description	Partial Result Supporting the Finding
F1.1	All staff in all institutions perceive that the Competence "A. Teachers' support" is one of the most well-developed in their institution. They associated initiatives related to training the trainers (mostly for supporting the digital transition) and activities for teachers' professional development. However, we noticed that, from all the roles analyzed (Administrative, Managers and Teaching/Academics staff), the managers were the ones giving the lowest values to this competence, while the Teaching/Academic staff in two universities (U1 and U4) evaluated them as the most well-developed.	[PR1.1] In all institutions, the Competence "A. Teachers' support" was valued as one of the most developed (Table A3 in Appendix D). [PR1.3] Participants from U2 and U4 evaluated the Competence "A. Teachers' support" as the most well-developed competence in the institution, and the Competence "E. Evidence-based practices" as the least developed (Table A3 in Appendix D). [PR1.4] In all institutions, the "Manager Staff" evaluates the Competence "A. Teachers' support" with the lowest values, together with the "Teaching Staff" from U3. However, "Teaching/Academic Staff" from U1 and U4 evaluated it as the most well-developed. (Table A3 in Appendix D) [PR2.1] To the Competence "A. Teachers' support", institutions associated initiatives for training the teachers. The types of trainings vary in frequency and format depending on the institution, including courses, workshops, seminars, and diplomas (a set of courses with several ECTS credits). Most of trainings focus on learning about digital tools. Participants also associate with these competences related to teaching recognition, teaching evaluation and the share of good practices (Table A4 in Appendix D).
F1.2	All staff in all institutions perceive that the Competence "B. Students' support" is one of the least developed. Participants associated initiatives such as online courses, video-tutorials as well as academic support or on the Learning Management Systems employed by the university. Participants also recognize that, in some cases, the Competence "Students' Support" is a bit poor.	Results in Table A3 in Appendix D. [PR2.2] To the Competence "B. Students' support" participants associated initiatives such as online courses, video-tutorials as well as academic support or on the Learning Management Systems employed by the university. Participants also recognize that, in some cases, the Competence "Students' Support" is a bit poor (Table A4 in Appendix D).
F1.3	Despite the Competence "C. Leadership, Culture and Transformation" was not perceived as one of the most well-developed competencies; participants were able to associate some institutional activities, mainly related with the development of the "sense of belonging" to the institution, self-assessment activities, cross-institutional initiatives, and digital transformation.	Results in Table A3 in Appendix D. [PR2.3] To the Competence "C. Leadership, Culture and Transformation" participants associated activities such as (1) programs for developing the sense of belonging to the institution and its culture; (2) instances for self-evaluation, and instances for interacting with other institutions through research international programs. They also mentioned activities addressed to teaching/academics and administration staff related to the digital transformation of institutional processes (Table A5 in Appendix D).
F1.4	Participants evaluated the Competence "D. Technology for Learning" as one of the most well-developed competencies and associated activities mainly related to training initiatives in the use of institutional platforms. Most of these initiatives were addressed to the teachers/academic staff, which indicates that these initiatives are closely related with Competence "A. Teachers' support".	Results in Table A3 in Appendix D. [PR2.4] To the Competence "D. Technology for Learning" participants associated initiatives such as training in the use of technological platforms (i.e., Moodle, Google Classroom) and tools (i.e., Google Suit) through online material, tutorials and courses (Table A5 in Appendix D).

Table 3. *Cont.*

Finding Code	Description	Partial Result Supporting the Finding
F1.5	The Competence "E. Evidence-based Practice" was perceived by all participants as the least developed competence in the institution. Participants associated to this competence initiatives related to the use of institutional data (Learning Analytics) for monitoring students' and teachers' progress and performance as well as activities related to the continuous improvement of the curriculum and benchmarking for studying initiatives in other institutions.	[PR1.2] All institutions, evaluated the Competence E "Evidence-based Practices" as the least developed. [PR2.5] To the competence "E. Evidence-based Practices", participants associated initiatives related to the use of institutional data. The refer to initiatives for monitoring teachers and students' performance. They also associated activities and initiatives related to the continuous curriculum improvement and benchmarking initiatives looking for other institutions practices as a reference.
F1.6	The use of the model as a self-assessment mechanism also shows that we can distinguish between those institutions with the highest and lowest competencies. In this case study, U2 was one of the institutions with the highest competencies, which is one of the institutions with more experience in the digital transformation of their teaching and learning processes.	[PR1.5] Institution U2 has reported the highest values in terms of competence dimensions and compared to the other institutions.

To address RQ2 about how the PROF-XXI framework helps with identifying the TLC competencies evolution before and during the pandemic lockdown, we analyzed the [Pre&Post-COVID Form] completed per the leader of the TLC at each institution. One of the implicated researchers organized in a table the different initiatives aligned with the competencies of the PROF-XXI framework and indicated whether the initiatives were created before or during the pandemic and if they have been still carried out by the institution (see Tables A3 and A4 in Appendix D). Table 4, together with a qualitative analysis of the initiatives described, was then generated by the three researchers to extract a consensual list of findings that explains the evolution of the initiatives in the different institutions before and during the pandemic lockdown.

Table 4. Findings from RQ2. Findings for RQ2 obtained from the analysis of [Pre&Post Pandemic Lockdown Forms]. The supporting data of these findings are in Tables A5 and A6 of Appendix E. Column period indicates the period (before, during or after the pandemic lockdown) referred in the findings.

Period	Finding Code	Description
Before the lockdown	F2.1	Before the pandemic, most of the institutions counted with long training programs for teachers (diplomas of several weeks, for example). These programs were designed for training the teachers in different areas (digital tools, pedagogical support, etc.) and are still maintained after the pandemic lockdown. However, any institution create new training programs of this type during the pandemic lockdown. Only short training programs, such as workshops for showing specific tools or training teaching methodologies, were created during this period. All these initiatives are related to Competencies A ("Teachers' Support") and D ("Technology for Learning") of the PROF-XXI framework.
	F2.2	Before the pandemic lockdown, the least developed competence from the PROF-XXI framework was the Competence "B. Students' Support" (5 initiatives out of the 16 existing initiatives before the pandemic lockdown), but the initiatives related to this competence augmented during the pandemic lockdown (8 out of the 15 originated during this period). The most well-developed were "A. Teachers' support" (12 out of 16) and "D. Technology for Learning".

Table 4. *Cont.*

Period	Finding Code	Description
During the lockdown	F2.3	During the pandemic lockdown, institutions invested most of their efforts in developing the Competencies "A. Teachers' support" (10 out of 15 initiatives were related to this competence) and "D. Technology for Learning" (12 out of 15 initiatives were related to this competence); investment in Competencies "E. Evidence-based Practices" decreased (from 6 initiatives related to this competence before the lockdown, only 3 were reported associated with this competence during this period).
	F2.4	The initiatives created by the TLC before the pandemic lockdown were related with the Competencies "A. Teachers' Support" (12 of the 16 initiatives existing in this period for all universities) and "D. Technology for Learning" (13 of the 16 in total of this period for all universities). Whereas, during the pandemic lockdown, initiatives related to "B. Students' Support" doubled (5 out of 16 before the lockdown and 6 out of 15 originated during this period).
	F2.5	During the pandemic, all institutions created courses and materials (such as guidelines or video tutorials) for teachers and administrative staff that they facilitated through their online institutional systems. Some of the universities organized these materials in the form of online programs (i.e., U2). All universities related these initiatives to the competencies "A. Teachers' Support", "B. Students' Support", and "D. Technology for Learning". Only U3 related the initiative created during the pandemic to all competencies of the framework.
	F2.6	During the pandemic, U1 and U2 initiated activities for supporting teachers in the use of digital tools. Examples of these activities are coaching for teachers, personalized support, etc. These institutions explicitly mentioned that they created these initiatives for promoting innovating in online assessment practices. For example, they installed Proctoring tools for facilitating online assessment. U2 related some of these initiatives to the Competence "C. Culture and Transformation". U1 also associated some of these initiatives with the Competencies "A. Teachers' Support" and "D. Technology for Learning" of the PROF-XXI framework.
Maintained after the lockdown	F2.7	Three out of the four universities (except U3) maintain the activities that were originated for facing the pandemic lockdown. In U4, two of these initiatives are still under study to see if they are maintained or not.
	F2.8	After the pandemic lockdown, U1, U2, U4 reported they started to use the institutional platforms (i.e., VLE, Simulators, videoconferencing, etc.) in a more systematic way. These initiatives were usually related to the Competencies "D. Technology for Learning", and to Competencies "A. Teachers' Support" and "B. Students' Support" for U2.
	F2.9	After the pandemic lockdown, the number of initiatives of the TLC increased (from 16 existing before the pandemic to 27 maintained today). Although the number of initiatives associated to the different competencies increased, the universities still relate the majority of their initiatives to competencies "A. Teachers' support" (15 out of the 27 initiatives are related to this competence) and "D. Technology for Learning" (18 out of the 27 initiatives are related to this competence), whereas Competencies "C. Leadership, Culture and Transformation" (9 out of 27) and "E. Evidence-based Practices" (9 out of 27) are still the least supported competencies.

5. Results and Discussion

This section presents the main findings of the study, after analyzing the different data sources. Section 5.1 presents the findings related to the research question RQ1 about how the PROF-XXI framework can be used for analyzing the competence level of the teaching and learning centers at the institution from the perspective of different teaching staff. Section 5.2 presents the findings related to the research question RQ2 about how the PROF-XXI framework helps with identifying the competencies developed by the TLCs before, during and after the pandemic lockdown.

5.1. The PROF-XXI Framework as a Tool for Analyzing Institutional Teaching and Learning Competencies Development

Six findings were obtained from analyzing the data gathered in the workshop with 83 participants (administrative, teaching/academics, and manager staff) from four different institutions (See Sections 3 and 4 in this paper). All findings suggest that the PROF-XXI framework is a good support for getting a holistic perspective of the competencies that the institution has put more effort in developing and those that are still under development. Table 4 summarizes all these findings and the analyzed data supporting them.

The first finding [F1.1] suggests that institutions invest significantly in developing "training the trainers" initiatives for developing the competencies of their teaching staff. Most of the initiatives consist of teaching programs that vary in time, frequency, and format (online, hybrid or traditional face-to-face) for training the teachers in particular competencies and promote the exchange of good practices. Therefore, the staff perceives that the competence "A. Teachers' support" is one of the most well-developed at an institutional level. However, we observe some differences depending on their role at the university. For example, the managers are the ones giving the lowest values at this competence, whereas in two universities (U1 and U4) they rated it as the most well-developed competence [PR1.4]. This suggests that certain initiatives have more impact on some institutional staff than on others. Another possible explanation relates to the fact that this competence does not limit itself to teacher training but refers also to pedagogical and technological advising to teachers.

While the competence "A. Teachers' support" is perceived as the most developed one, the competence "B. Students' support" is perceived as the least developed [F1.2]. This finding suggests institutions consider that having an impact on teachers' competencies will have a direct impact on students' performance. Recent literature indicates that these types of approaches can have, indeed, an effect on teaching practices that influence [11] students' perception of the learning experience [32]. However, as literature on distance education has demonstrated, learner support is paramount to cover the affective dimensions of the learning experience, along with the cognitive and systemic dimensions [14]. This explains why institutions perceive that the students' support is still neglected and should be reinforced. After the pandemic, some literature reported that students had difficulties in following the courses in online environments [32,33]. As a response, some of the universities in this study initiated certain activities for supporting students in these new scenarios, such as video lectures or manuals on the use of their Learning Management Systems.

The competence "C. Leadership, Culture and Transformation" was also perceived in three of the four universities analyzed (U1, U2 and U4) as one of the least developed competencies [F1.3]. The participants associated to this competence initiatives related with the development of the sense of belonging, with mechanisms for self-evaluating the institution, or with activities including exchanges with other institutions as a benchmarking effort for identifying good practices. We observed, however, that the participants associated a smaller number of initiatives with this competence compared to others, suggesting that the institutions are developing some of the aspects related to this competence indirectly through other initiatives targeting other objectives. Nevertheless, this possibility indicates that leadership in their institutions is not implementing holistic, strategically driven integrated approaches as is recommended by the literature and best practices.

The participants from the different institutions evaluated the competence "D. Technology for Learning" as one of the most well-developed ones [F1.4]. Most of the institutions associated with this competence, with training programs targeting especially teachers for the development of digital skills. Although some of these courses were designed as tutorials for learning about a particular tool, many institutions reported initiatives focused on training teachers to operate in their virtual learning environments (VLEs). This suggests that institutions already have some digital strategy including the use of a VLE and other digital support, but they still need courses for promoting its usage among the teaching/academic staff. Since the engagement of teaching/academic staff with technology has been a recurrent problem in higher education [33], institutions have focused on providing support to face the resistance to change. However, prior studies suggest that teachers respond better to change when "their beliefs and practices are integrated, negotiated and reconciled with the demands of a changed context" [34]. The pandemic lockdown completely changed the context and, consequently, the teachers' demand increased, making institutions react urgently with new initiatives that could have their effect beyond the pandemic lockdown.

Regarding the competence "E. Evidence-based practice", the associated data suggest that participants perceived this competence as one of the least developed [F1.5]. The participants did not report many initiatives for this competence, but they associated initiatives mainly related to the use of educational data such as, for example, monitoring students' performance and teachers' progress. Some of them talked about Learning Analytics initiatives for promoting continuous curriculum improvement. In some cases, the participants referred to initiatives of benchmarking as a mechanism of self-evaluation and a way for looking for new practices. This finding suggests that, although this is one of the least developed competence, institutions are starting to see in educational data a good potential for supporting decision making processes [35,36]. The capacity to collect data and evidence should be complemented though with a much tighter connection between reflective teaching practice, educational research and innovation.

Finally, data supporting finding [F1.6] suggest that the PROF-XXI framework is a good support for identifying the overall competence level of teaching and learning practices in an institution and comparing it with others in a benchmarking exercise. For example, in this study, we identified institution U2 as the most well-developed and as a potential leader in the region compared to others. In fact, U2 is one of the institutions which has a larger experience and a higher level of maturity in the use of technology for digital learning and in initiatives for promoting teaching and learning innovation.

5.2. The PROF-XXI Framework as a Tool for Analyzing and Understanding the Evolution of TLCs Strategy

Eight findings were obtained from analyzing the data from the four leaders of the TLCs of the universities participating in this study. The findings obtained from this analysis show how the PROF-XXI framework can be used to understand the evolution of the TLCs competencies. Table 4 summarizes all these findings and the analyzed data supporting them.

First, findings suggest that (1) institutions should benefit from the course and initiatives created during the pandemic for updating and re-adapting their institutional plans for training the teachers to include training in those competencies required during the lockdown; (2) these courses should be complemented with learning capsules delivered in flexible formats (such as small learning capsules or online courses) to facilitate their adaptation and consumption. Findings F2.1 and F2.2 indicate that, even if institutions already put lots of effort in developing teachers' competencies related to "A. Teachers Support" and "D. Technology for Learning" before the pandemic (especially in digital learning and quality), these competencies were not enough to face the lockdown challenges. Consequently, and as indicated by F2.3 and F2.5, institutions reinforced these two competencies during the lockdown through manuals and online courses for training teachers in particular tools. Moreover, F2.6 indicates that some institutions also implemented during this period tools such as proctoring tools for supporting new practices that they expect to maintain after this lockdown period (F2.7, F2.8 and F2.9).

These findings align with current literature, which emphasizes the importance of looking for models to adapt teachers' training to their personal needs [11] as the only way to promote actual changes in the institutions' culture and practices. Moreover, a recent publication shows that short online pedagogical training for university teachers has an effect on their interpretation of teaching–learning practices [33], suggesting that these types of courses could facilitate training teachers at scale in an effective manner. In addition, some authors show that the resources generated during this period can also be a mechanism to transform formal education [37]. That is, institutions have now the opportunity of benefiting from the resources developed during the pandemic lockdown to expand their training offer and effectively transform their traditional practices.

Second, findings show the importance of introducing, as part of the institutional strategy, initiatives dedicated to support students in the transition to digital education, especially in the digital competencies needed to succeed in online and hybrid practices. F2.3 indicates that initiatives related to "B. Students' Support" were one of the most neglected competencies before the pandemic, but that institutions doubled the number of initiatives related to this competence during the lockdown (F2.4). Moreover, F2.5 shows that some of the material produced during the pandemic for supporting teachers was also associated with students' support competence, suggesting that this material had a double purpose, to support both teachers and students.

Recent literature on the impact of the pandemic lockdown on students provides evidence that students faced various problems during this period [28,38]. Some of these problems were (1) the lack of in-home infrastructure for following online courses, especially in countries with higher socio-economical inequalities, and (2) an unfavorable study environment [32]. However, data collected after this first lockdown period show that students adapted well to the new forms of teaching and learning [32], but that institutions should still provide support for assuring the psychological well-being of students in these circumstances [32].

6. Conclusions and Implications

This paper presents the result of evaluating the PROF-XII framework as a tool for analyzing the institutional teaching and learning competencies development, as well as the evolution of the TLC strategy. The PROF-XXI framework advances the existing frameworks for supporting HEIs in their development for facing the challenges of the 21st century by providing a holistic vision of the competencies that institutions should support and develop. For the first time, a model integrates the critical interdependence between the institutional vision, strategy, and policies, on the one hand; infrastructure, processes, and organization, on the other; as well as faculty, educational practices, teacher and learner support; and, finally, quality and evaluation.

The evaluation was conducted with 83 participants from four Latin American universities (including managers, administrative and teaching/academic staff) and the four leaders of the TLCs of these institutions. The mixed-methods analysis of the collected data shows that the framework can be used as a self-assessment method for analyzing the actual development of the teaching innovation competencies at the institution through the perspective of the different stakeholders. In addition, the model was shown useful as a reference for classifying the different initiatives conducted by the TLCs and analyzing how they evolved across time, according to the emerging contextual needs. The results of this analysis indicate that the four analyzed institutions had already installed initiatives to train the teachers in digital skills and quality pedagogical practices before the pandemic, which were reinforced with short courses and learning capsules during the pandemic lockdown. These courses were created for both teachers and students since the competencies related to the support of these last ones were a bit neglected before this emergency period.

The presented work has several implications at different levels. At a theoretical level, this paper contributes with a new framework that discusses the competencies that TLCs should consider for adapting to the new societal needs and become the core of the sustainable innovative digital development at the institution. This framework proposes a holistic perspective of all the elements to be considered, which is aligned with recent policy models proposed for facing the emergency changes derived from the COVID-19 pandemic [6]. Moreover, as suggested by Alam et al. (2021) [39], HEIs are used to prioritize market-oriented outcomes, but nowadays more and more universities are focusing on promoting sustainability in their operations as well. The PROF-XXI framework offers a starting point to discuss about how operations related with teaching and learning innovation can become a daily practice in the long term.

At a practical level, the PROF-XXI framework can be used as a self-assessment tool to identify what the level of competency of a particular institution is and develop a strategy accordingly. This idea of self-effacement tool is in line with what is currently proposed by the UE platform "Digital Skills & Job Platform" [40], which proposes a test for teachers to self-evaluate their teaching digital skills. Similar solutions for evaluating what the level of competency of the institution is in terms of teaching and learning innovation could be offered using the PROF-XXI framework as a basis.

7. Limitations and Future Work

This is the first study that evaluates the use of the PROF-XXI framework as a reference tool for supporting the development of competencies in HEIs. Since this evaluation was conducted with four Latin American institutions, the main conclusions are limited to this sample. Although we included a sample of universities from two different countries and of different nature in terms of size and management, more studies with other universities could help extract further conclusions about the framework. These studies should also include students as part of the stakeholders' analysis, since recent students show that they play a key role when conducting institutional transformations [39]. In addition, we did not analyze weather the use of this framework will produce changes in the policies in the long term. This is another interesting aspect to explore in future studies.

Future work should include analysis of other institutions. To facilitate a large-scale evaluation of the framework with other institutions, we are currently working on a web-based dashboard to visualize data of the PROF-XXI framework. This tool will facilitate the distribution of the questionnaire about the competencies employed in this study among all the educational stakeholders, including students, and compare their perception about their innovation in teaching and learning competencies. We expect that both the framework and the results of this study could help HEIs in Latin America and beyond to understand how to improve their training programs and advance on those competencies that need to be addressed for anticipating the post-COVID-19 pandemic era in a sustainable way.

Supplementary Materials: Supplementary material could be found in https://osf.io/e742r/.

Author Contributions: M.P.-S. coordinated the preparation of this paper, designed the evaluation process, collected the data, leaded the data analysis, conducted the cross-analysis and leaded the writing process of the whole paper; I.K. participated in the data collection and analysis and contributed to the writing process about the teaching and learning centers models and Table A1; F.M. participated in the data analysis; A.T. and M.d.C.T.P. participated in the conception of the PROF-XXI framework and wrote all the section about the teaching and learning models history and evolution; J.B. supervised the evaluation design and reviewed the different versions of the paper; C.A.-H., Ó.J. and C.D.K., participated in the conception and funds of the project PROF-XXI, reviewed the different versions of the paper and contributed the introductory section; B.G. participated in getting the funds for writing this paper; M.M., M.S., L.M.O.-C. and A.H.G.L., participated in the data collection by recruiting the participants of the evaluation process and completed Table A7. All authors have read and agreed to the published version of the manuscript.

Funding: The authors acknowledge PROF-XXI, which is an Erasmus+ Capacity Building in the Field of Higher Education project funded by the European Commission (609767-EPP-1-2019-1-ES-EPPKA2-CBHE-JP). This publication reflects the views only of the authors and funders cannot be held responsible for any use which may be made of the information contained therein.

Informed Consent Statement: Informed consent was obtained from all subjects involved in the study.

Data Availability Statement: All data-gathering instruments and original data can be accessed here: https://osf.io/e742r/ (accessed on 21 December 2021).

Conflicts of Interest: The authors declare no conflict of interest.

Appendix A.

Table A1. Analysis of the different models for supporting competencies of Teaching and Learning Centers (TCLs) most referred in the literature.

Framework/Model	Description
Teachers' Focus	
European Framework for the Digital Competence of Educators (DigCompEdu)	DigCompEdu was published in late 2017 by the Joint Research Centre of the European Union (JRC) (Redecker & Punie, 2017). Its main objective is to align the European educational policies with such a reference framework. DigCompEdu is a digital competence model with six differentiated competence areas: Professional engagement, Digital resources, Teaching and Learning, Assessment, Empowering learners, and Facilitating learners' digital competence. Each area has a series of competencies that "teachers must have in order to promote effective, inclusive and innovative learning strategies, using digital tools" (Redecker y Punie, 2017, p. 4).
UNESCO ICT Competence Framework for Teachers (ICT-CFT)	This framework, developed by UNESCO, presents "a wide range of competencies that teachers need in order to integrate ICT in their professional practice" (Butcher, 2019, p. 2). It fosters practical knowledge of the advantages that ICT provides in education systems. Moreover, it suggests that teachers, apart from acquiring competencies related to ICT, must be able to use these to help their students to become collaborative, creative, innovative, committed, and decisive citizens (Rodríguez et al., 2018). This framework presents six fundamental areas or aspects of the professional teaching practice: Understanding ICT in the educational policies, Curriculum and evaluation, Pedagogy, Application of digital abilities, Organization and administration, and Professional learning.
Common Spanish Framework of Digital Competence for Teachers of the "Spanish Institute of Educational Technology and Teacher Training	The Spanish Ministry of Education, Culture and Sport launched a project in 2012 to define the Common Framework of Digital Competence for Teachers, updated four times (Instituto Nacional de Tecnologías Educativas y Formación del Profesorado, INTEF, 2017a, 2017b). It is based on the DigComp Framework of Digital Competence for Citizens (Carretero, Vuorikari, & Punie, 2017; Vuorikari, Punie, Carretero, & Van-Den-Brande, 2016). It is a generic digital competence model for educators. The competence areas (5) and competencies (21) are those of the DigComp framework.
British Framework of Digital Teaching	The British Framework of Digital Teaching was created by the Education and Teaching Foundation (ETF) in association with the JISC company (Education and Training Foundation, 2019). Its main objective is to increase the understanding of teachers in the use of digital technologies to enrich their teaching practices and improve their professional development (Pérez-Escoda et al., 2019). This framework consists of seven key areas, with three levels for each of them: exploration, adaptation, and leadership. The seven elements are Pedagogical Planning, Pedagogical Approach, Employability of the Students, Specific Teaching, Evaluation, Accessibility and Inclusion, and Self-development.
ICT Competencies and Standards for the Teaching profession of the Chilean Ministry of Education	The Education and Technology Centre of the Chilean Ministry of Education published this framework in the year 2011, as an updated version of a previous framework published in 2006 (Elliot, Gorichon, Irigoin, & Maurizi, 2011). It presents five dimensions aligned with the UNESCO Framework of ICT Competencies for Teachers (Butcher, 2019). All five dimensions work through descriptors, criteria and competencies. Moreover, each standard allows teachers to recognize how to use and integrate ICTs, identify their training needs, and define personalized training itineraries (Ríos, Gómez, & Rojas, 2018).

Table A1. *Cont.*

Framework/Model	Description
Teachers' Focus	
Framework of Implementing Collaborative Learning in the Classroom (ICLC)	The ICLC framework is based on the metacognitive framework of teacher practice by Artzt and Armour-Thomas (1998) that describes teaching in analogy to the cognitive process of solving a problem in three phases: a pre-active phase, an inter-active phase, and a post-active phase (cf. Jackson 1968). While the framework focuses on the teacher level, the student level is also presented in the framework, as the teacher's goal is to ensure a high quality of student interaction, on which the effectiveness of collaborative learning depends (Dillenbourg et al., 1996; Kobbe et al., 2007; Webb, 1989). The ICLC framework distinguishes between five teacher competencies that span across all implementation phases of collaborative learning: the ability to plan student interaction, monitor, support, and consolidate this interaction, and finally reflect upon it.
Students' Focus	
Framework of the "International Society for Technology in Education" (ISTE) for teachers	The International Society for Technology in Education develops this competence framework focusing on the needs of the students of the 21st century (Crompton, 2017). Its main objective is to delve into the teaching practice, promote student collaboration, rethink the traditional approaches, and boost autonomous learning (Crompton, 2017; ISTE, 2018; Pérez-Escoda, García-Ruiz, & Aguaded, 2019). The general teacher profile is characterized by being active and innovative in the teaching–learning process (Gutiérrez-Castillo, Cabero, Almenara, & Estrada-Vidal, 2017). Thus, the ISTE standards for teachers are divided into seven roles or profiles that an educator must develop along his/her professional career. Framework with seven differentiated competence areas: Learners, Leaders, Citizens, Collaborators, Designers, Facilitators, and Analysts.
Managers' Focus	
ICT Competencies for Teachers' Professional Development of Colombian Ministry of Education	The model proposed by the Colombian Ministry of Education aims to guide the professional development of teachers to improve educational innovation with ICT (Fernanda, Saavedra, Pilar, Barrios, & Zea, 2013). It is targeted at both designers of training programs and teachers interested in generating ICT-enriched environments: relevant, practical, established, collaborative and inspiring (Hernández-Suárez, 2016). This framework has five competencies that teachers must develop: Technological, Communicative, Pedagogical, Management, and Research.
Framework for the Center for Teaching Development and Innovation (Centro de Desarrollo e Innovación de La Docencia (CeDID) at the Universidad Católica de Temuco (UCT))	A framework for the evaluation of educational development programs in Chile. This framework was designed to support the diverse needs of different stakeholders: (1) faculty to make judgments about their teaching in their school and disciplinary context; (2) the learning center to evidence the impact of their educational development programs; (3) the university to inform its attainment of its planned strategic goals; and finally (4) the ministry on the effectiveness and impact of the programs that it has funded. The CeDID Evaluation Framework drew on Guskey's five-level model, which identifies where educational development programs can demonstrate impact (Chalmers & Gardiner, 2015). These are (1) Teachers' reaction to the development program; (2) Conceptual changes in teachers' thinking; (3) Behavioral changes in the way teachers use the knowledge, skills and techniques learners; (4) Changes in organizational culture, practices, and support; and (5) Changes in student learning, engagement, perception, study approaches.

Appendix B.

Document facilitated to the participants of the workshop for explaining the details of the PROF-XXI Framework.

Appendix B.1. Introduction

This document presents a first version of the PROF-XXI competency framework, a framework created to guide higher education institutions in the design and implementation of Teaching and Learning Centers for the 21st century.

This document presents only a first version of the framework to be revised and improved in two phases: (1) a first revision by the Latin American partners belonging to the project from activities linked to the project such as training workshops; (2) a second revision including assessments and comments from external project staff.

Appendix B.2. Context

The PROF-XXI framework is proposed to guide higher education institutions (HEIs) in the design and implementation of Teaching and Learning Centers (TLCs) for the 21st century. As framework reference, we mainly take the DigCompEdu [1], a conceptual framework defined by the European Union to support educational institutions or companies in the sector in thinking towards the systematic integration of technology-supported learning. The objectives of the DigCompEdu framework are (i) to encourage self-reflection and self-evaluation within educational organizations to support them in their commitment to digital learning and pedagogies; (ii) to enable policy makers (at local, regional, national, and international levels) to design, implement and evaluate programs, projects and policy interventions for the integration of digital learning technologies in education and training systems. Concretely, this framework proposes 7 distinct elements and 15 sub-elements that are common to all education sectors as well as 74 descriptors that help institutions to reflect on the key elements towards this integration of technology-supported learning.

Like the DigCompEdu framework, the PROF-XXI framework is aimed at leaders and managers of higher education institutions who need to design an institutional strategy for innovation in education and the use of technologies. However, unlike other frameworks, the PROF-XXI framework proposes a set of competencies that institutions can take as a reference to develop the actions and strategies of their teaching and learning centers. This strategy will have a direct impact on their teaching and learning staff and students and will support the deep transformation of the institutional strategy.

Appendix B.3. The PROF-XXI Framework

To guide higher education institutions (HEIs) in the design and implementation of Teaching and Learning Centers (TLCs) for the 21st century, the PROF-XXI framework proposes a set of competencies that these centers should be able to acquire. These competencies are organized into five levels and five interrelated dimensions.

Appendix B.3.1. Levels of Competence of the TLCs

The 5 levels of TLCs competence are organized from lowest to highest from 1 to 5, where 1 means the lowest level of competence and 5 means the highest. In addition, and in order to facilitate the understanding of these levels, each level is associated with a strategic objective of the TLC within the institution:

- **Level 1 or "Development":** This is the first level of competencies and defines the basic competencies that any TLC should have to start its activities in the institution. Institutions at this level are able to identify innovative teaching practices, needs of their students and other stakeholders, and systematize prior learning about their activity in digital education.
- **Level 2 or "Innovation"**: This is the second level of competencies and defines the competencies that TLCs must have in order to be able to generate and promote educational innovation in their institution. Institutions at this level are capable of installing new educational experiences of references, promoting the use of technologies and the most innovative teachers, as well as generating opportunities for training and exchange of good practices among the different actors in the institution.
- **Level 3 or "Value generation"**: This is the third level of competencies and defines the competencies that the TLCs must have in order to be able to generate value in their institutions, generating changes and promoting transformations that affect their culture. Institutions at this level are able to disseminate new models of training and excellence to promote change, increase the educational quality of the institution, contribute to the cultural transformation of the institution, promote the installation of good practices in the use of technology and generate evidence on new practices to support decision-making.
- **Level 4 or "New Challenges and Opportunities"**: This is the fourth level of competencies and defines the competencies that TLCs should have to identify new institutional challenges related to innovation and teaching quality. Institutions at this level must be able to identify and visualize new horizons on teaching practice and quality learning scenarios that enhance student learning, define indicators and metrics that allow for the evaluation of educational innovations, involve the institution's stakeholders at various levels and systematize these challenges from the information collected into concrete actions for the institutional strategy.
- **Level 5 or "Public accountability of impacts for continuous improvement"**: This is the fifth and highest level of competencies and defines the competencies that TLCs must have to be able to ensure the monitoring and transparency of the actions carried out by the TLC in order to assess their impact and make this impact visible through both internal and public reporting and research on these actions.

Appendix B.3.2. Competence Dimensions of TLCs

Each competence level is further organized into five dimensions. These dimensions refer to the institutional aspects that are affected by the different competencies developed by the TLC. Each of the levels of competence defined above is related to each of these five dimensions through different competencies, between one and three competencies depending on the level and the dimension. See details of the competencies associated with each dimension in Annex 1 of this document:

1. **Dimension A or "Support for teaching"**: Dimension A refers to those competencies of the TLC that are related to supporting teaching processes. Actions related to these competencies will have a direct effect on teachers in the institution. This dimension defines three competencies for level 1 (A1–A3), three for level 2 (A4–A6), two for level 3 (A7 and A8), one for level 4 (A9) and two for level 5 (A10 and A11).
2. **La Dimension B or "Student support"**: Dimension B refers to the competencies of the TLC that are related to student support. Actions related to these competencies will have a direct effect on the students of the institution. This dimension defines three competencies for level 1 (B1-B3), two for level 2 (B4 and B5), two for level 3 (B6 and B7), two for level 4 (B8 and B9) and two for level 5 (B10 and B11).

3. **Dimension C or "Leadership, Culture and Transformation"**: Dimension C refers to TLC competencies that are related to leadership initiatives that promote a cultural transformation of the institution towards educational innovation. Actions related to these competencies will have a direct effect on the internal processes of the institution, both in its practices and policies. This dimension defines three competencies for level 1 (C1–C3), two for level 2 (C4 and C5), two for level 3 (C6 and C7), two for level 4 (C8 and C9) and two for level 5 (C10 and C11).
4. **Dimension D or "Technology at the service of learning"**: Dimension D refers to the competencies of the TLC that are related to technological educational initiatives, both in terms of practices and infrastructures (tools, services...). Actions related to these competencies will have a direct effect on the development of the institution's technological infrastructures as well as its educational models, conditioned by these infrastructures. This dimension defines three competencies for level 1 (D1–D3), two for level 2 (D4 and D5), two for level 3 (D6 and D7), one for level 4 (D8) and one for level 5 (D9).
5. **Dimension E or "Evidence-based practice"**: Dimension D refers to the competencies of the TLC that are related to initiatives that aim to collect data and information to understand the effect of the transformations and initiatives carried out in education. Actions related to these competencies will have a direct effect on the evaluation of the institutional initiatives carried out, and the TLC itself may affect decision-making in the definition of concrete policies and initiatives. This dimension defines two competencies for level 1 (E1 and E2), two for level 2 (E3 and E4), two for level 3 (E5 and E6), one for level 4 (E7) and one for level 5 (E8).

Each of these dimensions is related to one or more of the key dimensions defined in the DigCompEdu framework. Specifically, the dimensions A ("Support for Teaching") and B ("Student Support") are related to the dimensions "Teaching and Learning Practices", "Assessment of Practices" and "Content and Curriculum". Dimension C ("Leadership, Culture and Transformation") is related to the DigCompEdu dimension "Leadership and Governance", dimension D ("Technologies for Learning") to the dimension "Infrastructures" and dimension E ("Evidence-based Practice") to the dimensions "Professional Development" and "Collaboration and Networks".

Appendix B.4. Use of the PROF-XXI Framework

The PROF-XXI framework can be used in different ways depending on the objective of the institution. In this paper we propose the two main ways in which institutions can make use of the framework. It is important to recall that the framework is primarily intended for managers of the institution (from rectors and deans to management professionals), as well as for practitioners of the TLCs (professors or professionals who will participate in TLC activities).

Appendix B.4.1. The PROF-XXI Framework as a Reference and a Form of Internal Assessment

The PROF-XXI framework can be used as a reference framework that institutions can use to make an internal assessment of the level of competence of their institution's TLCs or learning and teaching services.

By providing a list of competencies, institutions can assess what level of competence they are at. To do so, institutions can use a questionnaire that assesses the level of competence of each of the competencies associated with the different dimensions. For each competence in the framework, this questionnaire asks the institution to select a competence level. In order to simplify the assessment, for each competence, the level of competence is assessed in four grades organized from lowest to highest:

Table A2. Grades for evaluating the competences of the PROF-XX framework.

Grade 1 (Minor)	Grade 2	Grade 3	Grade 4 (Major)
My institution/center does not have this competence	My institution/center is moderately prepared in this competence	My institution/center is moderately prepared in this competency	My institution/center is prepared in this competency

That is to say, for the competencies related to Dimension "A. Teacher support", institutions should select the degree of competence (from 1 to 4) for each of the competencies in that dimension (from A1 to A11).

The result of answering this questionnaire are several quantitative indicators (numerical values) that institutions can use in different ways to better understand where they stand in terms of competencies of their TLCs or education services.

1. LEVEL INDICATOR: This numerical value is calculated by adding up all the degrees of competence of the competencies associated to a level and dividing this value by the number of competencies in this level. For example, to calculate level 1, all the degrees of competence of the different dimensions of level 1 (A1 + A2 + A3 + B1 + B2 + B3 + C1 + C2 + C3 + D1 + D2 + D3 + E1 + E2)/11 will be added up.
2. DIMENSION INDICATOR. This numerical value is calculated by adding up all the degrees of competence of the competencies associated with a dimension and dividing it by the number of competencies in this dimension. For example, to calculate the value of dimension A, all the degrees of competence of the different dimensions of level 1 (A1 + A2 + ...+ A11)/12 will be added together.

Figure A1 shows the results of the indicators for an institution where the degree of competence was the highest for all competencies per dimension and level.

	Dimensión											
Nivel	A. Apoyo al profesorado	Indicar Grado Competencia	B. Apoyo al estudiante	Indicar Grado Competencia	C. Liderazgo, Cultura y Transformación	Indicar Grado Competencia	D. Tecnologías al servicio del aprendizaje	Indicar Grado Competencia	E. Prácticas Basadas en la Evidencia	Indicar Grado Competencia	INDICADOR DE NIVEL	
	A1	4	B1	4	C1	4	D1	4	E1	4	4	Nivel1
	A2	4	B2	4	C2	4	D2	4	E2	4		
1	A3	4	B3	4	C3	4	D3	4				
	A4	4	B4	4	C4	4	D4	4	E3	4	4	Nivel 2
	A5	4	B5	4	C5	4	D5	4	E4	4		
2	A5	4										
	A6	4	B6	4	C6	4	D6	4	E5	4	4	Nivel 3
3	A7	4	B7	4	C7	4	D7	4	E6	4		
	A8	4	B8	4	C8	4	D8	4	E7	4	4	Nivel 4
4	A9	4	B9	4	C9	4						
	A10	4	B10	4	C10	4	D9	4	E8	4	4	Nivel 5
5	A11	4	B11	4	C11	4						
INDICADOR DE DIMENSIÓN	A	4	B	4	C	4	C	4	E	4		

Figure A1. Picture of the calculation of indicators for an institution where all values of competencies, by level and dimension, is maximum (grade 4).

Appendix B.4.2. The PROF-XXI Framework as a Reference for Strategic Planning

In addition to being used as an internal evaluation framework, the PROF-XXI framework can also be used as a guide for the creation of an institutional strategy for the development of educational innovation and the use of educational technologies.

Managers, practitioners, or teachers involved in the teaching/learning institution or service can refer to the competencies set out in the framework as a tool for strategic planning and projection. Each competence or set of competencies can be "the target to be achieved". From there, the institution can work on the implementation of training workshops or activities and processes related to these competencies and plan the time for their implementation.

The use of the framework in this case should be accompanied by collaborative workshops with different actors in the institution in order to create a strategy that is as inclusive as possible.

Appendix C.

The following poster, created with Lucid app., was used by the participants to classify the different activities and initiatives conducted by their institution within the PROF-XXI framework. This is a snapshot of the original version of the poster in Spanish that was filled in. It included all the dimensions and levels of the framework.

The results of the poster session are accessible for visualization at the following link: https://lucid.app/lucidspark/a0bbe847-def3-445e-b663-9b72abc0722d/edit?viewport_loc=-2275%2C99%2C6790%2C3474%2C0_0&invitationId=inv_9d5f3086-a63a-4187-9aae-c8512252fe6d (accessed on 21 December 2021). All the data collected in the poster were organized in an Excel file accessible here: https://osf.io/mfjtg/ (accessed on 21 December 2021).

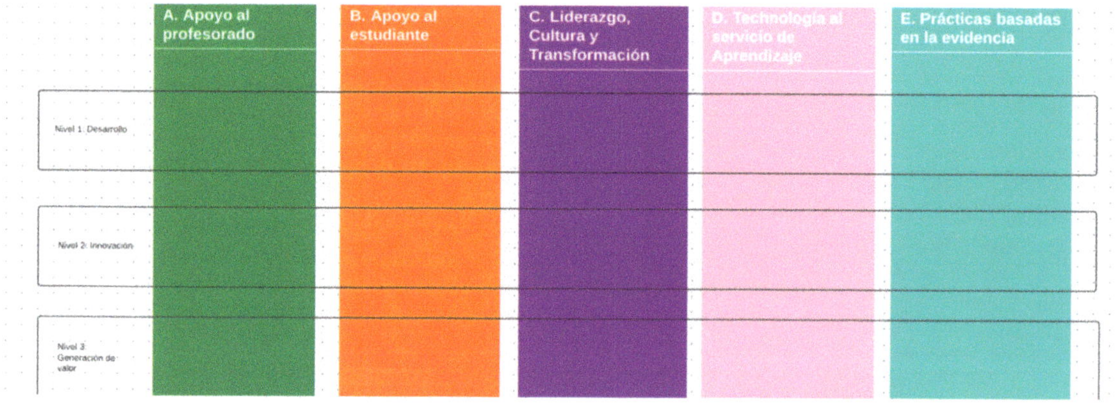

Figure A2. Poster used for the second activity of the first phase in the workshop with Administrators, Managers, and Teaching/Academic staff.

Appendix D.

Table A3. Data analysis of [Competencies Questionnaire]. We calculated the average value given for each competence dimension for the different stakeholders and per each institution. With **, we indicate the highest values for each competence per stakeholder in each institution. The values given by the participants are between 1 (minimum) and 4 (maximum).

Institution	Role in the University	Participants	Average of A. Teachers' Support	Average of B. Students' Support	Average of C. Leadership, Culture and Transformation	Average of D. Technology for Learning	Average of D. Evidence-Based Practice	Average per Competence per Role
U1 (U. San Carlos de Guatemala)	Administrative	11	2.51 (SD = 0.74)	2.44 (SD = 0.82)	2.58 (STD = 1)	2.61 (SD = 0.95)	2.63 (SD = 1.14) **	2.55
	Manager	6	2.45 (SD = 0.70)	2.45 (SD = 9.57)	2.55 (STD = 0.68) **	2.65 (SD = 0.38) **	2.58 (SD = 0.62)	2.54
	Teaching/Academic	9	2.69 (SD = 0.55) **	2.47 (SD = 0.75) **	2.55 (STD = 0.82)	2.52 (SD = 0.61)	2.22 (SD = 0.84)	2.11
	Total U1	**26**	**2.56 (SD = 0.65)**	**2.45 (SD = 0.72)**	**2.56 (STD = 0.84)**	**2.59 (SD = 0.72)**	**2.48 (SD = 0.93)**	**2.55**

Table A3. *Cont.*

Institution	Role in the University	Participants	Average of A. Teachers' Support	Average of B. Students' Support	Average of C. Leadership, Culture and Transformation	Average of D. Technology for Learning	Average of D. Evidence-Based Practice	Average per Competence per Role
U2 (U. Galileo)	Administrative	8	3.39 (SD = 0.57) **	3.13 (SD = 0.74) **	2.95 (SD = 0.85) **	3.25 (SD = 0.72) **	2.92 (SD = 1.05) **	3.13
	Manager	2	3.00 (SD = 0.64)	2.95 (SD = 0.71)	2.82 (SD = 0.64)	3.11 (SD = 0.47)	2.75 (SD = 0.71)	2.93
	Teaching/Academic	-	-	-	-	-	-	-
	Total U2	**10**	**3.32 (SD = 0.57)**	**3.09 (SD = 0.70)**	**2.93 (SD = 0.78)**	**3.22 (SD = 0.66)**	**2.89 (SD = 0.96)**	**3.09**
U3 (San Buenaventura)	Administrative	5	3.99 (SD = 0.65) **	2.93 (SD = 0.53) **	2.95 (SD = 0.82) **	2.76 (SD = 0.84)	2.85 (SD = 0.76)	2.41
	Manager	5	2.67 (SD = 0.88)	2.82 (SD = 0.71)	2.93 (SD = 0.64)	2.96 (SD = 0.58) **	2.88 (SD = 0.73) **	2.85
	Teaching/Academic	13	2.69 (SD = 0.46)	2.94 (SD = 0.56)	2.94 (SD = 0.42)	2.81 (SD = 0.58)	2.72 (SD = 0.67)	2.82
	Total U3	**23**	**2.75 (SD = 0.59)**	**2.91 (SD = 0.56)**	**2.94 (SD = 0.54)**	**2.83 (SD = 0.62)**	**2.78 (SD = 0.67)**	**2.84**
U4 (U. Cauca)	Administrative	1	3.00 **	2.45	3.36 **	3.22 **	3.13 **	3.03
	Manager	1	2.36	2.09	2.18	2.00	2.00	2.13
	Teaching/Academic	22	2.81 (SD = 0.67)	2.72 (SD = 0.72) **	2.57 (SD = 0.68)	2.52 (SD = 0.72)	2.32 (SD = 0.87)	2.59
	Total U4	**24**	**2.80 (SD = 0.67)**	**2.68 (SD = 0.70)**	**2.59 (SD = 0.67)**	**2.52 (SD = 0.71)**	**2.34 (SD = 0.85)**	**2.55**
	Total general	**83**	**2.77 (SD = 0.66) ****	**2.72 (SD = 0.69)**	**2.72 (SD = 0.72)**	**2.71 (SD = 0.71)**	**2.57 (SD = 0.85)**	-

Table A4. Partial results analysis of the data collected through the [Competencies Questionnaire], corresponding to the first activity of the first phase.

Partial Result Code	Description	Supporting Data Source (Tables A2 and A3)
PR1.1	In all institutions, the competence is Competence "A. Teacher support" was valued as one of the most developed.	Competence A is evaluated with the highest values (2.77; SD = 0.66), compared with other competencies B (2.72; SD = 0.69); C (2.72; SD = 0.72); D (2.71; SD = 0.71); and E (2.57; SD = 0.85) (Table A2)
PR1.2	All institutions, evaluated Competence E "Evidence-based practices" as the least developed	Competence E is evaluated with the lowest value (2.57; SD = 0.85), compared with other competencies A (2.77; SD = 0.66); B (2.72; SD = 0.69); C (2.72; SD = 0.72); and D (2.71; SD = 0.71) (Table A2)

Table A4. *Cont.*

Partial Result Code	Description	Supporting Data Source (Tables A2 and A3)
PR1.3	Participants from U2 and U4 evaluated the Competence "A. Teachers' support" as the most well-developed competence in the institution, and the Competence "E. Evidence-based practices" as the least developed.	Competence A in U2 is evaluated with the highest value (3.31; SD = 0.57), while Competence E (2.89; SD = 0.96) with the lowest, compared with other competencies B (3.09; SD = 0.70); C (2.93; SD = 0.78); D (3.22; SD = 0.66) (Table A2) Competence A in U4 is evaluated with the highest value (2.80; SD = 0.67), while Competence E (2.34; SD = 0.85) with the lowest, compared with other competencies B (2.68; SD = 0.70); C (2.57; SD = 0.67); D (2.52; SD = 0.71) (Table A2)
PR1.4	In all institutions, the "Manager Staff" evaluates the competence "A. Teachers' support" with the lowest values, together with the "Teaching Staff" from U3. However, "Teaching/Academic Staff" from U1 and U4 evaluated it as the most well-developed.	Values for Competence A for competencies and all stakeholders in the following order (Table A2): (1) Administrative: U1 (2.51; SD = 0.74) U2 (3.39; SD = 0.57); U3 (3.39; SD = 0.57); U4 (3, 00). (2) Manager: U1 (2.45; SD = 0.70) U2 (3.00; SD = 0.64); U3 (2.67; SD = 0.88); U4 (2.36). (3) Teaching/Academic: U1 (2.69; SD = 0.55) U2 (-); U3 (2.69; SD = 0.46); U4 (2.81; SD = 0.69).
PR1.5	Institution U2 has reported the highest values in terms of competence dimensions and compared with the other institutions.	Competence values in Table A2.

Table A5. Partial results analysis of the data collected through the [Poster Initiatives Classification], corresponding to the second activity of the first phase.

Partial Result Code	Description	Selected Supporting Data (Translated from the Original Data)
PR2.1	To the Competence "A. Teacher support", institutions associated initiatives for training the teachers. The types of trainings vary in frequency and format depending on the institution, including courses, workshops, seminars, and diplomas (a set of courses with several CETS credits). Most of trainings focus on learning about digital tools. Participants also associate to these competencies' initiatives related with teaching recognition, teaching evaluation and the share of good practices.	"Training courses for teachers in new digital tools" (U1) "Creation of a support and training unit to support teachers in virtual education. Training for teachers in the use of ICT. Workshops on good practices in Moodle, Meet, zoom, classroom and other tools" (U4) "Training for teachers" (U3) "Monthly training workshops on the use of the institutional educational platform" (U2) "Sharing and supporting teachers' successful experiences" (U4) "Learning from different experiences that led to good practices" (U4) "Evaluation on the teaching practice carried out" (U4) "Recognition of the teaching work" (U2)
PR2.2	To the Competence "B. Student support" participants associated initiatives such as online courses, video tutorials as well as academic support or on the Learning Management Systems employed by the university. Participants also recognize that, in some cases, the Competence "Student Support" is a bit poor.	"Facilitate technological tools for cooperative, collaborative and participatory work" (U1) "Only some help for internet connection" (U4) "Support to the student through a web page" (U2) "Video-lecture for the laboratory sessions" (U2)

Table A5. *Cont.*

Partial Result Code	Description	Selected Supporting Data (Translated from the Original Data)
PR2.3	To the competence "C. Leadership, Culture and Transformation" participants associated activities such as: (1) programs for developing the sense of belonging to the institution and its culture; (2) instances for self-evaluation, and instances for interacting with other institutions through research international programs. They also mentioned activities addressed to teaching/academics and administration staff related with the digital transformation of institutional processes.	" Institutional Membership Program" (to promote the sense of belonging to the institution) (U2) "Organizational culture program" (U2) "TLC project and organizational culture focused on innovation and presentation of results and indicators" (U3) "Summa Project: Impact of the university in its context, through continuing education programs" (U4) "Culture of continuous institutional and program self-evaluation" (U2) "Each department has a person in charge of digital education" (U4). "Group work between managers and teachers for the best choice of objectives and platforms for the new modalities of virtual teaching" (U4) "Institutional training plan in competencies oriented to Technology, Communication, pedagogy, management and research" (U1) "ICT training pathway for teachers" (U2) "Workshops on good practices: In what? Teacher support leaders" (U3)
PR2.4	To the competence "D. Technology for Learning" participants associated initiatives such as training in the use of technological platforms (i.e., Moodle, Google Classroom) and tools (i.e., Google Suit) through online material, tutorials, and courses	"Support materials and tutorials for the use of digital platforms and tools" (U2) "Training in visual and audiovisual technologies, for use in virtual classes" (U4) "New tools adapted to our own institutional platform, constant innovation" (U2) "Training courses, google classroom" (U3) "Training on the use of technology in the classroom" (U1) "Implementation of technologies and educational platforms for teaching, training of students and teachers" (U3)
PR2.5	To the competence "E. Evidence-based practices", participants associated initiatives related with the use of institutional data. The refer to initiatives for monitoring teachers and students' performance. They also associated activities and initiatives related with the continuous curriculum improvement and benchmarking initiatives looking for other institutions practices as a reference.	"Data analysis of educational data from online courses" (U4) "Learning analytics" (U4) "Curricular design based on students' performance" (U2) "Curricular updates at the end of the semester" (U1) "Evaluation of programs to determine innovation in teaching practice Preparation of a related semester report" (U3)

Appendix E.

Table A6. Original list of initiatives collected through the [Pre&Post Pandemic Lockdown Forms] translated to English and indicating the code we use to refer to them. U1, U2, U3 and U4 are the codes used to refer to each of the four institutions.

Code of the Initiative	Code of the Initiative
U1.1	Creation of the Distance Education in Virtual Environments Policy
U1.2	Creation of the Division of Distance Education in Virtual Environments
U1.3	Teacher training programs related to educational innovation
U1.4	Creation of the RADD (Digital Teacher Support Network)
U1.5	Enabling videoconferencing systems for the teachers in all academic units at the institution
U1.6	Creation of official accounts for the use of the videoconferencing system
U1.7	Creation of virtual classrooms with the Moodle platform for each academic unit
U1.8	Workshops for teachers and administrative staff, related to communication and technological innovation
U1.9	Diploma courses in digital teaching, virtual tutoring and instructional design
U1.10	Manual for quality in distance education
U1.11	Creation, in some academic units, a group for supporting distance learning
U1.12	Creation of the first online diploma "Bachelor's Degree in Criminology and Criminalistics"
U1.13	Design and creation of educational tutorials to support teaching
U1.14	Implementation of the remote supervision tool for online exams "proctorizer" in the School of Medicine and in the Bachelor's Degree in Criminology and Criminalistics
U2.1	Institutional implementation and management of an LMS: At the institutional level, the use of an LMS (Zoom, Meet) was standardized for the execution of synchronous and asynchronous sessions for academic continuity.
U2.2	Hybrid education: Academic programs currently have the particularity of being hybrid given the case that students can either attend their virtual classes or review the recording of the same.
U2.3	Use of tools for the improvement and quality of virtual classes: use of tools for the improvement and quality of the teaching-learning process and interactivity during the development of virtual classes.
U2.4	Supporting resources for teachers: Specialized resources available to all teachers (video tutorials, guides, podcasts, websites) were created for the process of academic continuity in the digital environment.
U2.5	Webinars for teachers: We implemented webinars on the different topics of our specialized programs.
U2.6	Personalized management advice and accompaniment: We decentralized the mentoring and coaching carried out by the project administration, with the objective of supporting teachers in the process from moving from a traditional teaching style to a virtual learning style.
U2.7	Automation of services: We conducted an automatization of certain existing processes to facilitate the access to university tools to all the educational community and assure its immediate use.
U2.8	Use of simulators, Learning Scenarios: The use of simulators is established with the objective of generating learning scenarios, to create a space for collaboration and practice for students.
U2.9	Formal assessment scenarios: The use of tools is implemented to strengthen the virtual teaching-learning process by creating a formal scenario for evaluation and assurance of academic integrity on the part of students.

Table A6. *Cont.*

Code of the Initiative	Code of the Initiative
U2.10	Continuous Learning Workshops for Teaching staff: The teacher training and education strategy was implemented on a continuous basis to achieve a development of Technological pedagogical competence.
U3.1	Diploma in Pedagogical Training
U3.2	Diploma in Design of Virtual Learning Environments
U3.3	ICT training plan for teachers
U3.4	Seminar-Workshop on e-Learning Activities
U3.5	TICatlón: An event to explain and show cases using ICT for educational practices
U3.6	Digital Competence Teacher Training Plan
U4.1	Diploma in Educational Innovations for Higher Education: training designed to encourage innovation in university teaching practice
U4.2	Diploma in University Teaching: training designed and offered to university professors who have recently joined the Institution
U4.3	Management of Teaching, Learning and Assessment Course: training designed and offered to university teachers in the context of the emergency remote teaching caused by the COVID-19 pandemic (it is a mini-course created from the Diploma in Educational Innovations for Higher Education, designed for a mass education environment)
U4.4	Visual and Auditory Narratives course: training designed and offered to university teachers in the context of the emergency remote teaching caused by the COVID-19 pandemic, focused on the production of multi-format educational materials (designed for a mass education environment)

Table A7. Original list of initiatives collected through the [Pre&Post Pandemic Lockdown Forms] translated to English. Columns of section "Periods": Before, the initiative existed before the pandemic lockdown; Originated, the initiative was originated during the pandemic lockdown; Continues, the initiative is maintained at the institution after the pandemic lockdown. We indicated under study those initiatives that the university is still studying whether to be maintained or not.

Initiatives	Period			Competence Dimensions of the PROF-XXI Framework				
	Before	Originated	Continues	A. Teacher Support	B. Students' Support	C. Leadership, Culture and Transformation	D. Technology for Learning	E. Evidence-Based Practices
U1.1	X		X			X		
U1.2	X		X	X	X	X	X	X
U1.3	X		X	X			X	X
U1.4		X	X	X			X	
U1.5		X	X				X	
U1.6		X	X				X	
U1.7		X	X				X	
U1.8		X	X	X			X	
U1.9	X		X	X			X	
U1.10			X			X		X
U1.11		X	X	X	X	X		

Table A7. *Cont.*

Initiatives	Period			Competence Dimensions of the PROF-XXI Framework				
	Before	Originated	Continues	A. Teacher Support	B. Students' Support	C. Leadership, Culture and Transformation	D. Technology for Learning	E. Evidence-Based Practices
U1.12			X		X	X		X
U1.13		X	X	X		X		
U1.14		X	X	X	X		X	X
U2.1		X	X				X	
U2.2		X	X					X
U2.3		X	X	X	X		X	
U2.4	X		X				X	
U2,5	X		X					
U2.6	X		X			X		X
U2.7	X		X	X	X		X	
U2.8	X		X	X	X		X	
U2.9		X	X	X	X		X	
U2.10	X		X				X	
U3.1	X			X		X	X	
U3.2	X			X		X	X	
U3.3	X			X		X	X	
U3.4	X			X		X	X	X
U3.5	X			X	X	X	X	X
U3.6		X	X	X	X	X	X	X
U4.1	X		X	X	X		X	X
U4.2	X		X	X				
U4.3		X	Under study	X			X	
U4.4		X	Under study	X	X		X	
TOTAL	16	15	27	**22**	**12**	**14**	**25**	**11**
Competencies Before				12	5	7	13	6
Competencies Originated				10	6	4	12	3
Competencies Continues				15	10	9	18	9

Appendix F.

This appendix contains all the information regarding the different institutions participating in the experience.

Table A8. Information from all the institutions participating in the evaluation.

U1	
Country	Guatemala
Type of administration	Public
Number of Students	235,212
Number of Academics	6856
Origins and mission	The founding of the Universidad de San Carlos de Guatemala (USAC) began with the management of the first bishop Francisco Marroquin to the King of Spain in his letter dated 1 August 1548, and after more than 120 years in which multiple projects were carried out to perfect the concept of a university based on the dream of a society that needed professionals to promote development, on 21 January 1676, was embodied in a Royal Charter the birth of the first university in Central America (USAC). Over time, it went through five eras where different names were established. It was with the revolution of 1944 that it was declared as a secular institution with a social orientation. USAC is the only state university; therefore, it is exclusively responsible for directing, organizing, and developing state higher education, as well as the dissemination of culture in all its manifestations. As part of its mission, it promotes research in all spheres of human knowledge, cooperating and solving national problems. USAC currently has an academic offer of more than 600 training programs that have allowed professional growth at the Central American level and the fulfillment of its motto "Go and teach everyone".
Existing Teaching and Learning Center	The Teaching and Learning Center (TLC) of the Division of Distance Education in Virtual Environments of the University of San Carlos de Guatemala "EDUMEDIA" has, as its mission, to implement and innovate educational practices through knowledge management and research, as well as learning in virtual environments using educational technologies as didactic-methodological resources, to achieve the purposes of the university and for this, it has proposed strategic actions framed in six objectives: (1) develop training and capacity building activities; (2) improve the generation of digital educational content; (3) promote educational innovation projects; (4) systematize experiences and good practices; (5) reinforce the use of virtual learning spaces; and (6) carry out technological surveillance for educational innovation. Among the services offered by the EDUMEDIA TLC are (a) advice on innovation projects for virtual education; (b) pedagogical-technological training; (c) space for the design of digital educational content; (d) technological and digital content production consulting; (e) University of San Carlos de Guatemala repository of learning objects; (f) systematization of good practices; (g) LMS installation and hosting service; (h) Google Workspace for teachers; (i) live streaming. To ensure the proper functioning of the TLC, an evaluation framework has been established with indicators that measure the development of the strategic objectives.
Link to TLC	https://youtu.be/ON7qZh0-SbU (accessed on 21 December 2021).

Table A8. *Cont.*

U2	
Country	Guatemala
Type of administration	Private
Number of Students	25,000
Number of Academics	1200
Origins and mission	Located in Guatemala, Galileo University is a higher education institution, the product of 40 years of constant work and effort of an elite group of academics and professionals, lead by Eduardo Suger Cofiño, Ph.D., founder and President. He has been able to put forward a completely innovative and non-traditional educational approach that Galileo calls "The revolution of education", which is also impelled by very a clear motto: "To educate is to change visions and transform lives." With thirty-eight years of successful experience, facing the rapid-changing times and the knowledge globalization, Galileo University has positioned itself as a relevant leader and a reference in the field of technology. This gives the University a very important role, not only in professional training, but also in the generation of knowledge, that responds to the needs of an increasingly competitive world, becoming an excellent choice for the education of the Guatemalan and Latin American new generations. Our mission is preparing professionals with world-class academic excellence, a high spirit of justice, human, and ethical values, at the service of our society by incorporating contemporary science and technology. We are committed to give everyone the opportunity to access university studies without distinction of race, social condition, or geographic location.
Existing Teaching and Learning Center	The Learning and Teaching Center (TLC) collaborates with the academic community at Galileo University to provide and promotes excellence in teaching and learning through different services and resources. TLC (Teaching and Learning Center) About Services Teaching support Student support Webinars Contact Us
Link to TLC	https://www.galileo.edu/page/cea/ (accessed on 20 December 2021).
U3	
Country	Colombia
Type of administration	Private
Number of Students	5000
Number of Academics	403
Origins and mission	The University of San Buenaventura in Colombia was founded by the Franciscan Order in 1688, named after the exalted doctor Saint Bonaventure. In 1973, the Colegio Mayor of San Buenaventura requested the change of its name to the University of San Buenaventura, an application that was accepted and ratified by Decree 1729 of 30 August 1973. In accordance with Article 19 of Law 30 of 1992, it retains its category of University and is based in the city of Santafé de Bogotá and sections in the cities of Medellín, Cali and Cartagena. The Cali campus was created on 24 August 1970, began academic work with the Bachelors of Law, Education and Accounting. The academic organization of the San Buenaventura Cali is made up of five faculties: Architecture, Art and Design; Economic and Administrative Sciences; Law and political science; Human and Social Sciences and Engineering, with 20 undergraduate programs, 21 face-to-face specializations, 4 virtual specializations, 24 masters, 5 PhD and 1 post-PhD, which guarantee their graduates and the general public to update and continue to advance in different fields of their professional development.

Table A8. *Cont.*

Existing Teaching and Learning Center	The Learning and Teaching Center (TLC) is a high-quality bet of San Buenaventura University as a fundamental element for the training processes and as a guarantor of high-quality strengths in higher education, promoting competitiveness, faculty development from research as the main source of generation and transfer of knowledge. From the infrastructure, there are multimedia rooms, sound laboratory and administrative office where the processes of innovation, teacher training and creation of didactic resources involving teachers and students of the educational community are centralized.
Link to TLC	Under construction
U4	
Country	Colombia
Type of administration	Public
Number of Students	16,562
Number of Academics	1309
Origins and mission	The Universidad del Cauca is an autonomous university entity of the national order, created by the Decree of April 24, 1827, issued by the President of the Republic Francisco de Paula Santander at Popayán (Cauca) Mission The Universidad del Cauca is an institution of higher education, public, autonomous, of national order, created in the origins of the Republic of Colombia. The Universidad del Cauca, founded on its tradition and historical legacy, is a cultural project that has a vital and permanent commitment to social development through critical, responsible and creative education. The University forms people with ethical integrity, relevance and professional suitability, democrats committed to the welfare of society in harmony with the environment. The Universidad del Cauca generates and socializes science, technology, art and culture in teaching, research and social projection.
Existing Teaching and Learning Center	The Teaching and Learning Center of the Universidad del Cauca is linked to the Center for Quality Management and Institutional Accreditation. It is in charge of organizing the Diploma in Educational Innovations, diagnosis and teacher training and the articulation of student orientation services of the Vice Rector's Office for Culture and Welfare.
Link to TLC	https://cgcai.unicauca.edu.co/innovacioneducativa/ (accessed on 20 December 2021).

References

1. Drüke, V. *Übergangsgeschichten: Von Kafka, Widmer, Kästner, Gass, Ondaatje, Auster und anderen Verwandlungskünstlern*; Athena-Verlag: Oberhausen, Germany, 2013.
2. UNESCO. Education: From Disruption to Recovery. Available online: https://en.unesco.org/covid19/educationresponse (accessed on 22 October 2021).
3. Singer, S.R. Learning and Teaching Centers: Hubs of Educational Reform. *New Dir. High. Educ.* **2002**, *119*, 59–64. [CrossRef]
4. Schumann, D.W.; Peters, J.; Olsen, T. Cocreating value in teaching and learning centers. *New Dir. Teach. Learn.* **2013**, *133*, 21–32. [CrossRef]
5. Holt, D.; Palmer, S.; Challis, D. Changing perspectives: Teaching and learning centres' strategic contributions to academic development in Australian higher education. *Int. J. Acad. Dev.* **2011**, *16*, 5–17. [CrossRef]
6. Alam, G.M.; Forhad, M.A. Clustering Secondary Education and the Focus on Science: Impacts on Higher Education and the Job Market in Bangladesh. *Comp. Educ. Rev.* **2021**, *65*, 310–331. [CrossRef]
7. Dondi, M.; Klier, J.; Panier, F.; Schubert, J. *Defining the Skills Citizens Will Need in the Future World of Work*; McKinsey & Company: Minato, Tokyo, 2021; Available online: https://www.mckinsey.com/industries/public-and-social-sector/our-insights/defining-the-skills-citizens-will-need-in-the-future-world-of-work (accessed on 22 October 2021).
8. Blaschke, P.; Demel, J.; Kotorov, I. Innovation Performance of Small, Medium-Sized, and Large Enterprises in Czechia and Finland. *ACC J.* **2021**, 27. [CrossRef]

9. Alam, G.M.; Asimiran, S. Online technology: Sustainable higher education or diploma disease for emerging society during emergency—Comparison between pre and during COVID-19. *Technol. Forecast. Soc. Chang.* **2021**, *172*, 121034. [CrossRef] [PubMed]
10. García-Morales, V.J.; Garrido-Moreno, A.; Martín-Rojas, R. The transformation of higher education after the COVID disruption: Emerging challenges in an online learning scenario. *Front. Psychol.* **2021**, *12*, 196. [CrossRef]
11. Jerez Yàñez, Ó.; Aranda Càceres, R.; Corvalán Canessa, F.; González Rojas, L.; Ramos Torres, A. A teaching accompaniment and development model: Possibilities and challenges for teaching and learning centers. *Int. J. Acad. Dev.* **2019**, *24*, 204–208. [CrossRef]
12. Cowan, J.; Harding, A.G. A Logical Model for Curriculum Development. *Br. J. Educ. Technol.* **1986**, *17*, 103–109. [CrossRef]
13. Tyler, R.W. *Basic Principles of Curriculum and Instruction*; University of Chicago Press: New York, NY, USA, 1949.
14. Tait, A. From Place to Virtual Space: Reconfiguring Student Support for Distance and E-Learning in the Digital Age. *Open Prax.* **2014**, *6*, 5–16. [CrossRef]
15. Tait, A. Planning Student Support for Open and Distance Learning Open Learn. *Open Learn. J. Open Distance e-Learn* **2000**, *15*, 287–299. [CrossRef]
16. Hénard, F.; Roseveare, D. *Fostering Quality Teaching in Higher Education: Policies and Practices*; OECD Publishing: Paris, France, 2012; p. 54. Available online: https://www.oecd.org/education/imhe/QT%20policies%20and%20practices.pdf (accessed on 21 December 2021).
17. Council of Australian Directors of Academic Development. Benchmarking Performance of Academic Development Units in Australian Universities. 2010. Available online: http://www.cadad.edu.au/wp-content/uploads/2017/02/Benchmarking_Report.pdf (accessed on 21 December 2021).
18. Moya, B.; Turra, H.; Chalmers, D. Developing and implementing a robust and flexible framework for the evaluation and impact of educational development in higher education in Chile. *Int. J. Acad. Dev.* **2019**, *24*, 163–177. [CrossRef]
19. Challis, D.; Holt, D.; Palmer, S. Teaching and learning centres: Towards maturation. *High Educ. Res. Dev.* **2009**, *28*, 371–383. [CrossRef]
20. Wright, M.C.; Lohe, D.R.; Little, D. The Role of a Center for Teaching and Learning in a De-Centered Educational World. *Chang. Mag. High. Learn.* **2018**, *50*, 38–44. [CrossRef]
21. Palmer, S.; Holt, D.; Challis, D. Strategic leadership of Teaching and Learning Centres: From reality to ideal. *High. Educ. Res. Dev.* **2011**, *30*, 807–821. [CrossRef]
22. Hurney, C.A.; Brantmeier, E.J.; Good, M.R.; Harrison, D.; Meixner, C. The faculty learning outcome assessment framework. *J. Fac. Dev.* **2016**, *30*, 69–77.
23. Genthon, M.M.K.E. Helping Teaching and Learning Centers Improve Teaching. In *Accent on Improving College Teaching and Learning. National Center for Research to Improve Postsecondary Teaching and Learning, 2400 SEB*; University of Michigan: Ann Arbor, MI, USA, 1989.
24. Behling, K.K.E. Collaborations between Centers for Teaching and Learning and Offices of Disability Services: Current Partnerships and Perceived Challenges. *J. Postsecond. Educ. Disabil.* **2017**, *30*, 5–15.
25. Brinthaupt, T.M.; Cruz, L.; Otto, S.; Pinter, M. A Framework for the Strategic Leveraging of Outside Resources to Enhance CTL Effectiveness. *Improv. Acad.* **2019**, *38*, 82–94. [CrossRef]
26. Wright, M.C.; Lohe, D.R.; Pinder-Grover, T.; Ortquist-Ahrens, L. The Four Rs: Guiding CTLs with Responsiveness, Relationships, Resources, and Research. *Improv. Acad.* **2018**, *37*, 271–286. [CrossRef]
27. Teixeira, A.M.; Bates, T.; Mota, J. What future(s) for distance education universities? Towards an open network-based approach. *RIED Rev. Iberoam. Educ. A Distancia* **2019**, *22*, 107–126. [CrossRef]
28. Alam, G.M.; Parvin, M. Can online higher education be an active agent for change?—Comparison of academic success and job-readiness before and during COVID-19. *Technol. Forecast. Soc. Chang.* **2021**, *172*, 121008. [CrossRef]
29. Delgado Kloos, C.D.; Alario-Hoyos, C.; Morales, M.; Rocael, H.R.; Jerez, O.; Perez-Sanagustin, M.; Kotorov, I.; Fernandez, S.A.R.; Oliva-Cordova, L.M.; Solarte, M.; et al. PROF-XXI: Teaching and Learning Centers to Support the 21st Century Professor. In Proceedings of the 2021 World Engineering Education Forum/Global Engineering Deans Council, WEEF-GEDC 2021, Madrid, Spain, 15–18 November 2021; IEEE: Piscataway, NJ, USA, 2021; pp. 448–455.
30. Johnson, R.B.; Onwuegbuzie, A.J. Mixed methods research: A research paradigm whose time has come. *Educ. Res.* **2004**, *33*, 14–26. [CrossRef]
31. Leech, N.L.; Anthony, J. Onwuegbuzie: A typology of mixed methods research designs. *Qual. Quant.* **2009**, *43*, 265–275. [CrossRef]
32. Kapasia, N.; Paul, P.; Roy, A.; Saha, J.; Zaveri, A.; Mallick, R.; Barman, B.; Das, P.; Chouhan, P. Impact of lockdown on learning status of undergraduate and postgraduate students during COVID-19 pandemic in West Bengal, India. *Child. Youth Serv. Rev.* **2020**, *116*, 105194. [CrossRef]
33. Blin, F.; Munro, M. Why hasn't technology disrupted academics' teaching practices? Understanding resistance to change through the lens of activity theory. *Comput. Educ.* **2008**, *50*, 475–490. [CrossRef]
34. Westberry, N.; McNaughton, S.; Billot, J.; Gaeta, H. Resituation or resistance? Higher education teachers' adaptations to technological change. *Technol. Pedagog. Educ.* **2015**, *24*, 101–116. [CrossRef]
35. Hilliger, I.; Ortiz-Rojas, M.; Pesántez-Cabrera, P.; Scheihing, E.; Tsai, Y.-S.; Muñoz-Merino, P.J.; Broos, T.; Whitelock-Wainwright, A.; Pérez-Sanagustín, M. Identifying needs for learning analytics adoption in Latin American universities: A mixed-methods approach. *Internet High. Educ.* **2020**, *45*, 100726. [CrossRef]

36. Viberg, O.; Hatakka, M.; Bälter, O.; Mavroudi, A. The current landscape of learning analytics in higher education. *Comput. Hum. Behav.* **2018**, *89*, 98–110. [CrossRef]
37. Radha, R.; Mahalakshmi, K.; Kumar, V.S.; Saravanakumar, A.R. E-Learning during lockdown of COVID-19 pandemic: A global perspective. *Int. J. Control Autom.* **2020**, *13*, 1088–1099.
38. Mishra, L.; Gupta, T.; Shree, A. Online teaching-learning in higher education during lockdown period of COVID-19 pandemic. *Int. J. Educ. Res. Open* **2020**, *1*, 100012. [CrossRef]
39. Alam, G.M.; Roslan, S.; Al-Amin, A.Q.; Leal Filho, W. Does GATS'Influence on Private University Sector's Growth Ensure ESD or Develop City 'Sustainability Crisis'—Policy Framework to Respond COP21. *Sustainability* **2021**, *13*, 4520. [CrossRef]
40. European Commission, Digital Skills & Jobs Platform 2021. Available online: https://digital-skills-jobs.europa.eu/en/inspiration/resources/digital-competence-test (accessed on 21 December 2021).

electronics

Article

'Should I Turn On My Video Camera?' The Students' Perceptions of the Use of Video Cameras in Synchronous Distant Learning

Živilė Sederevičiūtė-Pačiauskienė *, Ilona Valantinaitė and Vaida Asakavičiūtė

Faculty of Creative Industries, Vilnius Gediminas Technical University, Trakų St. 1, 00132 Vilnius, Lithuania; ilona.valantinaite@vilniustech.lt (I.V.); vaida.asakaviciute@vilniustech.lt (V.A.)
* Correspondence: zivile.sedereviciute-paciauskiene@vilniustech.lt

Abstract: One of the challenges teachers and students face in online synchronous learning is not turning on their video cameras. The reasons are multitasking, being concerned about the background, psychological barriers, and poor internet connection. In this study, social presence theory (SPT) was employed as the theoretical lens to understand the possible impacts of video cameras in synchronous online learning. Social presence allows individuals to make personal characteristics visible to the community. Students experience greater levels of trust and rapport because of verbal and nonverbal cues that occur when video cameras are turned on in video conferencing. The use of video cameras in synchronous distant learning creates intimacy and immediacy, leading to teacher–learner social presence, which leads to dialog. The phenomenographic study was carried out to analyze the students' perceptions of the phenomena. The eighty-two first-year undergraduate and doctoral students took part in the study. It showed that students perceive a video camera as a tool for cooperation, as well as for self-discipline and self-control. The students relate the use of video cameras with quality studies, the ability to interact, and to be a part of the process. They feel less inclined to participate when their cameras are off. That leads to the weaker student–teacher relationship, which is achieved with a higher social presence. It is essential to see one other to strengthen students' motivation, sense of belonging, and community in the courses for first-year students who are still developing learning habits and social networks.

Keywords: computer-mediated communication; teacher–student relation; E-learning; pandemic; social presence theory

Citation: Sederevičiūtė-Pačiauskienė, Ž.; Valantinaitė, I.; Asakavičiūtė, V. 'Should I Turn On My Video Camera?' The Students' Perceptions of the Use of Video Cameras in Synchronous Distant Learning. *Electronics* **2022**, *11*, 813. https://doi.org/10.3390/electronics11050813

Academic Editors: Boni García, Carlos Alario-Hoyos, Mar Pérez-Sanagustín, Miguel Morales and Oscar Jerez

Received: 22 January 2022
Accepted: 25 February 2022
Published: 4 March 2022

1. Introduction

Due to the pandemic, the education system has made a giant leap forward by moving to virtual learning. It was a challenge and provided many advantages: various computer-mediated platforms were tested, teachers mastered new technologies, and they adapted methodologies for distance communication. The sudden transition to distanced education mobilized by the educational community raised the need for supportive collaboration, self-control and discipline, and professional leadership [1]. The virtual environment allowed us to maintain education and at least partially maintain social relations. However, people lately complain about loneliness and lack of intimacy, even though technology has eliminated time differences and allowed direct and instant communication [2]. It is recognized that the affective domain (i.e., emotions, feelings, and moods) affects various aspects of the online experience, including distance learning [3–5]. However, how students interact with and respond to their online environment has been missed or overlooked in the specific literature on wellbeing [6]. Not only do students bring their own emotions to the online environment, but they also have to try and interpret and understand those of others (for example, tutors and peers) without the use of non-verbal cues, and deal with the emotions generated by this, which could potentially include distrust, isolation, and loneliness [7].

Teachers face no less stress. Many educators find themselves teaching remotely for the first time and facing a new set of challenges. One such challenge is not being able to see students during synchronous class meetings held via videoconferencing software because students do not have their video cameras turned on [8].

There are many reasons why students do not turn on their video cameras. These can be technical disturbances; either they do not have access to a private space or are embarrassed about their home environment [9]. Maybe they are self-conscious about their appearance and being seen by classmates [8,10,11], or they compromise their video presence by disabling the video display in their interface settings. It can also be a psychological barrier, as the COVID-19 pandemic has already increased college student anxiety and depression [12]. Scholars note that a mandate for camera use may add to that trauma [9]. During the initial transition to remote online learning, college students reported feelings of increased anxiety, fear, and depression consistent with responses to traumatic events [12,13].

F.R Castelli's and M.A Sarvary's (2020) [8] study with 312 undergraduate students revealed various reasons students do not turn on their video cameras during synchronous online class meetings. The reasons are that they are concerned about appearance (N = 113, 41% of students) and concerned about other people being seen in the background (N = 73, 26%). This reason was selected more frequently than the related reason of not wanting their physical location to be seen in the background. Of relatively moderate frequency were concerns about distracting their classmates or instructor. As for reasons related to technology, a very small number of students (N = 6, 2% of students) reported that their webcam was not working. However, a much larger number (N = 61, 22%) reported having a weak internet connection.

Thus, the reasons why students did not turn on their cameras were widely studied, [8,10,11,14] and the impact of video cameras on students' psychological statuses was investigated [9,12,13]. Given that not connecting the camera can be varied, the question arises as to how individuals perceive the use or non-use of video cameras in synchronous distant education. It is established that student–teacher relationships during video learning are crucial to students' academic success and satisfaction [15–17]. However, there is a lack of studies of the students' perceptions of computer-mediated communication during synchronous online learning not using video cameras. Do students see the use of video cameras as a prerequisite to the student–teacher relationship?

In this study, social presence theory (SPT) was employed as the theoretical lens to understand the possible impacts of video cameras in online learning. The research has shown that greater social presence within online classrooms increases students' satisfaction [18], motivation, and performance [19]. Much of this research was focused upon increasing social presence through the use of personal profiles, individualized video feedback, and one-on-one email communication [20–22]. However, there is limited research on fostering social presence perceptions among users [19], especially on using video cameras during synchronous distance learning and its relation to social presence theory.

The results can be used for further research on video camera (non) use to determine how it affects the specific social presence or wellbeing parameters.

2. Social Presence Theory

There are many reasons why video presence with cameras is crucial in distance education.

According to John Dewey's (1922) [23] philosophy, social interaction is central to education and is essential to creating a community of online learners. Garrison et al. (1999) [24] adapting Dewey's philosophy, said that three core elements must be present in the online space to facilitate learning: a social presence, a cognitive presence, and a teacher presence. The social learning theory states that behaviors result from people's social interaction and their environments [25]. "Virtually all learning phenomena resulting from direct experience occurs on a vicarious basis by observing other people's behavior and its consequences for them" [26] (pp. 11–12). Personal and environmental factors determine each other, and the influences are bi-directional. Social interaction between learners and role models is required for social learning to occur; no interaction, no learning [27]. Instructional dialogue is considered purposeful, bi-directional, and a constructive communication between the learner and teacher [28]. The quantity of dialogue within a course exists on a spectrum, from continuous dialogue between teachers and learners to a complete absence of communication. The lack of physical presence and the inadequate communication between teachers and learners in online learning could lead to students' frustration, dissatisfaction, less participation, or even higher dropout rates in online courses [29].

Social presence allows individuals to make personal characteristics visible to the community. Social presence, according to Garrison et al., is defined as "the ability of learners to project themselves socially and emotionally as 'real' people into a community of learners" and may facilitate the success of cognitive presence [24].

Social presence in computer-mediated communication is "the degree of feeling, perception, and reaction to another intellectual entity in the CMC [computer mediated communication] environment" [17] (p. 146). Social presence allows individuals to make personal characteristics visible to the community. According to Garrison et al., it is "the ability of learners to project themselves socially and emotionally as 'real' people into a community of learners." It may facilitate the success of cognitive presence [24] (p. 17).

The minimum level of social presence occurs when users feel that a form, behavior, or sensory experience indicates the presence of another intelligence. The amount of social presence is how a user feels access to the intelligence, intentions, and sensory impressions of another. Factors contributing to social presence are facial expression, the direction of gaze, posture, dress, and non-verbal and vocal cues [30]. For example, *intimacy* is a function of eye contact, proximity, the topic of conversation, etc. Communication with maintained eye contact, proximity, the body leaning forward, and smiling conveys greater intimacy [31]. Another dimension, *immediacy*, is the psychological distance between communicator and recipient [32]. Technological immediacy is achieved when the maximum amount of information is transmitted [33]. Social immediacy is conveyed through speech and verbal and non-verbal cues [25]. Birdwhistel (1970) [34] notes that these nonverbal cues perform two distinct functions. The first concerns itself directly with the passage of information from one individual to another; the second is the "integrational aspects" of the communication process. Integrational aspects include all the physical manifestations of information exchange that keep the conversation going, regulate the interaction process, particularly cross-reference messages to semantic meaning, and relate a specific context to larger contexts. Immediacy enhances social presence [27]. Because of the immediacy of response to verbal and nonverbal cues that occurs when video cameras are turned on in video conferencing, students can experience greater levels of trust and rapport [8]. In computer-mediated communication, the integrational activity is the dialogue that occurs between participants and the instructors/moderators/facilitators, and among participants [27]. The use of video cameras in synchronous distant learning creates intimacy and immediacy, leading to teacher–learner social presence, which leads to dialog. When instructors and students can see one other, there is an added layer of human connection, strengthening students' motivation, sense of belonging, and community in the course [11,35].

On the one hand, everything seems clear; cameras help build trust and connection, so students and teachers should use them in online synchronous classes. However, as Nowak et al. (2009) [36] note, people are cognitive and behavioral misers, and prefer doing a task using less effort than more effort. If students can participate in the synchronous distant learning classroom with an audio setting only, they probably would be keen to choose one. Therefore, it is interesting to find out how students perceive the use or non-use of video cameras in their synchronous distant learning. What does it mean for them to learn remotely via Zoom when no teacher is nearby?

3. Material and Methods

3.1. Procedure and Materials

The phenomenographic research approach was chosen to carry out the study. The study aims not to summarize the experience, but to present the variety of aspects that characterize the experience of video camera use [28,37–39]. The article does not examine distance learning itself as a phenomenon, but how education participants perceive the (non) use of video cameras in synchronous distant learning. Phenomenography describes the collective variation of experiences among the respondents. Phenomenography assumes some degree of transferability, as the descriptive categories in a phenomenographic study are abstracted to a collective level of experiences as concepts of the phenomenon [40]. The results of phenomenological research enable us to identify and describe individual and subjective peculiarities, and ways of perceiving the experienced phenomena, showing their interrelationship, and revealing the human relationship with the world [41,42]. The analysis focuses on identifying a small number of qualitatively distinct descriptive categories of how the subjects experience (or understand or conceptualize) the phenomena of interest [38,43].

3.2. Research Aim and Purpose

This study aimed to explore students' perceptions of the significance of video cameras' (non) use in synchronous distant learning. In this study, social presence theory (SPT) is employed as the theoretical lens to understand the impacts of video cameras in online learning. The study focuses on the teacher–learner relationship in synchronous distant education, which is affected by (non) use of video cameras.

The following research question guided data collection for this study: How do participants perceive the significance of (non) use of video cameras in synchronous distance learning?

3.3. Participants

Non-probability convenience sampling [44,45] was used in this study. One hundred ninety-eight Lithuanian students of the first year of undergraduate and doctoral social sciences and engineering studies were invited to participate in the research through a letter of invitation following their enrollment in the courses with access to the virtual classroom space. Forty-seven students from undergraduate studies and thirty-five from doctorate studies voluntarily agreed to participate in the study. The research participants were supposed to study face-to-face, but because of the COVID-19 pandemic, were switched to synchronous distance learning from the beginning of their studies. The study was carried out in one of the universities of Vilnius, in Lithuania.

The first-year students were chosen on purpose. The first-year students still develop learning habits and social networks, and the importance of seeing one other is higher because of the necessity to strengthen students' feelings of motivation, sense of belonging, and community in the course [11,35].

The survey was finished when the data saturation was achieved. If the number of informants was larger, the results might have been deepened, or another descriptive category discovered. Nevertheless, the descriptive category system of a phenomenography study is not definitive, as the results are derived from a limited number of sources. However, the variation of experiences within the investigated material can be described [40].

Confidentiality was assured as the persons were anonymized and assigned a code kept in safe custody.

3.4. Data Collection

The research was carried out 3–16 March 2021. Using the Zoom platform for remote synchronous meetings and video conferencing, the four research sub-questions were presented to the informants: (1) What is the significance of a video camera during distance learning? (2) What is the difference between a video camera on and off during the synchronous distant class? (3) What does it mean to you to learn remotely via Zoom when there is no teacher nearby? (4) What problems do you encounter during distance synchronous learning? Study participants provided synchronously written answers (all parties responded in real-time) using a Zoom chat tool. The researchers left students free to choose to provide an answer to the question to the researchers by sending a private message on Zoom or providing a visible answer to the group in the Zoom chat. Mobile messaging has been characterized as a unique semi-hybrid between spoken and written communication, bridging the spontaneity and informality of everyday conversation and the more edited, permanent nature of the written text [46]. There are some advantages and disadvantages of this data collection method. As a text-based medium, it allows one to structure and edit their remarks, and the text, once sent, is static and enduring [47]. However, people may be less willing to write long responses instead of a series of shorter comments supplemented with emojis and shorthand. One research point to mobile messaging is that is 'associated with less warmth and affection, and fewer expressed affiliation cues, and lower feelings of bonding' [48]. However, J. Chen and P. Neo (2019) [46] found that the virtual environment did not discourage the participants from writing longer answers, or from engaging with and responding to one another. We used additional strategies that helped increase the overall engagement levels among the focus group participants: we explained the importance of the research overall and the impact of students' participation on it. Students were free to choose when to quit responding, whether because they finished or they did not want to answer any further.

3.5. Data Coding

The phenomenographic research method was used to process the data. The research goal was not to generalize experiences but to present various aspects that characterize the experience of (non) use of video cameras in synchronous distance learning [37,39]. The results of phenomenographic research allowed for identifying and characterizing individual and subjective peculiarities and ways of perceiving the experienced phenomenon, showing their interrelations, and disclosing the relationship of an individual with the surrounding world [41,42]. The data were analyzed following seven steps: (1) familiarization (the text is read, content is learned, and technical mistakes are eliminated), (2) compilation (the most significant elements in the responses of every informant are identified), (3) reduction (the essence of every more extended response or dialogue is searched for), (4) grouping (responses are grouped according to categories), (5) preliminary comparison of categories (boundaries among categories are established), (6) naming (categories are named), and (7) contrastive comparison (categories are compared, and their differences are highlighted) [49]. The coder reliability check was conducted. Two researchers independently coded the transcribed texts of all the interviews and compared the received categories with each other. The dialogic reliability check was done through researchers' discussions [50]. The "critical friend" method was used to assure rigor, and focused on a reflective approach [51]. The role of the critical friend is to encourage reflection and explore multiple and alternative explanations and interpretations as these emerge from the data. [52]. Validation was done by verifying the findings during the research process with continual checks of the findings' credibility, plausibility, and trustworthiness [53]. Communicative validity checks were not carried out, as the study aimed not to capture a particular individual's understanding, but to capture the range of understandings within a particular group [50]. The authors followed the approach that the outcomes might then be judged regarding the insight they provide into more effective ways of operating in the world [54]. A pragmatic validity check was performed, providing the recommendations to the teachers and institutions' executives in the discussion sections. Ethical issues were carefully considered and addressed. All discussions were anonymous and voluntary to ensure that all respondents felt comfortable in this sensitive lockdown time. The whole process met the General Data Protection Requirements (GDPR).

The results of the phenomenographic study are descriptive categories and outcome space of the concept (Figure 1), expressed by a network of logically related, hierarchically arranged, and systematized categories [50]. The phemenographic research identified the categories describing the phenomenon under consideration and revealed hidden, unexpressed (tacit) depth meanings, and presented their interrelationships. Figure 1 presents the hierarchical categories singled out by the authors. Following the recommendations of F. Marton (1994) [41], the authors distinguished qualitatively different but interrelated categories describing how informants experience the phenomenon under consideration. Therefore, the analysis of the survey data did not seek to identify as many descriptive categories as possible. Descriptive categories have distinctive features that describe how a phenomenon or situation is experienced. Category descriptions mean that the phenomenon is experienced differently, collectively [50]. Descriptive categories must reflect the experiences of the study participants and cannot be given real-world meanings [55]. Analyzing individual experiences, the collective concept of the phenomenon is formed [56]. Therefore, the categories are described with a focus on the collective level [40]. Based on the methodological insights of F. Marton (1988) [37], the article selects a type of empirical descriptive category in which the description is based on the experiences of the study participants.

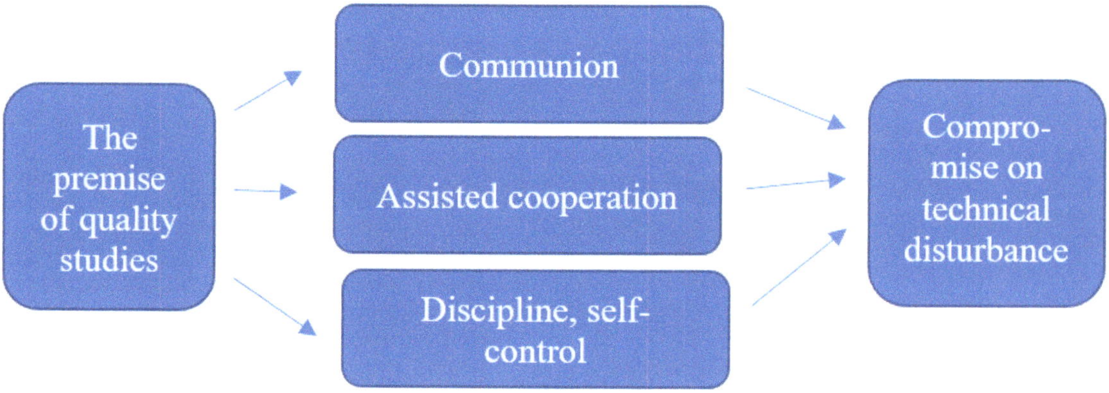

Figure 1. Relationships between research categories.

4. Results

The research data analysis allowed us to distinguish six descriptive categories that form hierarchically structured results (Figure 1). These categories reveal how students perceive and evaluate a video camera's (non) use during distance learning.

Prerequisites for a quality study process. Students point out that the (non) use of a video camera during distance learning can affect the quality of studies. Research participants noted that the video camera is associated with better communication, self-awareness, better information reception, understanding, student engagement during the session, collaboration, feedback, accountability, and academic integrity.

Learners' aspirations to obtain quality studies when studying remotely are related to the teacher–learner relationship, in which the real-time image of the teacher plays an important role. Observing the teacher's interpretation, following the change of emotions in the face creates conditions for a better understanding of the phenomenon discussed during the session. Textbooks' ideas and the themes of theories transform into a lively discussion here and now, during which existing knowledge can be combined with new information, and new conceptual constructs can be created. At the same time, the teacher can follow the growth of students' understanding of the topic and see the knowledge gaps issued in body language, and/or maintain a favorable atmosphere of learning and curiosity. Cognition, discovery, and learning become comfortable. Students indicate that it is good to feel heard and understood. It is good to receive information from all the senses given and to not just listen during a pandemic.

Discipline and self-control. Students in the study associate the use of a video camera with discipline. Learners claim to have experienced that it is challenging to speak for yourself when you see "only black boxes" on the screen. Many informants describe the feeling they experience as "talking to themselves." At the same time, however, they note that their video camera helps to focus on the topic of the activity, not to engage in extraneous activities, and to become even more actively involved in the organized activities or to listen to the teacher's explanation actively, and to take notes. In addition, a switched-on video camera makes you look good, in regard to clothes, hygiene issues, sitting at a table instead of lying down, and so on. Students also draw on the experience gained that active involvement in teacher-organized activities helps to understand the topic better. Passive listening to lectures does not allow one to get answers to the questions that have arisen or even to limit oneself to a superficial understanding of the subject being studied. The activated video camera compensates for the lack of will to constantly learn, actively listen, and even follow the work rhythm.

In summary, the video camera being used in distance learning disciplines students to look and act analogously to face-to-face lectures. As a result, even during a pandemic, it is possible to have quality free time and rest. Moreover, a balanced rhythm of work and rest creates conditions for good mental and physical wellbeing.

Communion. Students appreciate that one of the essential things in the study process is relationships. Establishing a connection between the teacher and the student allows the teacher to act qualitatively on the student's values as a future specialist. Establishing a connection between the student and the student allows them to be a part of the community, experience together, survive, and feel the study process. The connection of the participants in the educational process creates a sense of community for a common goal. Working together and solving problems in the community are based on mutual trust, respect, and recognition of the value of each member. Students name the use of a video camera as a show of respect for another. With the video camera turned on, the student broadcasts to the teacher to appreciate the teacher's efforts to overcome the pandemic's obstacles, and to organize the study process as well as possible. The video camera allows one to observe colleagues' faces, show emotions, and express respect and gratitude for being together, focusing on studies, on growth. The video camera helps to create and maintain communication between the participants of the study process. A sense of community helps overcome the obstacles caused by a pandemic and gives additional strength to the common goal of quality studies. The use of a video camera in distance learning can be seen as a statement of your determination to belong to a particular group. Turning on the video camera is a person's attitude to a particular community.

Assisted cooperation. Students point out that a video camera helps maintain a specific relationship with another person. The students who participated in the study associate successful communication and conversation with a video camera. If they can freely choose to use or not to use a video camera while listening to specific lectures, there is no choice left in discussions, working in groups, or pairs. Listening to another means not only hearing by ear, but also reading the signs of nonverbal language. It is also very important for the speaker to see the interlocutor to follow the change of emotions and reactions seen in the face while developing one or another topic. Visible reactions allow you to turn the analyzed issue in one direction or another. There is no room for monologues for yourself and yourself in the study process. Every conversation has its own goal that can be pursued or even achieved by another person or other people. The video camera enables one to act and even collaborate according to the reactions and experiences of the participants, sometimes responding only to the needs expressed in non-verbal language. Mutual assistance, support, and cooperation in the distance learning process are implemented more fully, in detail and with better quality, if the participants use video cameras.

Compromise on technical disturbances. Students pointed out the reason of turning their video of in the distance learning process with hardware and software or even the physical environment. Even students who tend to use a video camera regularly point out that using specific applications during the session disrupts the internet connection. You must not use a camcorder to maintain your login status. If a student has to choose between active participation in a lecture and the use of a video camera, in most cases, the student chooses to participate in the session actively. Another reason that can limit the use of a video camera during distance learning is the physical environment of the home: a few people working remotely in a small area (in which case, it can be challenging to use a microphone as well); young children in the background; incorrect background image, etc. Technical circumstances do not consider a person's wishes, needs, available experience, attitudes, goals, etc. The technical circumstances arise as a fact of the current minute, allowing you to choose only one of the two options: to participate in the session "as a dark box," or not to participate at all. In the long run, it is possible to make decisions according to the possibilities: to purchase an improved ISP service, to buy a new computer, to update the software, etc. However, in the initial stage of technical disruption, the situation here and now has to be addressed. After deciding to participate in a class without using a video camera, a student often experiences a feeling of discomfort, especially if his or her attitude is to always participate qualitatively using a video camera. Psychological discomfort limits a person's ability to act actively and creatively, and may even partially block the reception, understanding, and assimilation of information.

5. Discussion

The study showed that students perceive a video camera in synchronous distant learning as a tool for communion and cooperation, as well as for self-discipline and self-control (Figure 1). They describe these categories as the premise of quality studies. Students point out the technical disturbances as the reason for not using video cameras during the class: the trade-offs they make about not connecting a camcorder for better sound quality. Students have to compromise on a bed internet connection and personal visibility to the community, which they treat as a communion. Nevertheless, the students understand that without cameras, they lose the relationship with teachers and peers. A video camera means quality studies and students' involvement in active activities organized by the teacher. In the students' answers, a switched-on video camera was associated with interpersonal communication, relationships, interaction, communication, help, respect, and everyone's personal self-determination.

There is a consensus within the existing online learning literature that social presence or the ability to perceive others in a mediated environment is an important factor to cultivate within the online classroom [18]. It is established that *immediacy* enhances social presence [27]. Because of the immediacy when video cameras are turned on in video conferencing, students can experience greater levels of trust and rapport [8]. Not seeing a face or hearing a voice may pose the greatest obstacle to establishing the relationships that are so crucial in the success and satisfaction of students today [16]. Our students associate video cameras with the factors that influence immediacy. We can presume that they treat video cameras in online synchronous distant learning as a prerequisite for a greater social presence and, as a consequence, with quality studies. Students perceive it as the possibility to make connections with the teacher and peers; even they admit that sometimes they do not turn on their video cameras.

The social learning theory states that behaviors result from social interaction and environments [25,57]. Personal and environmental factors determine each other, and the influences are bi-directional. Social interaction between learners and role models is required for social learning to occur; no interaction, no learning [27]. Ch. H. Tu (2020) [58], describing the relationship between social learning and social presence theory, emphasizes that "social interaction on computer-mediated communication is affected by social presence. Learners must acknowledge and value the other person's social presence; otherwise, social interaction is absent and social learning will not occur" (u, 2020, p.4).

Our study confirmed that students relate video cameras with the community and assisted cooperation. The study showed that a video camera helps maintain a specific relationship with another person. Students indicate that it is good to feel heard and understood. Technically, it is unnecessary to see the other person to be heard and understood. However, students relate it with the teacher's visual presence, which creates the relationship. Philosopher Martine Buber (1998) [15] calls for a fundamental difference between I–You and I–It, which helps distinguish the opposite features of the study process: inclusion or exclusion. There is reciprocity, trust, and cooperation [15]. A mechanical or marketing dialogue in the second I–It determines that disconnection, distrust, and psychological tension prevail in the study process [59]. A mechanical dialog can be observed during synchronous distant learning. The transmission of information to students can be assured. However, educational growth occurs only when I–You is established.

Technological immediacy is achieved even when students do not use video cameras. However, social immediacy, conveyed through speech and associated verbal and non-verbal cues [25], is missing. That weakens the social presence of students because they feel more similar to passive listeners of the lecture than active participants. The increased social interaction and real-time collaboration among peers with video cameras turned on parallels the in-person learning experience and works to mitigate the negative cognitive consequences associated with loneliness [8]. Our students think that the video camera enables them to act and even collaborate according to the reactions and experiences of the participants, sometimes responding only to the needs expressed in non-verbal language. It is impossible to create and maintain the teacher–learner relationship without immediacy, which is essential in the educational process. Using video cameras in synchronous distant learning can create immediacy, leading to teacher–learner dialog.

The controversial results were shown by the Bradner and Mark (2001) [60] experiment, which discovered that visual feedback of a collaborating partner (or observer) is not necessary to create a sense of presence. When application sharing is used, a person's presence is salient, even when visible cues are not available to indicate their presence. Having in mind that there are many reasons why students and even teachers do not turn on their video cameras [8–11,14], and that a mandate for camera use may be, at some point, traumatic [9], there should be some balance in encouraging students to use it, and to let them stay online without their video cameras on. We argue that the relationship created by social presence is crucial for students' performances in any learning, and especially synchronous distant learning. We did not study the student's social presence measures or how visual feedback influences our students' social presences. According to our study, the students feel social distance if their peers turn the camera off, and they feel less likely to participate when they do not use their video cameras. The students associate video cameras with the community, integration, and cooperative assistance. Therefore, we presume that using video cameras during Zoom classes enhances the social presence of participants: students and teachers. It builds a stronger relationship. However, in the longer term, some compromises are possible in order not to lose students who want to take part in videoconferencing without a video camera. Andel et al. (2020) [19], found that even asynchronous comments enhance social presence in a video-centric online environment. They emphasized two personality traits (i.e., conscientiousness and extraversion) as moderators of the relationships between social presence perceptions and two outcomes: perceived learning and satisfaction. Castelli and Sarvary, 2021 [8] proposed strategies to encourage—without requiring—camera use, while promoting equity and inclusion. By explaining to students the rationale behind recommending camera use during synchronous class sessions, the instructor helps to set the norms for the course and maintains transparency about how camera use will enhance the learning experience [8]. We think it is essential to strengthen students' motivation, sense of belonging, and community. These can be achieved by encouraging students to use their cameras during synchronous distant classes. This is essential for first-year students who are still developing learning habits and social networks.

6. Conclusions

The study showed that students perceive video cameras during synchronous distant learning as a prerequisite to quality studies. The students understand that they lose the relationship with the teacher and peers without cameras. Video cameras were associated with interpersonal communication, relationships, interaction, communication, help, respect, and everyone's self-determination. Students perceive a video camera as a tool for cooperation, as well as for self-discipline and self-control. We can conclude that students' perceptions of video cameras are related to their social presence, and they understand the importance of the student–teacher relationship. Using video cameras in synchronous distant learning can create immediacy, leading to teacher–learner dialog. Computer-mediated communication with a video camera is a prerequisite to students' social presences. It enhances social presence and helps create communion within the group. This dialog is essential in instructional communication between the learner and teacher and creates an I–You relationship. The educational processes without the I–You relationship lose the essence and can be described as a process of knowledge transfer, which reduces the effectiveness of learning. However, in the longer term, some compromises are possible in order not to lose students who want to take part in videoconferencing without a video camera. The relationship with students can be maintained by explaining to students the benefits of social presence and using other tools for cooperation and communion.

The findings of this study, at a practical level, contribute to broader university decision making about the use of video cameras in synchronous distant learning. Perhaps more importantly, it is expected that findings will add to a more comprehensive understanding of if and how students are invited to switch on the cameras, and what arguments are for or against them. The generalization of this study depends on the context of synchronous online learning, which we suppose can be relatively the same for university studies beginners. Further clarification is still needed. There is a need for other instructors to gather information about their own student population to develop more tailored strategies that also promote communion and cooperation.

Limitations: As this was a qualitative descriptive phenomenographic study, causal links between social presence and the effect of the use of video cameras during synchronous distant learning were not concluded. The effect of video cameras on students' perceptions of social presence as an expression of emotions, humor, or self-disclosure was not investigated. We also did not investigate interactive categories of social presence (continuing thread, quoting from others' messages, referring explicitly to others' messages, asking questions, complimenting, expressing appreciation, and expressing agreement), nor cohesive categories of social presence (vocatives, addresses, or references to the group using inclusive pronouns, phatic speech, and salutations) [61]. Social presence is an important variable in perceived learning effectiveness and instructor satisfaction for typically developed populations; however, it can differ for learners with various disorders. However, these factors have not been studied. Lastly, it is unclear to what extent our findings would hold for long-term, computer-mediated communication.

To which settings could we generalize our findings? We assume that many teachers and students after being locked down during the COVID19 quarantine find them selfs in similar situations. Therefore, many can find useful insights that apply to their context. However, the various factors can be important while applying the results to the specific context.

Author Contributions: Conceptualization, Ž.S.-P., V.A. and I.V.; methodology, I.V. and Ž.S.-P.; validation, V.A.; formal analysis, I.V.; investigation, Ž.S.-P., V.A. and I.V.; resources, Ž.S.-P. and V.A.; data curation, I.V.; writing—original draft preparation Ž.S.-P., V.A. and I.V.; writing—review and editing, Ž.S.-P. and V.A.; visualization, Ž.S.-P. and V.A.; supervision, V.A. All authors have read and agreed to the published version of the manuscript.

Funding: This research received no external funding.

Informed Consent Statement: Informed consent was obtained from all subjects involved in the study.

Data Availability Statement: The data are not publicly. If you have any questions, please ask the contact author.

Conflicts of Interest: The authors declare no conflict of interest.

References

1. Sederevičiūtė-Pačiauskienė, Ž.; Valantinaitė, I.; Kliukas, R. Communion, care, and leadership in computer-mediated learning during the early stage of COVID-19. *Sustainability* **2021**, *13*, 4234. [CrossRef]
2. Chew, Q.H.; Wei, K.C.; Vasoo, S.; Chua, H.C.; Sim, K. Narrative synthesis of psychological and coping responses towards emerging infectious disease outbreaks in the general population: Practical considerations for the COVID19 pandemic. *Singap. Med. J.* **2020**, *61*, 350–356. [CrossRef]
3. Arguedas, M.; Daradoumis, T.; Xhafa, F. Analyzing the Effects of Emotion Management on Time and Self-Management in Computer-Based Learning. *Comput. Hum. Behav.* **2016**, *63*, 517–529. [CrossRef]
4. Cheshin, A.; Rafaeli, A.; Bos, N. Anger and Happiness in Virtual Teams: Emotional Influences of Text and Behavior on Others' Affect in the Absence of Non-Verbal Cues. *Organ. Behav. Hum. Decis. Processes* **2011**, *116*, 2–16. [CrossRef]
5. Fineman, S.; Maitli, S.; Panteli, N. Virtuality and Emotion. *Hum. Relat.* **2007**, *60*, 555–560. [CrossRef]
6. Jones, E.; Samra, R.; Lucassen, M. The world at their fingertips? The mental wellbeing of online distance-based law students. *Law Teach.* **2019**, *53*, 49–69. [CrossRef]
7. Ebner, N.; Zeleznikow, J. Fairness, Trust and Security in Online Dispute Resolution. *J. Public Law Policy* **2015**, *36*, 143–160. [CrossRef]
8. Castelli, F.R.; Sarvary, M.A. Why students do not turn on their video cameras during online classes and an equitable and inclusive plan to encourage them to do so. *Ecol. Evol.* **2021**, *11*, 3565–3576. [CrossRef]
9. Costa, K. Cameras be Damned. LinkedIn. 2020. Available online: https://www.linkedin.com/pulse/cameras-damned-karen-costa/ (accessed on 21 January 2022).
10. Jackson, T. COVID-19 and Videoclassism: Implicit Bias, Videojudgment, and Why I'm Terrified to Have You Look Over My Shoulder. (Posted 27 March 2020) LinkedIn. Available online: https://www.linkedin.com/pulse/covid-19-videoclassism-implicit-bias-videojudgment-why-jackson (accessed on 21 January 2022).
11. Nicandro, V.; Khandelwal, A.; Weitzman, A. Please, Let Students Turn Their Videos off in Class. *The Stanford Daily*. Available online: https://www.stanforddaily.com/2020/06/01/please-let-students-turn-their-videos-off-in-class/ (accessed on 24 November 2021).
12. Huckins, J.F.; DaSilva, A.W.; Wang, W.; Hedlund, E.; Rogers, C.; Nepal, S.K.; Wu, J.; Obuchi, M.; Murphy, E.I.; Meyer, M.L.; et al. Mental health and behavior of college students during the early phases of the COVID-19 pandemic: Longitudinal smartphone and ecological momentary assessment study. *J. Med. Internet Res.* **2020**, *22*, e20185. [CrossRef]
13. Kalman, R.; Esparza, M.M.; Weston, C. Student views of the online learning process during the COVID-19 pandemic: A comparison of upper-level and entry-level undergraduate perspectives. *J. Chem. Educ.* **2020**, *97*, 3353–3357. [CrossRef]
14. Reich, J.; Buttimer, C.J.; Coleman, D.; Colwell, R.; Faruqi, F.; Larke, L.R. What's Lost, What's Left, What's Next: Lessons Learned from the Lived Experiences of Teachers during the Pandemic. 2020. Available online: https://doi.org/10.35542/osf.io/8exp9 (accessed on 21 January 2022).
15. Buber, M. *Dialogo Principas I. AŠ ir TU. (The Principle of I-Thou)*; Katalikų Pasaulio Leidiniai: Vilnius, Lithuania, 1998.
16. Martin, J. Building Relationships and increasing engagement in the virtual classroom: Practical tools for the online instructor. *J. Educ. Online* **2019**, *16*, 8. Available online: https://eric.ed.gov/?id=EJ1204379 (accessed on 24 November 2021).
17. Tu, C.-H.; McIsaac, M. The relationship of social presence and interaction in online classes. *Am. J. Distance Educ.* **2002**, *16*, 131–150. [CrossRef]
18. Richardson, J.C.; Maeda, Y.; Caskurlu, J.; Lv, S. Social presence in relation to students' satisfaction and learning in the online environment: A meta-analysis. *Comput. Hum. Behav.* **2017**, *71*, 402–417. [CrossRef]
19. Andel, S.A.; de Vreede, T.; Spector, P.E.; Padmanabhan, B.; Singh, V.K.; de Vreede, G.J. Do social features help in video-centric online learning platforms? A social presence perspective. *Comput. Hum. Behav.* **2020**, *113*, 106505. [CrossRef]
20. Dunlap, J.; Lowenthal, P. Defeating the kobayashi maru: Supporting student retention by balancing the needs of the many and the one. 2010. Available online: https://er.educause.edu/articles/2010/12/defeating-the-kobayashi-maru-supporting-student-retention-by-balancing-the-needs-of-the-many-and-the-one (accessed on 21 January 2022).
21. Kear, K.; Chetwynd, F.; Jefferis, H. Social presence in online learning communities: The role of personal profiles. *Res. Learn. Technol.* **2014**, *22*, 1–15. [CrossRef]
22. Lowenthal, P.R.; Dunlap, J. Investigating students' perceptions of various instructional strategies to establish social presence. In Proceedings of the Annual Meeting of the American Educational Research Association, New Orleans, LA, USA, 8–12 April 2011.
23. Dewey, J. *Democracy and Education: An Introduction to the Philosophy of Education*; Macmillan: Basingstoke, UK, 1922.
24. Garrison, D.R.; Anderson, T.; Archer, W. Critical inquiry in a text-based environment: Computer conferencing in higher education. *Internet High Educ.* **1999**, *2*, 87–105. [CrossRef]
25. Walther, J.B. Interpersonal effects in computer-mediated interaction: A relational perspective. *Commun. Res.* **1992**, *19*, 52–90. [CrossRef]
26. Bandura, A. *Social Learning Theory. Englewood Cliffs*; Prentice-Hall: Hooboken, NJ, USA, 1977.
27. Gunawardena, C.N. Social presence theory and implications for interaction and collaborative learning in computer conferences. *Int. J. Educ. Telecommun.* **1995**, *1*, 147–166.
28. Moore, M.G. Theory of transactional distance. In *Theoretical Principles of Distance Education*; Keegan, D., Ed.; Routledge: New York, NY, USA, 1993; pp. 22–38.
29. Reio, T.G.; Crim, S.J. The Emergence of Social Presence as an Overlooked Factor in Asynchronous Online Learning. 2007. Available online: https://eric.ed.gov/?id=ED492785 (accessed on 23 March 2021).

30. Biocca, F. The Cyborg's dilemma: Embodiment in virtual environments. In Proceedings of the Second International Conference on Cognitive Technology Humanizing the Information Age, Aizu-Wakamatsu, Japan, 25–28 August 1997.
31. Burgoon, J.K.; Buller, D.B.; Hale, J.L.; de Turck, M. Relational messages associated with nonverbal behaviors. *Hum. Commun. Res.* **1984**, *10*, 351–378. [CrossRef]
32. Wiener, M.; Mehrabian, A. *Language within Language: Immediacy, a Channel in Verbal Communication*; Appleton-Century-Crofts: New York, NJ, USA, 1968.
33. Heilbronn, M.; Libby, W.L. Comparative effects of technological and social immediacy upon performance and perceptions during a two-person game. In Proceedings of the 81st Annual Convention of the American Psychological Association, Montreal, QC, Canada, 27–31 August 1973.
34. Birdwhistell, R.L. The Idealized Model of the American Family. *Soc. Casework* **1970**, *51*, 195–198. [CrossRef]
35. Lin, X.; Gao, L. Students' sense of community and perspectives of taking synchronous and asynchronous online courses. *Asian J. Distance Educ.* **2020**, *15*, 169–179.
36. Nowak, K.; Watt, J.H.; Walther, J.B. Computer-mediated teamwork and the efficiency framework: Exploring the influence of synchrony and cues on media satisfaction and outcome success. *Comput. Hum. Behav.* **2009**, *25*, 1108–1119. [CrossRef]
37. Marton, F. Phenomenography: A research approach to investigating different understandings of reality. *J. Thought.* **1986**, *1*, 28–49.
38. Yin, R.K. *Case Study Research: Design and Methods*; Sage: London, UK, 1984.
39. Walker, C. Learning to learn, phenomenography, and children's learning. *Educ. Child Psychol.* **1998**, *15*, 25–33.
40. Marton, F.; Booth, S. *Learning and Awareness*; Lawrence Erlbaum Associates, Inc.: New York, NY, USA, 1997.
41. Marton, F. Phenomenography. In *International Encyclopaedia of Education*; Husen, T., Postlethwaite, T.N., Eds.; Pergamon Press: Oxford, UK, 1994; pp. 4424–4429.
42. Stenfors-Hayes, T.; Hult, H.; Dahlgren, M.A. A phenomenographic approach to research in medical education. *Med. Educ.* **2013**, *47*, 261–270. [CrossRef] [PubMed]
43. Booth, S. On Phenomenography, Learning and Teaching. *High. Educ. Res. Dev.* **1997**, *2*, 135–158. [CrossRef]
44. Etikan, I.; Musa, S.A.; Alkassim, R.S. Comparison of convenience sampling and purposive sampling. *Am. J. Theor. Appl. Stat.* **2016**, *5*, 1–4. [CrossRef]
45. Yang, Z.; Liu, Q. Research and development of Web-based virtual online-e classroom. *Comput. E5 Educ.* **2007**, *48*, 171–184. [CrossRef]
46. Chen, J.; Neo, P. Texting the waters: An assessment of focus groups conducted via the WhatsApp smartphone messaging application. *Methodol. Innov.* **2019**, *12*, 1–10. [CrossRef]
47. Jesenská, P. How Electronic Communication Changes English Language. *J. Lang. Cult. Educ.* **2015**, *3*, 132–141.
48. Nesi, J.; Widman, L.; Choukas-Bradley, S.; Prinstein, M. Technology-based communication and the development of interpersonal competencies within adolescent romantic relationships: A preliminary investigation. *J. Res. Adolesc.* **2017**, *27*, 471–477. [CrossRef]
49. Dahlgren, L.; Fallsberg, M. Phenomenography as a qualitative approach in social pharmacy research. *J. Soc. Adm. Pharm.* **1991**, *8*, 150–156.
50. Åkerlind, G.S. Variation and commonality in phenomenographic research methods. *High. Educ. Res. Dev.* **2005**, *24*, 321–334. [CrossRef]
51. Bevington, G.; Wolcott, H.F. Transforming Qualitative Data: Description, Analysis, and Interpretation. *Mod. Lang. J.* **1996**, *80*, 405. [CrossRef]
52. Smith, B.; McGannon, K.R. Developing rigor in qualitative research: Problems and opportunities within sport and exercise psychology. *Int. Rev. Sport Exerc. Psychol.* **2018**, *11*, 101–121. [CrossRef]
53. Entwistle, N. Introduction: Phenomenography in Higher Education. *High. Educ. Res. Dev.* **1997**, *16*, 127–134. [CrossRef]
54. Kvale, S. Ten standard Objections to Qualitative Research Interviews. *J. Phenomenol. Psychol.* **1994**, *25*, 147–173. [CrossRef]
55. Bowden, J.A. Experience of phenomenographic research: A personal account. In *Phenomenography*; Bowden, J.A., Walsh, E., Eds.; RMIT University: Melbourne, Australia, 2000; pp. 47–61.
56. Sandberg, J. Are phenomenographic results reliable? *High. Educ. Res. Dev.* **1997**, *16*, 203–212. [CrossRef]
57. Bruce, C. *The Seven Faces of Information Literacy*; Auslib Press: Adelaide, Australia, 1997.
58. Tu, C.-H. Online learning migration: From social learning theory to social presence theory in a CMC environment. *J. Netw. Comput. Appl.* **2020**, *23*, 27–37. [CrossRef]
59. Laverty, M. Dialogue as Philosophical Inquiry in the Teaching of Tolerance and Sympathy. *Learn. Enq.* **2007**, *1*, 125–132. [CrossRef]
60. Bradner, E.; Mark, G. Social presence with video and application sharing. In Proceedings of the 2001 International ACM SIGGROUP Conference on Supporting Group Work, Boulder, CO, USA, 30 September–3 October 2001; pp. 154–161. [CrossRef]
61. Lomicka, L.; Lord, G. Social presence in virtual communities of foreign language (FL) teachers. *System* **2007**, *35*, 208–228. [CrossRef]

electronics

MDPI

Article

Smart Mobile Learning Success Model for Higher Educational Institutions in the Context of the COVID-19 Pandemic

Mohammed Amin Almaiah [1,*], Sarra Ayouni [2], Fahima Hajjej [2], Abdalwali Lutfi [3], Omar Almomani [4] and Ali Bani Awad [5]

1 College of Computer Sciences and Information Technology, King Faisal University, Al-Ahsa 31982, Saudi Arabia
2 Department of Information Systems, College of Computer and Information Sciences, Princess Nourah bint Abdulrahman University, P.O. Box 84428, Riyadh 11671, Saudi Arabia; Saayounij@pnu.edu.sa (S.A.); Fshajjej@pnu.edu.sa (F.H.)
3 College of Business, King Faisal University, Al-Ahsa 31982, Saudi Arabia; aalkhasawneh@kfu.edu.sa
4 Computer Network and Information Systems Department, The World Islamic Sciences and Education University, Amman 11947, Jordan; omar.almomani@wise.edu.sa
5 Deanship of Development and Quality Assurance, King Faisal University, Al-Ahsa 31982, Saudi Arabia; abanyawad@kfu.edu.sa
* Correspondence: malmaiah@kfu.edu.sa

Abstract: Smart mobile learning (M-learning) applications have shown several new benefits for higher educational institutions during the COVID-19 pandemic, during which such applications were used to support distance learning. Therefore, this study aims to examine the most important drivers influencing the adoption of M-learning by using the technology acceptance model (TAM). The structural equation modelling (SEM) method was used to test the hypotheses in the proposed model. Data were collected via online questionnaires from 520 undergraduate and postgraduate students at four universities in Saudi Arabia. Partial least squares (PLS)–SEM was used to analyse the data. The findings indicated that M-learning acceptance is influenced by three main factors, namely, awareness, IT infrastructure (ITI), and top management support. This research contributes to the body of knowledge on M-learning acceptance practices. Likewise, it may help to facilitate and promote the acceptance of M-learning among students in Saudi universities.

Keywords: mobile learning; distance learning; COVID-19; Saudi Arabia

Citation: Almaiah, M.A.; Ayouni, S.; Hajjej, F.; Lutfi, A.; Almomani, O.; Awad, A.B. Smart Mobile Learning Success Model for Higher Educational Institutions in the Context of the COVID-19 Pandemic. *Electronics* **2022**, *11*, 1278. https://doi.org/10.3390/electronics11081278

Academic Editors: Boni García, Carlos Alario-Hoyos, Mar Pérez-Sanagustín, Miguel Morales and Oscar Jerez

Received: 2 March 2022
Accepted: 13 April 2022
Published: 18 April 2022

1. Introduction

Universities throughout the world are attempting to develop electronic learning (e-learning) and mobile learning (M-learning) systems in order to merge these educational systems with the traditional ones currently used [1]. E-learning and M-learning systems have become critical in order to accomplish learning objectives in a successful way during the COVID-19 pandemic. It is critical for universities to take full advantage of e-learning in order to remain competitive in the globalised 21st century [2–4]. The recent revolution in information and communication technology (ICT) has resulted in a move away from face-to-face learning, toward e-learning. During the COVID-19 pandemic, this technology has helped universities on a global scale to ensure the continuity of learning processes [5]. It has also altered how students communicate and interact with instructors. This technological advancement has transformed the learning process through the use of e-learning systems and M-learning applications to achieve sustainable education [6–8].

The real implementation of M-learning systems in Saudi universities can offer some effective solutions for distance learning during the COVID-19 pandemic [9]. It can help Saudi students to continue their learning process, and offers them the autonomy to learn according to their own learning styles and attitudes [10–12]. In addition, online teaching sessions through smartphones will assist them in interacting with teachers at any time and place during the COVID-19 pandemic. M-learning has offered scope for students to undertake distance and online learning from their homes, and will help to decrease the spread of COVID-19 among students [13].

In fact, the application of M-learning systems for teaching and learning during the COVID-19 pandemic has been considered an excellent choice for both students and teachers. Despite the benefits of M-learning listed above, its widespread and effective application in the teaching and learning process remains very low among Saudi students [14–16]. One of the main issues in relation to the usage of new technology in the learning and teaching process is the acceptance of technology among students [17]. To this end, this study aims to answer the following research question:

What are the important drivers that would lead to the adoption of M-learning among Saudi students?

2. Literature Review and Research Background

2.1. Related Works on Mobile Learning Acceptance

With the increasing number of features offered by smartphones, there has been tremendous growth in interest in using mobile devices in higher education [18]. Hence, research interest in factors that affect M-learning acceptance among students has also increased. Several studies have been undertaken to explain the main drivers for the adoption of M-learning in different contexts [19–22]. According to previous studies [23–35], students' acceptance of M-learning is an essential step in guaranteeing the full usage of this system. To achieve such acceptance, the main aspects and factors of students' adoption of M-learning applications should be properly understood [36]. In addition, students' needs and requirements should be correctly determined by mobile service providers and designers from the outset. Several studies have addressed this issue. For example, For example, [19] found that information technology infrastructure (ITI) is one necessary component of M-learning acceptance. As a result, the ITI in Saudi universities requires extensive analysis. Providing adequate ITI is necessary when introducing new technologies such as M-learning applications, as insufficient IT resources and infrastructure can impede the acceptance and usage of any new technology [20]. Therefore, this study adopted this factor to investigate its effect on M-learning success in Saudi universities.

In addition, previous studies [21–24] have confirmed that university management support is vital to the development of M-learning system adoption and, thus, reflects positively on the actual use (AU) and acceptance of M-learning by students. According to [24], support by university management is associated with their willingness to provide all necessary resources to ensure the development success of the M-learning project. In other words, a positive attitude among top management towards an M-learning project is a real indicator that a university will support the adoption of M-learning. Therefore, our study included this factor in the proposed model to investigate its effect on M-learning success in Saudi universities.

On the other hand, university culture could play a crucial role in how universities adopt M-learning systems. Information system researchers have found that university culture is predictive of technology adoption, including M-learning adoption [25]. According to [26], public culture development is qualitatively distinct from physical infrastructure development. The COVID-19 pandemic has led to cultural shifts in attitudes towards distance learning technologies, as well as possible resistance from students to the use of these new technologies [27–30]. Hence, this study investigates the effect of university culture on M-learning success in Saudi universities.

Moreover, several researchers have indicated that one of the main issues that should be addressed to increase the involvement and use of M-learning applications is inadequate awareness of the technology's existence [31–35]. Prior studies have shown that aware ness is crucial in the adoption of M-learning systems. Therefore, this study adopted this factor to investigate its effect on M-learning success in Saudi universities.

Finally, in our proposed model, we adopted two main constructs of the TAM model as predictors of acceptance and usage of M-learning, namely, perceived ease of use (EUS), and perceived usefulness (PUS). Previous studies [36–40] have confirmed that these two factors could play a crucial role in the success, usage, and acceptance of M-learning systems. In general, users do not like to use systems that require high levels of skill or are highly complex. Several studies have supported the belief that EUS influences users' intention to use a particular technology [40–43]. Similarly, previous research has indicated that users find M-learning technology useful and productive if its use does not require much time and effort [44–48]. Hence, it can be argued that students are more likely to use M-learning services if they find that doing so is not complicated. Similarly, the success of an M-learning system would increase if users realised the importance of such a system in improving their performance.

2.2. Overview of Mobile Learning as a Distance Learning Tool

During the COVID-19 pandemic, many universities around the world started to use distance learning platforms, such as M-learning platforms, Blackboard, and others [49]. For example, many universities in Saudi Arabia—such as King Faisal University (KFU)—used the Blackboard and M-learning systems as distance learning tools, as a result of the decision by the Saudi government to close all universities during the COVID-19 pandemic [50]. M-learning systems served as online classrooms in which smartphones could be used by instructors and their students to continue the learning process during the COVID-19 pandemic. An M-learning system enables instructors to upload all learning materials, learning activities, assignments, and quizzes. On the other side, students can access online classrooms and interact with instructors through online classes, download learning materials, and submit homework using the M-learning system. An M-learning platform is a distance learning platform with many features that support the learning and teaching processes for all education levels in universities. It also contributes to ensuring that lesson plans are carried out and the educational goals of the curriculum are met [50–53]. An M-learning system features a package of educational tools to support the teaching and learning processes. It is a virtual classroom that enables learners and their teachers to meet simultaneously via virtual meetings, or at any convenient time through recorded lessons [54–57]. In addition, the platform includes excellent features for ease of communication between students and teachers, such as email service, Microsoft Teams, and a variety of channels for communication between students, teachers, and parents [58].

2.3. Factors Affecting the Mobile Learning Success Context

Several recent studies have examined a number of factors that could influence the acceptance, adoption, usage, and implementation of M-learning. For example, [59–61] found that information technology infrastructure (ITI) is one necessary component of M-learning acceptance. As a result, the ITI in Saudi universities requires extensive analysis. Providing adequate ITI is necessary when introducing new technologies such as M-learning applications, as insufficient IT resources and infrastructure can impede the acceptance and usage of any new technology [62]. Therefore, this study adopted this factor to investigate its effect on M-learning success in Saudi universities.

In addition, previous studies [63–67] have confirmed that university management support is vital to the development of M-learning system adoption and, thus, reflects positively on the actual use (AU) and acceptance of M-learning by students. According to [54], support by university management is associated with their willingness to provide all necessary resources to ensure the development success of the M-learning project. In other words, a positive attitude among top management towards an M-learning project is a real indicator that a university will support the adoption of M-learning. Therefore, our study included this factor in the proposed model to investigate its effect on M-learning success in Saudi universities.

On the other hand, university culture could play a crucial role in how universities adopt M-learning systems. Information system researchers have found that university culture is predictive of technology adoption, including M-learning adoption [68]. According to [69], public culture development is qualitatively distinct from physical infrastructure development. The COVID-19 pandemic has led to cultural shifts in attitudes towards distance learning technologies, as well as possible resistance from students to the use of these new technologies. Hence, this study investigates the effect of university culture on M-learning success in Saudi universities.

Moreover, several researchers have indicated that one of the main issues that should be addressed to increase the involvement and use of M-learning applications is inadequate awareness of the technology's existence [70]. Prior studies have shown that aware ness is crucial in the adoption of M-learning systems. Therefore, this study adopted this factor to investigate its effect on M-learning success in Saudi universities.

Finally, in our proposed model, we adopted two main constructs of the TAM model as predictors of acceptance and usage of M-learning, namely, perceived ease of use (EUS), and perceived usefulness (PUS). Previous studies [71–73] have confirmed that these two factors could play a crucial role in the success, usage, and acceptance of M-learning systems. In general, users do not like to use systems that require high levels of skill or are highly complex. Several studies have supported the belief that EUS influences users' intention to use a particular technology [74–77]. Similarly, previous research has indicated that users find M-learning technology useful and productive if its use does not require much time and effort [77–80]. Hence, it can be argued that students are more likely to use M-learning services if they find that doing so is not complicated. Similarly, the success of an M-learning system would increase if users realised the importance of such a system in improving their performance.

Based on the above discussion, we adopted ITI, university management support, university culture, awareness, EUS, and PUS in our proposed model, in order to examine their roles in increasing the acceptance of M-learning in Saudi universities.

3. Hypotheses of the Proposed Model

In this work, our proposed model was established by using the TAM model plus five external factors to determine the main drivers influencing M-learning acceptance in Saudi universities. According to previous studies [40–43], the TAM model is one of the most suitable models to measure the acceptance of M-learning among users. This model comprises five main factors—namely, AU, intention to use (IU), attitude (AT), EUS, and PUS—according to Davis [44–46]. These factors are considered to be predictors of the acceptance and usage of M-learning, according to previous studies [46–48]. Figure 1 presents the pictorial presentation of the TAM model constructs.

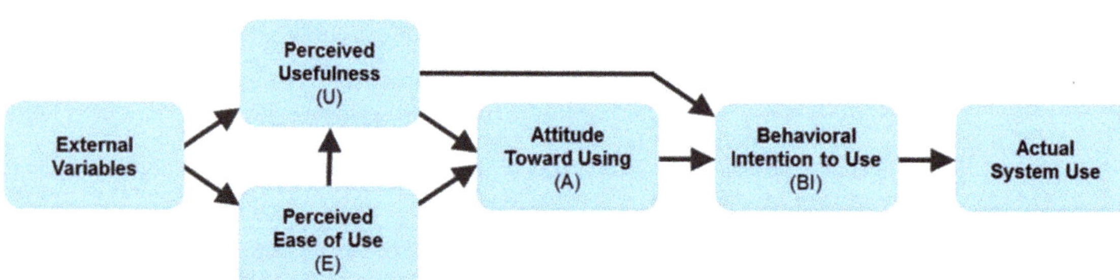

Figure 1. Technology acceptance model.

ITI refers to the combination of hardware, software communication networks, and software applications that should be offered by universities to enable students to access their online learning systems. Providing adequate ITI is necessary to introduce new technologies such as M-learning applications. Insufficient ITI resources can impede the acceptance and usage of any new technology [49]. Previous studies [50–53] have indicated that ITI is one necessary component of M-learning acceptance as shown in Figure 2. As a result, ITI in Saudi universities requires extensive analysis. Hence, this study proposes the following hypothesis:

Hypothesis 1 (H1). *ITI positively affects M-learning acceptance among students.*

According to [54], support from university management is associated with their willingness to provide all of the necessary resources to ensure the development success of an M-learning project. In other words, the positive attitude of top management towards an M-learning project is a real indicator of a university's support for the adoption of M-learning. Previous studies [55,56] have confirmed that university management support is vital to the development of M-learning system adoption and, thus, reflects positively on student AU and acceptance of M-learning. Hence, this study proposes the following hypothesis:

Hypothesis 2 (H2). *University management support positively affects M-learning acceptance among students.*

Universities should be prepared for any emergency conditions, such as the COVID-19 pandemic, and pursue alternative strategies to implement distance learning, such as adopting M-learning applications. Therefore, university culture issues can significantly affect the adoption of M-learning systems. According to [57], public culture development is qualitatively distinct from physical infrastructure development. The COVID-19 pandemic led to cultural shifts towards distance learning technologies, as well as possible resistance from students to the use of these new technologies. Thus, university culture could play a crucial role in how universities adopt M-learning systems. Information systems researchers have found that university culture is predictive of technology adoption, including M-learning adoption [58]. Hence, this study proposes the following hypothesis:

Hypothesis 3 (H3). *University culture positively affects M-learning acceptance among students.*

Student awareness of new technology, such as M-learning applications, is still limited [58]. Therefore, universities should increase the awareness among those of their students who lack adequate and essential technical information. According to [59], awareness has a strong impact on M-learning acceptance among users. Moreover, several researchers have indicated that one of the main issues that should be addressed in order to increase the involvement and use of M-learning applications is inadequate awareness of the technology's existence [60]. As prior studies have shown that awareness is crucial in adopting M-learning systems, the following hypothesis is proposed:

Hypothesis 4 (H4). *Awareness positively affects M-learning acceptance among students.*

Based on the TAM model, AU can be determined directly from one construct, namely, IU. In addition, AU can be determined indirectly from three constructs—namely, AT, EUS, and PUS—by the moderation of IU. Within the context of our study, the TAM model hypothesises that the three key constructs for AU of M-learning are students' perceptions of EUS, PUS, and IU.

In the M-learning context, we used the two main constructs of the TAM model as predictors of acceptance and usage of M-learning, namely, perceived EUS and PUS. EUS can be defined as the degree to which users perceive that using an M-learning system will be free of effort. PUS refers to the extent to which users perceive that using an M-learning system will improve their learning performance. In the TAM model, EUS and PUS affect AT toward using M-learning. PUS and AT affect IU, and IU affects AU of M-learning systems.

In general, users do not like to use systems that require high levels of skill or are complex. Several studies have supported the belief that EUS influences users' intentions to use a particular technology [31–35]. Similarly, prior literature has indicated that users find M-learning technology useful and productive if its use does not require much time and effort [36–38]. Hence, it can be argued that students are more likely to use M-learning services if they find that using such services is not complicated. Similarly, the success of an M-learning system would be increased if students realised the importance of such a system in improving their performance. Hence, we formulated the following hypotheses in our proposed model:

Hypothesis 5 (H5). *EUS positively affects M-learning acceptance among students.*

Hypothesis 6 (H6). *PUS positively affects M-learning acceptance among students.*

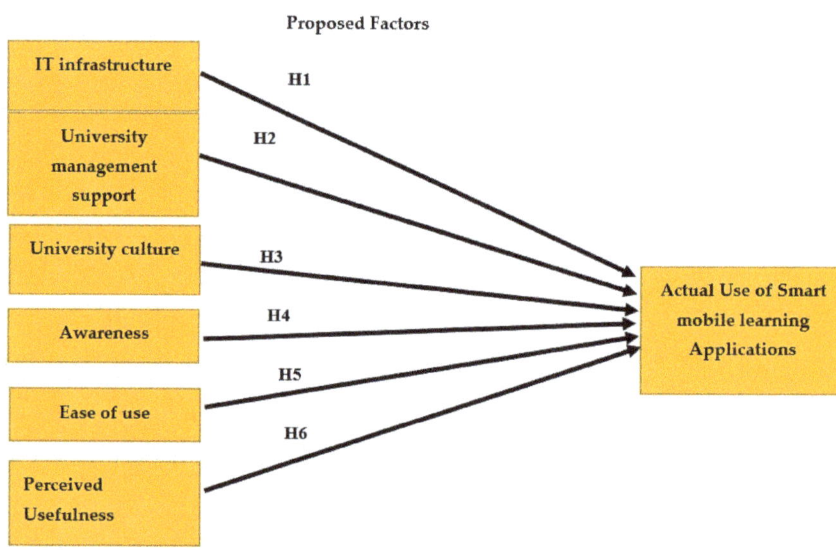

Figure 2. The proposed model for explaining the acceptance of an M-learning system.

4. Methodology

In this work, to test the hypotheses in the proposed model, the SEM method was used to determine the main drivers influencing M-learning acceptance in Saudi universities. In the context of M-learning acceptance, several previous studies have used the SEM method to examine students' acceptance [61]. These studies have indicated that the SEM method is capable of effectively discovering relationships between variables. In addition, it is an advantageous data analysis method, and tests the structural model in TAMs [62]. Consequently, we used the SEM method to test the hypotheses in the proposed model.

4.1. Data Collection and Participants

This study is based on an empirical examination of Saudi universities that are engaged in the adoption of distance learning activities, such as M-learning systems. The study collected data from four universities in Saudi Arabia, namely, KFU, King Saud University (KSU), King Khalid University (KKU), and Al-Dammam University (DU). The study gathered data from 520 respondents from the four universities as follows: KFU = 215, KSU = 105, KKU = 110, and DU = 90. To collect the data from respondents, we distributed the online questionnaire to all respondents during online classrooms, with assistance from their instructors.

The participants were 520 students enrolled in the Information System Analysis and Design course at undergraduate level and the Advanced Analysis and Design course at postgraduate level at four universities in Saudi Arabia (see Table 1). The courses, offered through the College of Information Technology, were taught online via the Blackboard system. The majority of respondents were in their sophomore or junior year. There were many more female students (69.2%, $n = 360$) than male students (30.8%, $n = 160$). Most were aged between 21 and 25 (78.8%, $n = 410$). About 88.4% reported owning an iPhone, and 92.3% had had the experience of using an iPhone in the learning process while attending an online class during COVID-19.

Table 1. Analysis of demographic information.

Characteristic		Sample (*n*)	Frequency (%)
Gender	Male	160	30.8%
	Female	360	69.2%
Age	18–20	30	5.7%
	21–25	410	78.8%
	Over 25	80	15.3%
Level	Undergraduate	395	75.9%
	Postgraduate	125	24.0%
Mobile Owner	Android	60	11.6%
	iPhone	460	88.4%
Prior Experience with Mobile Learning Apps	Yes	480	92.3%
	No	40	7.7%
Universities	KFU	215	41.3%
	KSU	105	20.1%
	KKU	110	21.1%
	DU	90	17.3%
Total	Total	520	100%

4.2. Research Measurements

To ensure that the items in the online questionnaire were measured in a valid and reliable manner, validated scales from prior studies were used for all constructs in our study. For instance, the items for the IT infrastructure construct were adopted from [3–5], items for university management support and university culture from [7–9], and items for awareness from [10]. Items for PUS and EUS were taken from [12]. Finally, the AU of M-learning was measured using items from [13], which assess the extent to which universities have adopted an M-learning system. All variables were quantified using a scale with poles ranging from strongly disagree (1) to strongly agree (5).

To ensure that the questionnaire items were valid and clear, we sent the questionnaire to seven faculty members with experience in the M-learning field, so that they could check the appropriateness and clarity of all questions and the appropriateness of each item for each construct. Based on the expert feedback, we corrected all comments and then re-sent the questions to them. The expert results indicated that all items were clear and appropriate for each construct.

5. Data Analysis and Results

5.1. Reliability Analysis

In this study, reliability analysis was conducted using Cronbach's alpha on the data collected to measure the internal consistency of each construct. Table 2 presents the values of Cronbach's alpha for all constructs. The results indicate that all values were higher than 0.70, which is acceptable according to [63]. This means that the reliability values for all constructs were accepted for further analysis.

Table 2. Reliability and convergent validity analyses.

Constructs	Cronbach's Alpha	(AVE > 0.5)
IT Infrastructure	0.792	0.937
University Management Support	0.873	0.918
University Culture	0.821	0.829
Awareness	0.890	0.811
Ease of Use	0.905	0.850
Perceived Usefulness	0.897	0.882
Actual Use	0.852	0.912

5.2. Convergent and Discriminant Validity Analysis

To conduct the validity analysis, convergent and discriminant validity were analysed. As shown by the results in Table 3, AVE values were greater than the threshold of correlation values between two variables, indicating that these values were acceptable according to [65].

Table 3. Discriminant validity analysis.

	ITI	UMS	UC	AW	EUS	PUS	AU
ITI	**0.945**						
UMS	0.797	**0.921**					
UC	0.630	0.758	**0.894**				
AW	0.646	0.684	0.545	**0.882**			
EUS	0.759	0.769	0.563	0.689	**0.901**		
PUS	0.769	0.792	0.643	0.707	0.790	**0.879**	
AU	0.530	0.623	0.506	0.643	0.527	0.614	**0.974**

5.3. Structural Model Analysis

According to the findings of the SEM modelling test, as shown in Table 4, all six proposed hypotheses in the research model were accepted. The findings revealed that the ITI factor influenced the AU of M-learning positively (β-value = 0.357, $p < 0.001$), with this result supporting H1. The findings also indicated that the factor of university management support had a significant influence on the AU of M-learning (β-value = 0.378, $p < 0.001$); this result means that H2 is accepted. In addition, H3 was supported by the present study's findings, indicating that university culture has a negative influence on the AU of M-learning (β-value = −0.395, $p < 0.001$); this result means that H3 is accepted. The results also indicated that awareness has a significant influence on the AU of M-learning (β-value = 0.327, $p < 0.001$). Thus, the results indicated that H4 was supported. Finally, we also found that EUS and PUS had significant influence on the AU of M-learning (β-value = 0.351, $p < 0.001$ and β-value = 0.342, $p < 0.001$, respectively). Thus, the results indicated that H5 and H6 were supported.

Table 4. Results of structural equation modelling analysis.

Hypotheses	Path	Impact	β	SE	*t*-Value	Results
H1	ITI→AU	Positive (+)	0.357	0.051	4.733	Supported
H2	PEU→AU	Positive (+)	0.378	0.042	4.137	Supported
H3	PU→AU	Negative (−)	−0.395	−0.075	−1.331	Supported
H4	BI→AU	Positive (+)	0.327	0.044	3.471	Supported
H5	CQ→AU	Positive (+)	0.351	0.091	3.114	Supported
H6	SYQ→AU	Positive (+)	0.342	0.06687	5.108	Supported

6. Discussion

The results indicated that the TAM model constructs with four external factors (ITI, university management support, university culture, awareness, EUS, and PU) play a primary role in increasing the acceptance of M-learning in Saudi universities. The findings indicate that all factors have a significant influence on M-learning acceptance among students.

Based on the results, the ITI factor influences the AU of M-learning positively, because it enables students to access information, increases the utility of learning activities, increases interactivity with instructors, and improves the efficiency of the learning and teaching processes. On the other hand, the ITI in Saudi Arabia is very strong, with good internet penetration and high usage among students. Based on the findings, the present study concluded that providing high ITI specifications would lead to greater acceptance of M-learning among students. This finding is consistent with prior studies [66,67].

In addition, the findings indicate that the factor of university management support has a significant influence on the AU of M-learning among students in Saudi universities. In other words, universities should prioritise top management support when deciding to implement new technologies such as M-learning systems, so as to ensure the successful development process of such systems from the first step to the last. University management support includes top managers' pledges and commitments to embrace M-learning systems, provide the necessary resources and financial support, and ensure that a high-quality system is offered in order to ensure effective student usage of M-learning. Saudi universities have high levels of management capabilities that could help in implementing M-learning systems successfully. Our findings are consistent with those of prior studies [66,67]. The present study concluded that university management support is one of the most significant factors influencing the acceptance of M-learning among students in Saudi universities.

Our findings indicate that university culture has a negative influence on the AU of M-learning among students in Saudi universities. This implies that an incorrect university culture could impede the implementation of M-learning in Saudi universities. This study found that university culture had a negative impact on students' willingness to accept M-learning. The main reason for this result is the cultural differences between students, which may affect their acceptance. For example, the COVID-19 pandemic led to cultural shifts towards distance learning technologies, as well as possible resistance from students to the use of such technologies. Based on these findings, university culture could play a crucial role in how universities adopt M-learning systems. Previous studies have found that university culture is predictive of technology adoption, including M-learning adoption [68–72].

Furthermore, our findings indicate that there is a significant and positive correlation between awareness and AU of M-learning. The study found that a majority of Saudi students are unfamiliar with mobile applications and how to use them. This study found that when the percentage of awareness among students of how to use M-learning systems is very low, there is a decrease in their acceptance of such systems. This result is consistent with prior research [72–77].

Finally, this study found that both EUS and PU have a significant and positive correlation with AU of M-learning. This indicates that when students find an M-learning system to be user-friendly, simple, easy to use, clear, and useful for the learning process, this could encourage them to use it effectively; thus, this will reflect on their opinions about accepting the M-learning system. Based on this result, this study recommends that designers and developers take those two factors into consideration in the development of an M-learning system. Previous studies [77–79] have confirmed that EUS and usefulness are among the primary dimensions that motivate users to accept any type of educational technology. Our findings are consistent with those of prior studies [80].

Research Contributions

This study makes both theoretical and practical contributions. Focusing on the theoretical contribution, this study contributes to the body of knowledge on M-learning acceptance by providing a new model that captures the most significant drivers of such acceptance among students in public Saudi universities. Second, this study clarifies that important factors—such as ITI, awareness, university management support, and university culture—played a key role in increasing the acceptance of M-learning systems during COVID-19-pandemic, and will ensure the continuity of the learning process through the use of this distance learning tool. Third, this study confirms that the TAM model is suitable for analysis of the factors influencing students' acceptance of M-learning. With regard to practical contributions, the findings of this study can help Saudi universities to better understand the process of M-learning project implementation. The universities should consider important factors related to ITI, awareness, university management support, and university culture in order to improve the acceptance of M-learning systems among students. Finally, this study's findings will benefit decision makers, designers, and developers in universities to ensure that students participate actively in using M-learning systems during the COVID-19 pandemic.

7. Conclusions, Limitations, and Future Work

Despite the major benefits of M-learning, the widespread and effective application of such technologies in the teaching and learning processes has remained low among Saudi students during the COVID-19 pandemic.

To address this issue, this study developed six direct hypotheses using the TAM model to explain the main drivers influencing M-learning acceptance. The SEM method was used to test the hypotheses in the proposed model. Data were collected via online questionnaires from 520 undergraduate and postgraduate students at four Saudi universities. PLS–SEM was used to analyse the data. The findings indicated that TAM model constructs with four external factors—namely, ITI, university management support, university culture, awareness, EUS, and PUS—play a primary role in increasing the acceptance of M-learning in Saudi universities. The findings indicated that all of the studied factors have a significant influence on M-learning acceptance among students. This research contributes to the body of knowledge on M-learning acceptance practices. Likewise, it may help to facilitate and promote the acceptance of M-learning among students in Saudi universities.

Although this work makes several interesting contributions, several limitations should be covered in future work. First, further investigation into the drivers of M-learning acceptance among students is needed. Second, future work could explore teachers' perceptions and needs in relation to adopting M-learning systems. Finally, there is a need to investigate other important factors related to system quality and usability factors, as well as their effects on students' acceptance of M-learning systems.

Author Contributions: Conceptualisation, M.A.A., S.A. and F.H.; methodology, A.L.; software, O.A.; validation, A.B.A., M.A.A. and M.A.A.; formal analysis, M.A.A.; investigation, M.A.A.; resources, F.H.; writing—original draft preparation, O.A.; writing—review and editing, O.A.; visualisation, M.A.A.; supervision, M.A.A.; project administration, M.A.A.; funding acquisition, M.A.A. All authors have read and agreed to the published version of the manuscript.

Funding: This work was funded by King Faisal University and Princess Nourah bint Abdulrahman University.

Institutional Review Board Statement: Not applicable.

Informed Consent Statement: Not applicable.

Data Availability Statement: The data presented in this study are available upon request from the corresponding author. The data are not publicly available for reasons of privacy.

Acknowledgments: This work was supported through the Annual Funding track by the Deanship of Scientific Research, Vice Presidency for Graduate Studies and Scientific Research, King Faisal University, Saudi Arabia (Project No. GRANT 368); and Princess Nourah bint Abdulrahman University Researchers (Supporting Project Number PNURSP2022R236), Princess Nourah bint Abdulrahman University, Riyadh, Saudi Arabia.

Conflicts of Interest: The authors declare no conflict of interest.

References

1. Almaiah, M.A. Acceptance and usage of a mobile information system services in University of Jordan. *Educ. Inf. Technol.* **2018**, *23*, 1873–1895. [CrossRef]
2. Almaiah, M.A.; Jalil, M.M.A.; Man, M. Empirical investigation to explore factors that achieve high quality of mobile learning system based on students' perspectives. *Eng. Sci. Technol. Int. J.* **2016**, *19*, 1314–1320. [CrossRef]
3. Aburub, F.; Alnawas, I. A new integrated model to explore factors that influence adoption of mobile learning in higher education: An empirical investigation. *Educ. Inf. Technol.* **2019**, *24*, 2145–2158. [CrossRef]
4. Al-Adwan, A.S.; Albelbisi, N.A.; Hujran, O.; Al-Rahmi, W.M.; Alkhalifah, A. Developing a Holistic Success Model for Sustainable E-Learning: A Structural Equation Modeling Approach. *Sustainability* **2021**, *13*, 9453. [CrossRef]
5. Pimmer, C.; Mateescu, M.; Gröhbiel, U. Mobile and ubiquitous learning in higher education settings. A systematic review of empirical studies. *Comput. Hum. Behav.* **2016**, *63*, 490–501. [CrossRef]
6. Shawai, Y.G.; Almaiah, M.A. Malay language mobile learning system (MLMLS) using NFC technology. *Int. J. Educ. Manag. Eng.* **2018**, *8*, 1.
7. Almaiah, M.A.; Al-Khasawneh, A.; Althunibat, A.; Almomani, O. Exploring the main determinants of mobile learning application usage during COVID-19 pandemic in Jordanian universities. In *Emerging Technologies during the Era of COVID-19 Pandemic*; Springer: Cham, Switzerland, 2021; pp. 275–290.
8. AlJordansbeh, M.; Abuhelaleh, M.; Almaiah, M. Towards a model of quality features for mobile social networks apps in learning environments: An extended information system success model. *Int. J. Interact. Mob. Technol.* **2019**, *13*, 75–93. [CrossRef]
9. Khan, M.A.; Nabi, M.K.; Khojah, M.; Tahir, M. Students' perception towards e-learning during COVID-19 pandemic in India: An empirical study. *Sustainability* **2020**, *13*, 57. [CrossRef]
10. Almaiah, M.A.; Al-Khasawneh, A. Investigating the main determinants of mobile cloud computing adoption in university campus. *Educ. Inf. Technol.* **2020**, *25*, 3087–3107. [CrossRef]
11. Nikolopoulou, K.; Gialamas, V.; Lavidas, K.; Komis, V. Teachers' readiness to adopt mobile learning in classrooms: A study in Greece. *Technol. Knowl. Learn.* **2021**, *26*, 53–77. [CrossRef]
12. Aini, Q.; Budiarto, M.; Putra PO, H.; Rahardja, U. Exploring e-learning challenges during the global COVID-19 pandemic: A review. *J. Sist. Inf.* **2020**, *16*, 57–65. [CrossRef]
13. Almaiah, M.A.; Almulhem, A. A conceptual framework for determining the success factors of e-learning system implementation using Delphi technique. *J. Theor. Appl. Inf. Technol.* **2018**, *96*, 5962–5976.
14. Almaiah, M.A.; Nasereddin, Y. Factors influencing the adoption of e-government services among Jordanian citizens. *Electron. Gov. Int. J.* **2020**, *16*, 236–259.
15. Alghazi, S.S.; Kamsin, A.; Almaiah, M.A.; Wong, S.Y.; Shuib, L. For Sustainable Application of Mobile Learning: An Extended UTAUT Model to Examine the Effect of Technical Factors on the Usage of Mobile Devices as a Learning Tool. *Sustainability* **2021**, *13*, 1856. [CrossRef]
16. Al-Emran, M.; Mezhuyev, V.; Kamaludin, A. Technology Acceptance Model in M-learning context: A systematic review. *Comput. Educ.* **2018**, *125*, 389–412. [CrossRef]
17. Cicha, K.; Rizun, M.; Rutecka, P.; Strzelecki, A. COVID-19 and higher education: First-year students' expectations toward distance learning. *Sustainability* **2021**, *13*, 1889. [CrossRef]
18. Salloum, S.A.; Alhamad, A.Q.; Al-Emran, M.; Monem, A.A.; Shaalan, K. Exploring students' acceptance of e-learning through the development of a comprehensive technology acceptance model. *IEEE Access* **2019**, *7*, 128445–128462. [CrossRef]
19. Mulhem, A.A.; Almaiah, M.A. A Conceptual Model to Investigate the Role of Mobile Game Applications in Education during the COVID-19 Pandemic. *Electronics* **2021**, *10*, 2106. [CrossRef]
20. Almaiah, M.A.; Almomani, O.; Al-Khasawneh, A.; Althunibat, A. Predicting the Acceptance of Mobile Learning Applica-tions During COVID-19 Using Machine Learning Prediction Algorithms. In *Emerging Technologies during the Era of COVID-19 Pandemic*; Springer: Cham, Switzerland, 2021; Volume 348, p. 319.
21. Davis, F.D. Perceived usefulness, perceived ease of use, and user acceptance of information technology. *MIS Q.* **1989**, *13*, 319–340. [CrossRef]
22. Aldossry, B. Evaluating the madrasati platform for the virtual classroom in Saudi arabian education during the time of COVID-19 Pandemic. *Eur. J. Open Educ. e-Learn. Stud.* **2021**, *6*, 89–99. [CrossRef]
23. Madrasati. Ministry of Education: Riyadh, Saudi Arabia. 2021. Available online: https://backtoschool.sa/ (accessed on 19 June 2021).
24. Khalife, W. The effectiveness of using active learning based on distance learning on improving academic achievement and developing the trend towards distance learning for the learning technologies course for mathematics education students. *Sci. J. Phys. Educ. Sports Sci.* **2019**, *87*, 321–334.
25. Khanfar, A.R. Distance-learning entrepreneurship education in the time of corona virus—COVID-19 challenges & solution. *J. Entrep. Educ.* **2020**, *23*, 1–17.
26. MOE. Ministry of Education: Riyadh, Saudi Arabia. 2020. Available online: https://www.moe.gov.sa/ar/mediacenter/MOEnews/Pages/MR2-2020-453.aspx (accessed on 21 June 2021).
27. Almaiah, M.A.; Alamri, M.M. Proposing a new technical quality requirements for mobile learning applications. *J. Theor. Appl. Inf. Technol.* **2018**, *96*, 6955–6968.

28. Almaiah, M.A. Thematic analysis for classifying the main challenges and factors influencing the successful implementation of e-learning system using Nvivo. *Int. J. Adv. Trends Comput. Sci. Eng.* **2020**, *9*, 142–152. [CrossRef]
29. Lutfi, A.; Alsyouf, A.; Almaiah, M.A.; Alrawad, M.; Abdo, A.A.; Al-Khasawneh, A.L.; Ibrahim, N.; Saad, M. Factors Influencing the Adoption of Big Data Analytics in the Digital Transformation Era: Case Study of Jordanian SMEs. *Sustainability* **2022**, *14*, 1802. [CrossRef]
30. Althunibat, A.; Almaiah, M.A.; Altarawneh, F. Examining the Factors Influencing the Mobile Learning Applications Usage in Higher Education during the COVID-19 Pandemic. *Electronics* **2021**, *10*, 2676. [CrossRef]
31. Al-Emran, M.; Arpaci, I.; Salloum, S.A. An empirical examination of continuous intention to use m-learning: An integrated model. *Educ. Inf. Technol.* **2020**, *25*, 2899–2918. [CrossRef]
32. Hoi, V.N. Understanding higher education learners' acceptance and use of mobile devices for language learning: A Rasch-based path modeling approach. *Comput. Educ.* **2020**, *146*, 103761. [CrossRef]
33. Gharaibeh, M.K.; Gharaibeh, N.K. An empirical study on factors influencing the intention to use mobile learning. *Adv. Sci. Technol. Eng. Syst. J.* **2020**, *5*, 1261–1265. [CrossRef]
34. Al-Nawayseh, M.K.; Baarah, A.H.; Al-Masaeed, S.A.; Alnabhan, M.M. Mobile learning adoption in Jordan: Technology influencing factors. *Int. J. Netw. Virtual Organ.* **2019**, *20*, 400–417. [CrossRef]
35. Badwelan, A.; Bahaddad, A.A. Functional Requirements to Increase Acceptance of M-Learning Applications among University Students in the Kingdom of Saudi Arabia (KSA). *Int. J. Comput. Sci. Netw. Secur.* **2021**, *21*, 21–39.
36. Fauziati, C. Faktor-faktor yang Memengaruhi Penerimaan Mobile Learning Applications (Apps)(Acceptance Factors of Mobile Learning Applications (Apps)). *J. Iptekkom* **2020**, *22*, 45–57.
37. Lutfi, A. Factors Influencing the Continuance Intention to Use Accounting Information System in Jordanian SMEs from the Perspectives of UTAUT: Top Management Support and Self-Efficacy as Predictor Factors. *Economies* **2022**, *10*, 75. [CrossRef]
38. Wahab, R.A. Student's acceptance of digital educational tools: A case study. In Proceedings of the 2020 Sixth International Conference on e-Learning (econf), Sakheer, Bahrain, 6–7 December 2020; pp. 81–86.
39. Badwelan, A.; Bahaddad, A.A. Towards the Acceptance of Functional Requirements in M-Learning Application for KSA University Students. *Int. J. Comput. Sci. Netw. Secur.* **2021**, *21*, 145–166.
40. Wei, N.; Li, Z. Telepresence and Interactivity in Mobile Learning System: Its Relation with Open Innovation. *J. Open Innov. Technol. Mark. Complex.* **2021**, *7*, 78. [CrossRef]
41. Al-Emran, M.; Mezhuyev, V.; Kamaludin, A. An innovative approach of applying knowledge management in M-learning application development: A pilot study. *Int. J. Inf. Commun. Technol. Educ.* **2019**, *15*, 94–112. [CrossRef]
42. Zhang, M.; Chen, Y.; Zhang, S.; Zhang, W.; Li, Y.; Yang, S. Understanding mobile learning continuance from an online-cum-offline learning perspective: A SEM-neural network method. *Int. J. Mob. Commun.* **2022**, *20*, 105–127. [CrossRef]
43. Kumar, J.A.; Osman, S.; Sanmugam, M.; Rasappan, R. Mobile Learning Acceptance Post Pandemic: A Behavioural Shift among Engineering Undergraduates. *Sustainability* **2022**, *14*, 3197. [CrossRef]
44. Alturki, U.; Aldraiweesh, A. Students' Perceptions of the Actual Use of Mobile Learning during COVID-19 Pandemic in Higher Education. *Sustainability* **2022**, *14*, 1125. [CrossRef]
45. Mingyue, F.; Ndavi, J.W.; Qalati, S.A.; Huang, L.; Zhengjia, P. Applying the time continuum model of motivation to explain how major factors affect mobile learning motivation: A comparison of SEM and fsQCA. *Online Inf. Rev.* **2022**, 32. [CrossRef]
46. Chibisa, A.; Mutambara, D. Determinants of High School Learners' Continuous Use of Mobile Learning during the COVID-19 Pandemic. *Int. J. Learn. Teach. Educ. Res.* **2022**, *21*, 20–33. [CrossRef]
47. Bacolod, D.B. Mobile Learning as a Solution for Restricted Learning during the COVID-19 Pandemic. *J. Digit. Educ. Technol.* **2022**, *2*, ep2203. [CrossRef]
48. Lin, H.C.; Ho, C.F.; Yang, H. Understanding adoption of artificial intelligence-enabled language e-learning system: An empirical study of UTAUT model. *Int. J. Mob. Learn. Organ.* **2022**, *16*, 74–94. [CrossRef]
49. Pishtari, G.; Rodríguez-Triana, M.J. An Analysis of Mobile Learning Tools in Terms of Pedagogical Affordances and Support to the Learning Activity Life Cycle. In *Hybrid Learning Spaces*; Springer: Cham, Switzerland, 2022; pp. 167–183.
50. Hartley, K.; Andújar, A. Smartphones and Learning: An Extension of M-Learning or a Distinct Area of Inquiry. *Educ. Sci.* **2022**, *12*, 50. [CrossRef]
51. Bai, H. Students' Use of Learning Management System in Hybrid Learning: Mobile or Not. *Int. J. E-Learn.* **2022**, *21*, 5–23.
52. Okuonghae, O.; Igbinovia, M.O.; Adebayo, J.O. Technological readiness and computer self-efficacy as predictors of E-learning adoption by LIS students in Nigeria. *Libri* **2022**, *72*, 13–25. [CrossRef]
53. Sayaf, A.M.; Alamri, M.M.; Alqahtani, M.A.; Alrahmi, W.M. Factors Influencing University Students' Adoption of Digital Learning Technology in Teaching and Learning. *Sustainability* **2022**, *14*, 493. [CrossRef]
54. Hsu, H.T.; Lin, C.C. Extending the technology acceptance model of college learners' mobile-assisted language learning by incorporating psychological constructs. *Br. J. Educ. Technol.* **2022**, *53*, 286–306. [CrossRef]
55. Mehra, A.; Rajput, S.; Paul, J. Determinants of adoption of latest version smartphones: Theory and evidence. mboxemphTechnol. Forecast. Soc. Chang. **2022**, *175*, 121410. [CrossRef]
56. Ismail, S.N.; Omar, M.N.; Don, Y.; Purnomo, Y.W.; Dzahir, M. Teachers' acceptance of mobile technology use towards innovative teaching in Malaysian secondary schools. *Int. J. Eval. Res. Educ.* **2021**, *11*, 120–127. [CrossRef]

57. Khlaif, Z.N.; Sanmugam, M.; Ayyoub, A. Impact of Technostress on Continuance Intentions to Use Mobile Technology. *Asia-Pac. Educ. Res.* **2022**, *12*, 15–26. [CrossRef]
58. Alsharida, R.; Hammood, M.; Al-Emran, M. Mobile Learning Adoption: A Systematic Review of the Technology Acceptance Model from 2017 to 2020. *Int. J. Emerg. Technol. Learn.* **2021**, *16*, 147–162. [CrossRef]
59. Moya, S.; Camacho, M. Identifying the key success factors for the adoption of mobile learning. *Educ. Inf. Technol.* **2021**, *26*, 3917–3945. [CrossRef]
60. Gupta, Y.; Khan, F.M.; Agarwal, S. Exploring Factors Influencing Mobile Learning in Higher Education—A Systematic Review. *Int. J. Interact. Mob. Technol.* **2021**, *15*, 140–157. [CrossRef]
61. Alowayr, A.; Al-Azawei, A. Predicting mobile learning acceptance: An integrated model and empirical study based on higher education students' perceptions. *Australas. J. Educ. Technol.* **2021**, *37*, 38–55. [CrossRef]
62. Criollo-C, S.; Guerrero-Arias, A.; Jaramillo-Alcázar, Á.; Luján-Mora, S. Mobile learning technologies for education: Benefits and pending issues. *Appl. Sci.* **2021**, *11*, 4111. [CrossRef]
63. Alhumaid, K.; Habes, M.; Salloum, S.A. Examining the factors influencing the mobile learning usage during COVID-19 Pandemic: An Integrated SEM-ANN Method. *IEEE Access* **2021**, *9*, 102567–102578. [CrossRef]
64. Sungur-Gül, K.; Ateş, H. Understanding pre-service teachers' mobile learning readiness using theory of planned behavior. *Educ. Technol. Soc.* **2021**, *24*, 44–57.
65. Buabeng-Andoh, C. Exploring University students' intention to use mobile learning: A research model approach. *Educ. Inf. Technol.* **2021**, *26*, 241–256. [CrossRef]
66. Al-Hamad, M.; Mbaidin, H.; AlHamad, A.; Alshurideh, M.; Kurdi, B.; Al-Hamad, N. Investigating students' behavioral intention to use mobile learning in higher education in UAE during Coronavirus-19 pandemic. *Int. J. Data Netw. Sci.* **2021**, *5*, 321–330. [CrossRef]
67. Ibrahim, M.; Khairudin, N.; Salleh, D. Unwinding Environment in the Lecture Room Using Mobile Learning (M-Learning) Adoption among the Millennial Generation. In *Journal of Physics: Conference Series*; IOP Publishing: Bristol, UK, 2021; Volume 1874, p. 012033.
68. Saikat, S.; Dhillon, J.S.; Wan Ahmad, W.F.; Jamaluddin, R. A systematic review of the benefits and challenges of mobile learning during the COVID-19 pandemic. *Educ. Sci.* **2021**, *11*, 459. [CrossRef]
69. Antee, A. Student perceptions and mobile technology adoption: Implications for lower-income students shifting to digital. *Educ. Technol. Res. Dev.* **2021**, *69*, 191–194. [CrossRef] [PubMed]
70. Loh, X.K.; Lee, V.H.; Loh, X.M.; Tan, G.W.; Ooi, K.B.; Dwivedi, Y.K. The dark side of mobile learning via social media: How bad can it get? *Inf. Syst. Front.* **2021**, *22*, 22–35. [CrossRef] [PubMed]
71. Yuan, Y.P.; Tan, G.W.; Ooi, K.B.; Lim, W.L. Can COVID-19 pandemic influence experience response in mobile learning? *Telemat. Inform.* **2021**, *64*, 101676. [CrossRef]
72. Vallejo-Correa, P.; Monsalve-Pulido, J.; Tabares-Betancur, M. A systematic mapping review of context-aware analysis and its approach to mobile learning and ubiquitous learning processes. *Comput. Sci. Rev.* **2021**, *39*, 100335. [CrossRef]
73. Okai-Ugbaje, S. Towards a pedagogical and sociotechnical framework for the strategic integration of mobile learning in higher education in low and middle income countries. *High. Educ. Res. Dev.* **2021**, *40*, 581–598. [CrossRef]
74. Criollo-C, S.; Lema, M.; Gonzalez, M.S.; Jaramillo-Alcázar, A.; Guerrero-Arias, A.; Luján-Mora, S. Exploring the technological acceptance of a mobile learning tool used in the teaching of an indigenous language. *PeerJ Comput. Sci.* **2021**, *7*, e550. [CrossRef]
75. Qashou, A. Influencing factors in M-learning adoption in higher education. *Educ. Inf. Technol.* **2021**, *26*, 1755–1785. [CrossRef]
76. Zhang, X.; Lo, P.; So, S.; Chiu, D.K.; Leung, T.N.; Ho, K.K.; Stark, A. Medical students' attitudes and perceptions towards the effectiveness of mobile learning: A comparative information-need perspective. *J. Librariansh. Inf. Sci.* **2021**, *53*, 116–129. [CrossRef]
77. Sophonhiranrak, S. Features, barriers, and influencing factors of mobile learning in higher education: A systematic review. *Heliyon* **2021**, *7*, e06696. [CrossRef]
78. Dorouka, P.; Papadakis, S.; Kalogiannakis, M. Nanotechnology and mobile learning: Perspectives and opportunities in young children's education. *Int. J. Technol. Enhanc. Learn.* **2021**, *13*, 237–252. [CrossRef]
79. Shodipe, T.O.; Ohanu, I.B. Electrical/electronics technology education teachers attitude, engagement, and disposition towards actual usage of Mobile learning in higher institutions. *Educ. Inf. Technol.* **2021**, *26*, 1023–1042. [CrossRef]
80. Hoi, V.N.; Mu, G.M. Perceived teacher support and students' acceptance of mobile-assisted language learning: Evidence from Vietnamese higher education context. *Br. J. Educ. Technol.* **2021**, *52*, 879–898. [CrossRef]

electronics

MDPI

Article

Administrators and Students on E-Learning: The Benefits and Impacts of Proper Implementation in Nigeria

Esen Sucuoğlu and Azubike Umunze Andrew *

Department of Educational Administration and supervision, Near East University, Nicosia 99138, Cyprus; esen.sucuoglu@neu.edu.tr
* Correspondence: princeazubike82@gmail.com

Abstract: The quest for better education and knowledge acquisition has triggered the introduction, acceptance and incorporation of e-learning into Nigerian learning. The introduction of the concept of e-learning to Nigerian learning can be dated back to the 1980s, when reputable Nigerians enrolled in several universities in London. In addition, the introduction of e-learning to a premier university in Nigeria, rooted in the college of Ibadan, led to greater interest, causing locals to seek extramural work and other studies at Oxford University. This study examines the impacts that proper educational administration, policy making and implementation, as well as the adoption of e-learning, can have to fix the dilapidated Nigerian educational structure. A quantitative method of data collection was used, through well-structured questionnaires for both administrators and students issued to the four universities sampled in this study. A total of 240 questionnaires were issued to respondents, with 60 each to the different universities and with 30 each for both students and administrators. A total of 180 were retrieved, and descriptive analysis was carried out with SPSS (23). Internal consistency was determined with Cronbach's alpha, having an internal consistency of 0.78. The findings show that all the administrators were graduates with a minimum of a Bachelor's degree. It was revealed that 32 (17.8%) of the students possessed smartphones as gadgets for e-learning and that administrators contributed to the enhancement of student performance, hence creating impacts in their examination grades, with a mean of 2.66, being rated 'Good' for their performance. Unfavorable government policies and unprofessionalism of administrators in e-learning implementations were the major constraints, with a mean of 4.6. The cost of the procurement of the needed resources (data) for e-learning also impacts e-learning. Internet resources used by the students contributed to huge success in e-learning for 28 (24.6%) and 24 (21.9%) students. Although the constraints limit the effectiveness of e-learning in Nigeria, it also impacts student advancement compared with the face-to-face learning process. The government's proactive measures will improve e-learning.

Keywords: e-learning; administrators; proper implementations; well-structured questionnaires; un-professionalism

Citation: Sucuoğlu, E.; Andrew, A.U. Administrators and Students on E-Learning: The Benefits and Impacts of Proper Implementation in Nigeria. *Electronics* **2022**, *11*, 1650. https://doi.org/10.3390/electronics11101650

Academic Editor: Christos J. Bouras

Received: 4 April 2022
Accepted: 18 May 2022
Published: 22 May 2022

1. Introduction

Education is an investment that needs to be made by every nation and all people aspiring for rapid and robust growth in their economic, social and even political aspects of the nation's needs, in order for them to attain the height and level of productivity they need [1]. In recent times, the quest for more advanced security and a more positively impacted educational system, capable of providing solutions to learning and teaching environments, has been a problem in the field of education. Several countries have sourced means to resolve this issue but with little or no achievement made in the past [2].

Schools across the globe have been shut down amidst the emergence of the COVID-19 pandemic, which originated from Wuhan, China, thus causing the authorities of the world to initiate lockdown as an emergency measure to manage the spread of COVID-19, including the closure of schools [3]. This development has brought about the acquirement and implementation of e-learning in most universities [4]. The cost of obtaining required instruments has allowed the continuation of learning and teaching and has also allowed for advancement from normal face-to-face learning patterns, with excellent results and grades [5]. The success of e-learning in the COVID-19 era has had to deal with the satisfaction that e-learning has brought to the end users and with the system that e-learning has to offer [6,7].

The Nigerian educational system has observed this, and it has arranged for the implementation and proper administration of e-learning [8].

E-learning is relevant for any country that seeks positive impacts in the educational sector and hence is widely used in this sector, yielding adequate and appropriate benefits [9].

Administration and administrators in schools and in educational spheres deal with different social processes that allow the people or administrators to identify, maintain, control and stimulate both formal and informal human resources within a given and well-organized system, such as schools [10,11]. The administration of good e-learning is entrusted into the hands of people at different apex positions in the educational sector, which spawn from the teachers, head of departments, deans and council members who make sure that the administration and implementation of e-learning are achieved along with the desired results [12].

In other to be a good teacher or administrator who is able to impact a learner's interest, an individual needs to possess some attributes and unique features which involve the internet and multimedia in order to effectively pass the information they want to give to timid students. These resources are readily available but require proper utilization for optimal output [13,14]. The implementation of e-learning across different academic levels, from the primary, secondary and tertiary institutions, is accompanied by several gaps which need to be addressed, and these gaps require the attention of both new and old schools for proper administration to be achieved [12]. The Nigerian educational system is advancing in its quest for well-organized and standard educational service delivery to help students make use of the resources which are readily available to them, hence the need for good administration and implementation of e-learning facilities in most academic spheres [15]. This administration and implementation require the principal and concerned members (the Head of Departments, Deans of Faculties and the council members of the universities) to come together and adopt the latest internet, technological, multimedia and other relevant tools that make learning possible and that allow the desired aim to be achieved. [16], highlighted e-learning to be a learning procedure that deals with the learning and use of computers. This is a result of the age of greater advances in the learning and education sector. E-learning entails the process of creating an enabling environment for students or their employers to be effectively trained in the process of continuous learning and teaching [17].

E-learning provides several opportunities such as financial benefits mostly in cases where there is lesser or limited provision of the required basic amenities. The need to address the task is being set up, thereby allowing the utilization of a cyber café in achieving the target, hence generating income and financial assistance [18].

The effectiveness of e-learning is solely dependent on the aggregation of more relevant and important technologies that play vital roles in making sure that the use of e-learning is easy and simple to understand [19].

An enabling environment plays a vital role in the acceptance, implementation and actualization of e-learning in the educational sector in order to drive and achieve the success that it has to offer [20]. The use of technologies such as multimedia, with the most important being informational tools, makes it possible to achieve flexibility in the understanding and usage of e-learning facilities in schools and institutions of higher learning, which is beneficial to both administrators and students [21,22]. Administrators of the Nigerian educational system have identified e-learning as being advantageous to both students and administrators, hence advocating for good quality and relevant gadgets to further enhance learning benefits for students [23]. The implementation of e-learning in the Nigerian educational system, compared with other developed nations, has suffered major setbacks with respect to meeting standards and acquiring needed materials and manpower, and the provision of basic amenities, such as light, gadgets, ICT, etc., could not be met [24].

The educational sector in Nigeria is faced with several challenges amid the novel COVID-19 pandemic, which has ravaged the educational sector as well as other financial institutions [25].

According to [26], the utilization of e-learning is accompanied firstly by its adoption before students can use it in both private HEIs and in public schools.

2. Literature Review

2.1. Concept of School Management

The optimal administration and incorporation of the best trends in educational sectors provide a good chance of a higher performance in schools, including either universities or secondary schools, where they are implemented [27]. The introduction of e-learning to a premier university in Nigeria, which has its roots in the college of Ibadan, has led to greater interest, which has caused locals to seek extramural work and several other studies at Oxford University [28].

The smooth running of education and e-learning in the Nigerian educational sector is hampered by several limitations and setbacks, including the curriculum, which is not adequately prepared, the lack of adequate electricity, the unprofessionalism of the staff and other factors [29,30].

Ref. [31], in his view, asserted that education needs resources and proper mobilization by the parents and guardians of the students in order to meet demands and what is expected of schools regarding academic success. It is a process of mobilizing school resources toward the achievement of desirable educational goals. School administration is a process of activation that requires expertise and training in educational principles and practices in order to ensure proper management to achieve results in education [11].

2.2. Concept of E-Learning

E-learning, also known as "electronic learning", is associated with the use of internet resources and other such tools to gain knowledge, with the maximum utilization of multimedia, technologies and ICT in making sure that the desired aim is achieved [9]. It is a system that helps in the effective passing and receiving of information from one source to another, i.e., from teachers to students [9,32].

According to [18], e-learning is an educational method for which popularity is on the rise in all aspects and endeavors of life, and as such, it has gained a high level of recognition and acceptability in the 21st century. In most formal school systems, however, e-learning is faced with many expectations, such as conforming to educational curriculum. E-learning entails the process of creating an enabling environment for the students or their employers to be effectively trained in the process of continuing to teach [17]. Facilitators and instructors of e-learning make learning more interesting and create more benefits; therefore, further exposure is seen in the learning process [33].

Ref. [2] acknowledged that, following many advances that have been observed, the rapid change in technological progress and the most recent globalization trend in the quest for higher educational performance, as well as the elimination of boundaries among students, have led to the adoption of new methods and perspectives in educational practice, such as e-learning. According to the Federal Republic of Nigeria [34], formal education in Nigeria is controlled by a series of policies.

2.3. Benefits of E-learning in the Educational Sector

One of the benefits of e-learning is the fact that e-learning saves 50% of time and 40–60% of costs when used well [35]. It also provides the following:

- **Location flexibility:** E-learning provides learners with flexibility, and the same is also provided for teachers. This makes learning easier for both parties notwithstanding the location.
- **Good access to instructors:** Learners can acquire a better understanding of the concepts being taught, as the materials that are provided can be re-accessed along with knowledge from instructors.
- **Reliable learning skills:** Certain circumstances have been observed by researchers to be of great benefit to learners using online learning, compared with the classroom learning process. Improvements in reliable learning skills are seen online, compared with normal face-to-face learning that is adopted in most countries' educational systems [35].

With the inception of e-learning, there are no barriers, as online learning provides a global perspective because it connects users beyond the boundaries of their location.

2.4. E-Learning in Nigerian Universities

Universities or higher institutions are seen as well-organized educational settings above the secondary level of learning, which offer degrees, diplomas or other relevant certificates that are widely recognized and accepted [28]. Higher institutions provide humans with opportunities to run the economy, which hence provides the economy with greater transformation in the society in which it is practiced [31]. [36] posited that e-learning has, in recent times, brought about a change in the relationship between teachers and students with respect to learning, which has transformed the learning process from the traditional education system. The benefits of implementing e-learning in the country are necessitated by the government through the Federal Ministry of Education to initiate policies that enable growth and stability but that are yet to be fully implemented for optimal performance [37]. Despite the educational benefits of e-learning, there is also a financial benefit, but this is always taken for granted by both students and administrators [18]. Continual advancements in information technology have provided the educational sector with the necessity to advance their teaching, learning and research methods, since these proactive measures are needed to effectively perform the duties assigned to them for optimization and for running smoothly [9,38].

2.5. Administrators of E-Learning in Nigerian Universities

Different principal officers play key roles in the administration and implementation of policies in organizations where they serve and hold principal offices. The effects are widely felt in the educational aspect, where administrators seek the most suitable methods of improving the educational performance of their students, mostly in higher institutions of learning [15]. The administrators of universities in Nigeria have roles to play in ensuring that e-learning and other relevant teaching strategies are for the benefit of the students [39].

AFRIHUB, which is responsible for providing ICTC in most Nigerian universities, agrees with Mr. Chukwuemerie Nnamdi that over (18) tertiary institutions in Nigeria utilize the help and assistance from them to unleash the "power of ICTs in almost all Nigeria and other African countries for the development of human capacity building and also to enhance economic empowerment, not only in the educational sectors" [40]. These tertiary institutions where programs are currently in operation are the Michael Okpara University of Agriculture, also known as Umudike; the Federal University of Technology Owerri (FUTO); Nnamdi Azikiwe University; the Awka University of Nigeria Nsukka:- and Enugu Campuses (UNN, UNEC); the University of Abuja; the University of Benin; Minna (FUTA); the Federal College of Education, Technical, Omoku; and the University of Calabar, amongst others in the country. With these establishments, plans are also in preparation to start ICT operations in more tertiary institutions in the country under a public–private partnership (PPP) [40]. The administrators of different universities in the country have started putting things in place to help standardize the internet and e-learning programs, which are seen to be of great importance to the academic environment and to the economy of the nation as a whole.

According to [12] in their study of the role of educational administrators in the implementation e-learning programs, carried out a study focused on the challenges faced by education administrators in e-learning implementation and on ways to remedy them. The study was implemented using mixed research methods (pragmatism research philosophy) for data collection, and analysis was conducted by [41], with primary administrators and school heads (principals used in the study) using well-structured questionnaires and with random data collection techniques. The findings of the research show that most of the school heads are degree holders and inspectors with a Master's degree in educational administration, but none of them had a relevant degree equivalent in the ICTC spheres to help them be more effective in teaching and e-learning implementation [27]. The study concluded that no optimal progress can be achieved in e-learning in schools without well-trained ICTC personnel being brought into the educational sector of Zimbabwe, despite the government's efforts to avert the problems faced.

3. Methodology

Research Design: A survey research design was adopted that uses a randomized standardized sampling technique, with valid, well-structured questionnaires aimed at considering each respondent's attitude towards and perception of e-learning. A quantitative method was used for data collection from administrators and students.

Instrumentation and data collecting tools: The collection of data was carried out by administering 240 well-structured questionnaires that addressed the problems of interest to the participants, who were both students and administrators. A total of 60 questionnaires were administered in different universities, with 30 each for both students and administrators. This data collecting tool was administered to both the administrators and the students to collect their ideas according to the information provided in the questionnaire. Randomized sampling and quantitative data methods were used to gather the opinions of administrators on their understanding and perception of e-learning and on the benefits that the learning procedure can offer to them.

Method of questionnaire administration: The data collecting tool (questionnaire) with a closed-ended pattern of questions to address the situation was administered by the researcher and through self-administration by the respondents. The paper and pen method was used, hence offering the respondents the same pattern of questions and allowing the easy retrieval of the filled questionnaires.

Determining a student's performance: The performances of students were determined based on overall performances put in during examinations, and different grade levels were designated as Poor, Average, Good, Very Good, etc. These performances of students were provided by the administrators and other staff who assumed roles in different universities, with those who used these grade levels as criteria for the performances of students.

Subject Sampling: The study was carried out in the southern part of Nigeria, and four public higher institutions were included, which were carefully selected from three states (Rivers State, Bayelsa State and Akwa-Ibom State), with two universities being from Rivers State (University of Port Harcourt and National Open University), the University of Uyo being from Akwa Ibom State and the Federal University Otueke being from Bayelsa State. The coordinates of the University of Port Harcourt are located at longitude 4°90′69″ N and latitude 6°91′70″ E; for the National Open University (NOUN) Port Harcourt Study Center, the coordinates are 4°86′81″ N and 6°95′66″ E; for the University of Uyo, the coordinates are 5°04′08″ N and 7°91′98″ E; and for the Federal University Otueke, the coordinates are 4°47′27″ N, and 7°91′98″ E.

Data analysis procedure: The data were retrieved from both student and administrator respondents regarding e-learning in the various universities. The administrators were the deans of faculties and the heads of departments of the universities. The data were coded into Statistical package for Social Sciences SPSS Statistics 23.0, and a statistical analysis was performed using descriptive statistics and frequencies, with a significance level of 0.5. The reliability of the instrument was determined and tested using Cronbach's alpha (α) formula: α = K/(K−1) [1−Σ (S2 y)/(S2 x)] [42]

where K is the n = Number of test items;

ΣS2 y = is the sum of item variance;

and S2 x = is the variance of the total score.

4. Results

Findings of the Study Based on Research Questions

Research Question 1: Who are the principal administrators, technical personnel and others that facilitate e-learning in tertiary institutions, with impacts on student performance?

Table 1 below gives a summary of the educational qualifications of the administrators and respondents from the different universities used in the study. It can be observed from the table that, out of the 180 respondents, all of the administrators were graduates with relevant qualifications, which offered them the privilege of impacting students and being able to effectively utilize e-learning strategies in teaching students. The lecturers had all obtained degrees, from B.Sc./B.Ed. degrees held by 16 (8.9%) lecturers to Ph.D. degrees held by 48 (26.7%) lecturers. Additionally, the heads of departments (HOD) who responded had a frequency of 8 (4.4%) and 12 (6.7%), and the deans of different faculties that engaged in the study had frequencies of 39 (21.7%) and 21 (11.6%) for B.Ed., PGD and Ph.D. degrees, educational background of the administrators are shown in Figure 1 below.

Table 1. Principal administrators and technical staff that impact e-learning in the tertiary institutions used in the study.

S/N	Degree	Position	No. of Respondents	Percentage (%)
1	B.Sc./B.Ed.	Technicians	16	8.9
2	M.Sc./M.Ed.	Lecturer	36	20.0
3	Ph.D.	Lecturer	48	26.7
1	B.Sc./B.Ed., M.Ed.	HOD	8	4.4
2	B.Sc./B.Ed., M.Ed. and Ph.D.	HOD	12	6.7
1	B.Sc./B.Ed., M.Ed. and Ph.D.	Dean	39	21.7
2	B.Sc., PGD, M.Ed. and Ph.D.	Dean and other council members	21	11.6
Total			**180**	**100%**

Research Question 2: What tools and strategies are required to necessitate effective e-learning in tertiary institutions in Nigeria?

Table 2 below gives an overview of the tools and strategies that are employed by the administrators of different universities in Nigeria in order to achieve smooth running of e-learning and in order for students to make the best of it. The different gadgets and tools used by these students range from gadgets in their possession, such as smartphones and personal computers (PCs), to projectors and other tools used. Additionally, the responses from the students indicate that 32 (17.8%) of the respondents agreed that smartphones play a role in education and are predominantly used, whereas 9 (5%) disagreed with that. With regard to PCs and desktops, 17 (9.4%) and 25(13.9%) agreed and disagreed, respectively, with the availability of resources and tools and the use of them; 8 (4.4%) and 35 (19.4%) agreed and disagreed, respectively, with projector usage; and 22 (12.2%) and 32 (17.8%) agreed and disagreed, respectively, with the presence of functional e-library accessibility to students.

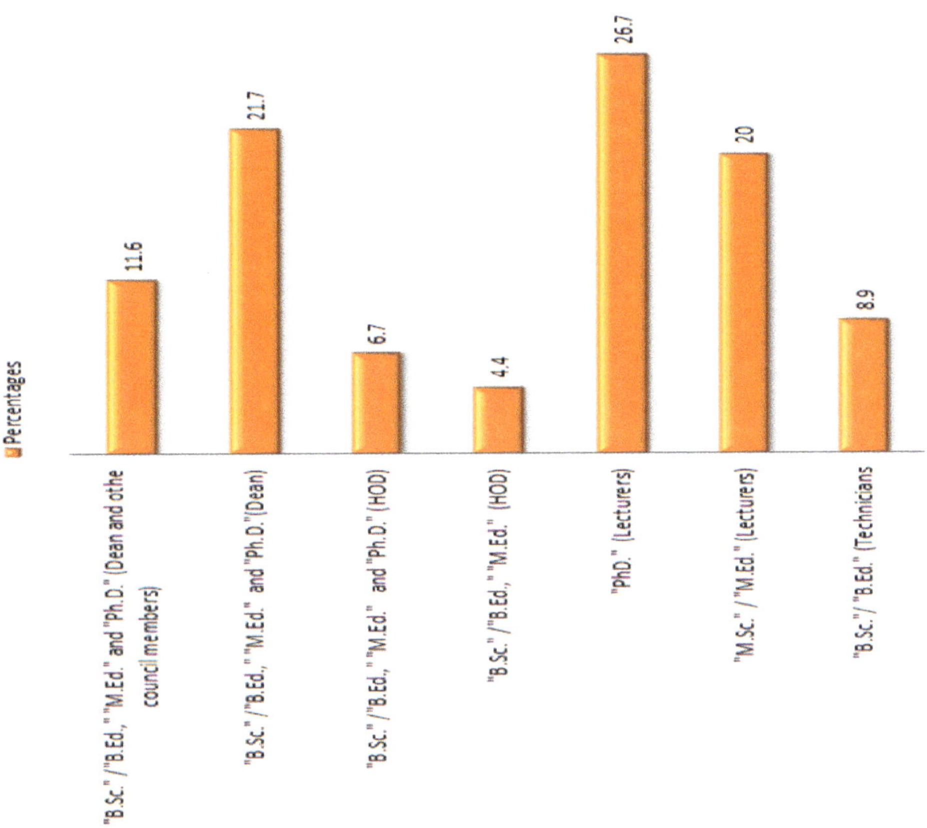

Figure 1. Educational background of administrators and other technical personnel.

Table 2. Tools and strategies for effective e-learning Implementation.

S/N	Tools Used, Strategies and Effectiveness	Responses		Percentage (%)	
		Yes	No	Yes	No
1	The use of smart and Android phones in learning	32	9	17.8	5.0
2	Personal computers and desktops are made available for teaching and learning	17	25	9.4	13.9
3	Projectors are used in learning	8	35	4.4	19.4
4	Functional e-library and open access for the students for learning purposes	22	32	12.2	17.8
	Total	**180**		**43.8%**	**56.1%**

Research Question 3: Is there any relationship between the administrators of e-learning implementation and the performance level of students compared with face-to-face learning?

Table 3 below shows that the way situations are handled and the overall contributions of administrators towards the success of e-learning in Nigerian universities have an impact on the performances of students, either in a positive or in a negative way. These impacts give a complete explanation of the way forward, and headway has been observed in the educational sector. The study shows that the formation of good policies by administrators has a positive impact (2.86) on the output of students during e-learning; allocating tasks to be carried out by students has an impact (2.80); student perceptions of the openness of administrators to them is a factor that has not strongly impacted them positively (2.26); and frequent references to online resources by administrators to the students significantly impact them and their performances (2.72). Additionally, the performance rates of students were graded as being poor, good and very good.

Table 3. The mean performance and remarks on the contributions of administrators towards performance compared with face-to-face learning.

S/N	Contributions of Administrators Toward E-Learning Implementation for Students, with Regard to Student Performance with E-Learning Compared with Face-To-Face Learning	Scores with Attitude Testing	Mean Performance of Students	Remarks
1	Formulation of relevant and positive policies to aid e-learning for students and lecturers.	60–90%	2.86	Very Good
2	Constant allocation of tasks to students to boost their overall performance.	60–80%	2.80	Very Good
3	Being open to students expressing the challenges they are facing.	20–40%	2.26	Poor
4	Frequent references to online resources for effective e-learning and optimum performance.	35–50%	2.72	Good
	Grand Mean		**2.66**	

Research Question 4: What impact has the implementation of e-learning had on the receivers and providers, i.e., students and administrators?

From Table 4 below, it can be seen that the impacts of e-learning on both students and administrators are enormous. Moreover, as they cannot be ignored, the assessment of the impacts of e-learning on both students and teachers or administrators was examined and rated below a <50% and >70% scale, respectively, and the percentages of the outcomes based on the respondents were determined.

Table 4. Frequencies and percentages of the tools and strategies for effective e-learning implementation.

S/N	Impacts of E-learning on Student Performance	Frequency <50	(%)	Frequency >70	(%)
1	Examination grades of e-learning compared with face-to-face learning	11	16.7	28	24.6
2	Impacts on exposure to and effective use of computer resources	9	13.6	19	16.7
3	Impact on the rate of examination malpractice	14	21.2	24	21.0
4	Impacts when compared with face-to-face learning	7	10.6	12	10.5
5	Contribution to the exposure of knowledge and self-development of e-learning compared with face-to-face learning.	25	37.9	31	27.2
	Total	**66**	**100%**	**114**	**100%**

Thus, most aspects of the learning and expression of understanding with respect to success, examinational misconduct, comparison with face-to-face learning and cost-effectiveness were examined. The results show that the success ratio was high, according to 28 (24.6%) participants, and that some factors that bring about optimal performance had a positive impact (19 responses (16.7%)). The impacts of e-learning on examination malpractice and misconduct were lower (14 responses (21.2%)). The impacts of e-learning compared with face-to-face learning in a university show a frequency of 7 responses (10.6%), and the cost of engaging with and implementing e-learning for the benefit of the students had little impact, with a frequency of 25 responses (37.9%).

5. Discussion of Findings

5.1. Educational Background of Administrators

The demographic and educational status of the administrators that participated in the study show that all of the administrators had passed through one level of education or another before they attained an administrative rank in their respective universities, and they had obtained a minimum of a B.Sc. and B.Ed. in their respective fields of study, potentially up to the Ph.D. level. The findings reveal that 26.7% of the administrators had a Ph.D. degree, and 21.7% also possessed a B.Sc./Ph.D in order for them to attain the position of a dean in their respective schools, thus having been exposed to the importance of e-learning. They should, therefore, be able to teach their students about the necessity of implementing this study method in Nigerian educational systems. This study has some similarities with the research of [14], who recorded that 35% of teachers and administrators held a Ph.D.

5.2. Needed Gadgets for E-Learning

For optimal impacts on the e-learning system, certain tools and different materials are needed for reflective and impactful learning, but these are not always available to students, except for smartphones, which serve as major tools, according to 32 (17.8%) of the participants. Additionally, other tools such as projectors are not often used for clearer explanation and for better understanding of the subject matter, thus making the identification of the process of ideas seem more complicated. These findings are similar to those of [39], who identified that CD-ROMs, the internet and other computer software packages are gadgets needed for e-learning to be effective. E-learning requires functional ICT and the accessibility of e-libraries for students, which are not always available to them, thus making it hard for students who do not have smartphones to have access to the provided materials for the learning process [37].

5.3. Contributions of Administrators

With regard to the contributions of administrators in the implementation of e-learning, and for them to ensure that the system succeeds despite facing some challenges, much is expected of them regarding resolving problems. Apart from them having an integral role to play in administration and implementation, they are placed in the position of counseling students in order to help build the right approach, and this is achieved by the formulation of policies, giving tasks to students, encouraging them to utilize internet resources for more understanding, etc. The findings from this study (2.66) are similar to those of [8], who observed a mean total of 2.77, necessitating the need for administrators to be active in proper e-learning implementation.

This research has shown many limitations, and the effective implementation of e-learning has suffered some setbacks, needing not only adjustment, but also an overhaul of the whole system, as the identified factors and limitations are faced by everyone in the nation; as such, they affect all sectors. The limitations include the lack of constant power supply, unprofessional attitudes of administrators, the lack of needed facilities to enhance e-learning and to achieve optimal performance and insufficient funds to invest in the educational sector.

This research would have adopted both the qualitative method and direct face-to-face data collection, but due to the COVID-19 situation, it was limited to the quantitative method, which concerns people's opinions and thoughts about the concept of e-learning in Nigeria's educational system.

Further research to address the hindrances associated with e-learning in Nigeria's educational system is worth looking at. The current research seeks to provide solutions to the impacts of e-learning and to expand upon the benefits, but little or no impact has been made regarding the factors that cause deficiency in e-learning; hence, further research needs to be carried out on factors that negatively impact e-learning in Nigerian educational sectors.

6. Conclusions

The impacts of e-learning on the performance of students work in line with the efforts of administrators and technical staff in universities who make sure that proper implementation is carried out for students to achieve optimal performance in the learning procedure, which, overall, means that it differs from face-face learning. This study, which focuses on e-learning and the key players who influence and implement them, has shown great advancement and impartation on students with respect to good grades, exposure to several and relevant technological skills, etc. The electronic learning process in other Nigerian educational systems will be of immense and great benefit to the drowning educational system that has been practiced in the nation in the past.

Recommendations for Further Study

The need to understand various roles played by administrators in ensuring electronic learning is achieved; however, the current study is centered on students and the benefits they derive from e-learning through their attitudes. Hence, more detailed studies that show the strategies which administrators use for effective learning and teaching strategies should be incorporated and studied.

Author Contributions: Conceptualization, E.S. and A.U.A.; methodology, E.S. and A.U.A.; software, E.S. and A.U.A.; validation, E.S. and A.U.A.; formal analysis, E.S. and A.U.A.; investigation, E.S. and A.U.A.; resources, E.S. and A.U.A.; data curation, E.S. and A.U.A.; writing—original draft preparation, E.S. and A.U.A.; writing—review and editing, E.S. and A.U.A.; visualization, E.S. and A.U.A.; supervision, E.S. and A.U.A.; project administration, E.S. and A.U.A.; funding acquisition, E.S. and A.U.A. All authors have read and agreed to the published version of the manuscript.

Funding: The research was fully funded by me the researcher and also the co-researcher also, advises, corrections and instructions were some of the contributions made by the co-researcher of the research work.

Institutional Review Board Statement: The study was conducted in accordance with the Declaration of Helsinki, and approved by the Institutional Ethics Committee of NEAR EAST UNIVERSITY, CYPRUS.

Informed Consent Statement: The concent of the respondents was been sought for in cause of the study with written statement on the data collection tools (Questionnaire) stating clearly that the data was for the sake of the study and that their data will be handled with confidentially.

Data Availability Statement: The data used in the research could only be released on request from corresponding authors (the reason for this was the ethical guidelines that was employed while taking the data with promise of a confidential handling of the data).

Conflicts of Interest: The authors declare no conflict of interest.

References

1. Usman, Y.D. Educational Resources: An Integral Component for Effective School Administration in Nigeria. *Res. Humanit. Soc. Sci.* **2016**, *6*, 27–37.
2. Mohammed, M.N.H.J. The Challenges and Prospects of Using E-learning among EFL Students in Bisha University. *Arab World Engl. J. (AWEJ)* **2020**, *11*, 124–137. [CrossRef]
3. Shahzad, A.; Rohail, H.; Adejare, Y.; Aremu, A.; Hussain, R.; Nawaz, L. Effects of COVID-19 in E-learning on higher education institution students: The group comparison between male and female. *Qual. Quant.* **2021**, *55*, 805–826. [CrossRef] [PubMed]
4. Sahu, P. Closure of universities due to Coronavirus Disease 2019 (COVID-19): Impact on education and mental health of students and academic staff. *Cureus* **2020**, *12*, e7541. [CrossRef]
5. Alvino, F.; Di Vaio, A.; Hassan, R.; Palladino, R. Intellectual capital and sustainable development: A systematic literature review. *J. Intellect. Cap.* **2020**, *22*, 76–94. [CrossRef]
6. Cidral, W.A.; Oliveira, T.; Di Felice, M.; Aparicio, M. E-learning success determinants: Brazilian empirical study. *Comput. Educ.* **2018**, *122*, 273–290. [CrossRef]
7. Selvaraj, C. Success of e-learning systems in management education in chennai city-using user's satisfaction approach. *Online J. Distance Educ. E-Learn.* **2019**, *7*, 124.
8. Nwajioha, P.N.; Chukwu, T. The Place of Educational Administrators in Promoting E-Learning in Nigerian Universities for Global Competitiveness. *Int. J. Educ. Res.* **2021**, *8*, 22–32.
9. Eze, S.C.; Chinedu-Eze, V.; Bello, A.O. The utilization of e-learning facilities in the educational delivery system of Nigeria: A study of M-University. *Int. J. Educ. Technol. High. Educ.* **2018**, *15*, 34. [CrossRef]

10. National Open University of Nigeria (NOUN). *Supervision of Instruction in Education*; Press Craft, Production Company: Lagos, Nigeria, 2006.
11. Adebayo, A. *Principles and Practice of Public Administration in Nigeria*; Spectrum Books Ltd.: Ibadan, Nigeria, 2001.
12. Albert, D.; Borniface, C.C. The Role of Educational Administrators on E-Learning Programmes Implementation: A Case Study of Goromonzi District Primary Schools. *Int. J. Res. Stud. Comput. Sci. Eng. (IJRSCSE)* **2021**, *8*, 43–52.
13. Ogwo, B.A. Modern Instructional Techniques and their applications in Technical Vocational Education (TVE) programmes of Polytechnics and Monotechnics. *A Comm. Pap. Presented Capacit. Build. Workshop Lect. Polytech. Monotechnics Niger. By Educ. Trust. Fund (ETF) Held Mushood Abiola Polytech. August* **2005**, *24*, 2005.
14. Olelewe, C.J.; Nzeadibe, C.A.; Nzeadibe, C.T. Availability and Utilization of Educational Resources in Selected Rural Communities of Enugu State: Implications for Achieving Universal Primary Education of Millennium Development Goals (MDGs) in Nigeria. *Educ. Res. Int.* **2014**, *3*, 15–24.
15. Leonard, C.I.; Collins, C.N. Optimizing E-Learning in Nigerian Universities for Sustainable Development. *J. Inf. Eng. Appl.* **2014**, *4*, 96–104. Available online: www.iiste.org (accessed on 29 November 2021).
16. Olaniyi, S.S. E-Learning Technology: The Nigeria Experience. In Proceedings of the Shape the Change XXIII FIG Congress, Munich, Germany, 8–13 October 2006.
17. Akbar, M.S.U. E-learning in developing countries: Challenges and opportunities: Bangladesh perspective. In Proceedings of the Second International Conference on E-Learning for Knowledge-Based Society, Bangkok, Thailand, 4–7 August 2015.
18. Chinyere, O.A. E-Learning in the Formal Education System: The Enablers, The Challenges Additionally, The Cost Implication. In *ADECT 2019 Proceedings*; Oklahoma State University Libraries: Stillwater, OK, USA, 2019.
19. Abdelaziz, A.M.; Riad, M.A.; Senousy, M.B. Challenges and Issues in Building Virtual Reality-Basede. *Int. J. E-Educ. E-Bus. E-Learn.* **2014**, *4*, 320–328.
20. Winarno, J.S. Penerapan Sistem E-Learning pada Komunitas Pendidikan Sekolah Rumah (Home Schooling). *Ultim. InfoSys* **2013**, *4*, 45–51. [CrossRef]
21. Sujana, J.G.; Dan Yuyu, Y. *Per-Kembangan Perpustakaan di Indonesia*; IPB Press: Bogor, Indonesia, 2005.
22. Tjokro, S.L. *Presentasi Yang Men-Cekam*; Elex Media Komputindo: Jakarta, Indonesia, 2009.
23. Okigbo, C.; Ndolo, I. Optimizing e-learning opportunities for effective educational service delivery in Nigeria. In *Optimizing E-Learning Opportunities for Effective Educational Service Delivery*; Onyegegbu, N., Eze, U., Eds.; Timex Enterprises for Institute of Education, University of Nigeria Nsukka: Nsukka, Nigeria, 2011.
24. Oye, N.D.; Mazleena, S.; Iahad, N.A. Challenges of E-learning in Nigerian University Education Based on the Experience of Developed Countries. *Int. J. Manag. Inf. Technol.* **2011**, *3*, 39–48. [CrossRef]
25. UNESCO. COVID-19 Educational Disruption and Response. 2020. Available online: https://en.unesco.org/news/covid-19-educational-disruption-and-response (accessed on 11 November 2021).
26. Eze, S.C.; Vera, C.C.; Adenike, B. Factors influencing the use of e-learning facilities by students in a private Higher Education Institution (HEI) in a developing economy. *Humanit. Soc. Sci. Commun.* **2020**, *7*, 133. [CrossRef]
27. Akudolu, L.R. Exploring Appropriate Pedagogical Designs for Optimizing ELearning in Higher Education in Nigeria. *Unizik Orient J. Educ.* **2010**, *6*, 105–119.
28. Peter, J.K.; Esezi, I.O. Higher education for all in need through the National Open University of Nigeria: A paradox in policy practice. *Merit Res. J. Educ. Rev.* **2013**, *1*, 172–180.
29. Sina, M.A.; Adeel, M.A. Assessment of Stand-Alone Photovoltaic System and Mini-Grid Solar System as Solutions to Electrification of Remote Villages in Afghanistan. *Int. J. Innov. Res. Sci. Stud.* **2021**, *62*, 179. [CrossRef]
30. Smiley. Lockdown: Is E-Learning a Good Idea for Tertiary Institution in Nigeria. 2021. Available online: https://ng.opera.news/ng/en/education (accessed on 11 November 2021).
31. Okebukola, P.A.O. Clipping the wings of degree mills in Nigeria. *Int. High. Educ.* **2008**, *43*, 12–15. [CrossRef]
32. Olojo, O.J.; Adewumi, M.G.; Ajisola, K.T. E-learning and its effects on teaching and learning in a global age. *Int. J. Acad. Res. Bus. Soc. Sci.* **2012**, *2*, 203–210.
33. Falana, F.T. *Prospects and Challenges of E-Learning in Nigerian University Education Using National Open University of Nigeria*; Akure Study Centre, Department of Science and Technical Education (Computer Education Unit) Adekunle Ajasin University: Akungba, Nigeria, 2015.
34. Federal Republic of Nigeria. *National Policy on Education*; NERDC Press: Lagos, Nigeria, 2014.
35. Alu, N.C. Utilizing e-learning in science and technology education: Problems and prospects. In *Optimizing E-Learning Opportunities for Effective Educational Service Delivery*; Nyegegbu, N.O., Eze, U., Eds.; Times Enterprises for Institute of Education University of Nigeria: Nsukka, Nigeria, 2011.
36. Darcy, S. Disability, Access, and Inclusion in the Event Industry: A Call for Inclusive Event Research. *Event Manag.* **2012**, *16*, 259–265.
37. Islam, M.T. Educational Technology for 21st century. *Observer Magazine*, 9 May 1997; 3–4.
38. Agostini, L.; Nosella, A. The adoption of Industry 4.0 technologies in SMEs: Results of an international study. *Manag. Decis.* **2020**, *58*, 625–643. [CrossRef]

39. Babatola, J.E.T. The Role of Nigerian University Professional Administrators in a Globally Competitive University Environment. Available online: https://www.academia.edu/33306384/ROLE_OF_NIGERIAN_UNIVERSITY_ADMINISTRATORS_IN_GLOBALLY_COMPETITIVE_UNIVERSITY_ENVIRONMENT (accessed on 1 November 2021).
40. Nwogu, U.F.; Oguejiofor, C.S. Professional competencies of secretaries for managing new office technologies in Nigeria. *Niger. J. Bus. Educ.* **2014**, *1*, 114–120.
41. El-Seoud, S. The Effect of E-learning on Learner's Motivation: A Case Study on Evaluating E-Learning and its Effect on Egyptian Higher Education. In Proceedings of the International Conference on E-Learning in the Workplace (ICELW 2013), New York, NY, USA, 12–14 June 2013.
42. Cronbach, L.J. Coefficient alpha and the internal structure of tests. *Psychometrika* **1951**, *16*, 297–334. [CrossRef]

electronics

Article

The Effect of Graded-Reading Websites/Applications on EFL Undergraduates' Reading Comprehension during COVID-19 Pandemic

Talal Musaed Alghizzi [1,*] and Tariq Elyas [2]

1 Department of English Language and Literature, College of Languages and Translation, Imam Mohammad Ibn Saud Islamic University, Riyadh 11432, Saudi Arabia
2 Department of Foreign Languages and Literature, Faculty of Arts and Humanities, King Abdulaziz University, Jeddah 21589, Saudi Arabia; telyas@kau.edu.sa
* Correspondence: tmalghizzi@imamu.edu.sa

Abstract: The COVID-19 pandemic has resulted in many educational changes, especially the shift towards the use of technology in all subjects. This longitudinal study was conducted to investigate the effect of learning environments—blended and online, alone and with graded-reading websites/applications—on the reading comprehension of Saudi undergraduates majoring in English during COVID-19 pandemic. In this study, 130 participants were selected (control: male [N = 21], female [N = 54]; or experimental: male [N = 21], female [N = 34]). Although the four gender-based groups were exposed to the same learning environments—first blended and later online, which were either partially or dependent on technology—only the male and female experimental groups were required to use graded-reading websites/applications for approximately 10 months during the COVID-19 school lockdowns. All participants took four tests (pretest, posttest 1, posttest 2, and delayed posttest). Using the SPSS program, the results indicated that the learning environments alone had a limited positive effect on the control groups' reading comprehension in the short term, which either decreased significantly (male control group) or remained unchanged (female control group) in the long term. There were significant differences between all control groups and experimental groups across all tests ($p < 0.000$). However, the experimental male group outperformed their male counterpart across all posttests except for the second posttest: experimental male group mean was 15.43 whereas it was 16.19 for the control male group. However, combining learning environments and graded-reading websites/applications yielded gradual positive effects on the reading comprehension of the experimental groups in the short term, which continued into the long term for the male experimental group. The experimental groups outperformed the control groups on at least two out of three posttests. The study concluded that the effect of technology on the reading comprehension of Saudi male and female undergraduates is bounded by the type of specialized technology (i.e., reading websites/applications) and the applied learning environments (i.e., blended and online). Additionally, the study indicated that there is a need to investigate other important factors related to technology used in Saudi institutes, as well as its effects on students' learning processes in ongoing changes in the education sector in Saudi Arabia.

Keywords: graded reading; reading websites/applications; technology; EFL; COVID-19

Citation: Alghizzi, T.M.; Elyas, T. The Effect of Graded-Reading Websites/Applications on EFL Undergraduates' Reading Comprehension during COVID-19 Pandemic. *Electronics* **2022**, *11*, 1751. https://doi.org/10.3390/electronics11111751

Academic Editors: Boni García, Carlos Alario-Hoyos, Mar Pérez-Sanagustín and Miguel Morales

Received: 29 April 2022
Accepted: 30 May 2022
Published: 31 May 2022

1. Introduction

While countries are at different points in their COVID-19 infection rates, there are currently more than 186 countries affected by school closure due to the pandemic [1–3]. Therefore, as a result of COVID-19, education has changed dramatically, with a distinctive rise in e-learning, whereby teaching is undertaken remotely or virtually and on digital platforms. The closure of all educational institutions in Saudi Arabia, due to the COVID-19 outbreak, has caused an unplanned rapid shift from the customary "traditional" learning approach [2–4] to the new government-endorsed approach, namely, online learning. Such changes from traditional face-to-face instruction to use of technology in classrooms have yield issues in the effectiveness of use of technology for all subjects. Teachers have found the learning experience challenging and consider that it failed to meet the needs of the students. Since the beginning of the period of school suspension in Saudi Arabia, the MoE has worked hard to efficiently adapt the educational system to distance learning [2,3]. According to Al-Bogami and Elyas, regarding the current educational shift "the Ministry of Education was in a dilemma. Remarkably, nonetheless they managed to control the situation by introducing the new official way of learning and creative way of online learning" [5]. In our paper, we intended not only to investigate English reading comprehension by using different online websites but also to examine how such websites have impacted learning English for Saudi ungraduated students.

Improving reading comprehension in the English as a foreign language (EFL) field is becoming an increasingly important goal in applied linguistics. Recent technological developments have heightened the need to investigate recent and advanced online applications to improve reading comprehension skills. Studies in the field of reading comprehension and the use of technology in improving such skills are considered essential. These studies compare and identify appropriate applications for improving reading comprehension in specific learning environments [6]. In addition, researchers have specifically focused on difficulties in the understanding of the text during reading, as EFL learners tend to face difficulties in this area [7].

Although many studies have investigated the role of technology in developing reading comprehension, the issue of developing EFL learners' reading comprehension has not been fully examined in the context of Saudi Arabia. In addition, exposing EFL learners to texts with different levels of difficulty is a practical way to develop their reading comprehension abilities. Therefore, using websites/applications that provide EFL learners with this exposure may efficiently improve their abilities through digital scaffolding that allows them to reach higher levels of comprehension [8]. According to Rydland et al. [9], when learners are provided with texts of different levels of difficulty, they master different reading skills gradually by connecting previous knowledge to new knowledge while activating their schemata. However, to the best of our knowledge, no study has investigated the role of graded-reading websites/applications in improving EFL learners' reading comprehension, especially in the Saudi context.

2. Literature Review

2.1. Theoretical Background

Researchers have proposed various theories to understand the process of reading comprehension, thereby creating valuable opportunities for EFL reading research. One prominent theory in reading comprehension is the schema theory, which states that the reader goes through a complex internal process during reading. The process involves interactions between the reader, writer, text, and the reader's previous knowledge, which leads to an accurate understanding of the information presented in the text [9]. Therefore, according to this theory, EFL teachers should activate their students' previous knowledge and connect it with new knowledge to obtain desirable outcomes.

Wood et al. presented the concept of scaffolding [10], which refers to any teaching technique used by the teacher to help the learner understand the presented material through the zone of proximal development proposed by Vygotsky [11]. The zone of proximal development refers to the gap between what a learner can execute without help and what they can execute with the guidance of an adult. Thus, the term "proximal" refers to those skills that the learner is "close" to achieving [12]. In this study, digital scaffolding was provided to EFL learners to help them read texts with varying levels of difficulty to improve their reading skills. Ardeshiri [8] asserted that digital scaffolding helps learners improve their reading comprehension skills independently. He confirmed that digital scaffolding refers to a variety of digital support that assists a language learner in developing new skills (e.g., reading) within the zone of proximal development with the goal to moving toward greater self-regulation to complete a given task.

2.2. Using Technology in Learning

The second half of the last century has witnessed a significant improvement in computers, phone applications, and learning programs for fostering the learning process. Researchers and educators in the field of teaching and learning, along with technology experts, are constantly inventing new technological tools to support learners and teachers as well as the learning process [13]. However, Golonka et al., who reviewed over 350 studies addressing the effectiveness of said technologies in foreign language learning (EFL) and teaching and who compared new educational technologies with more traditional methods and materials, concluded that only a few of these studies were well-designed and can empirically substantiate their positive impact on learning [14]. In addition, the use of such tools for learning requires the teachers' expertise and sometimes training as well [15]. Research in this field has reported promising results regarding the use of technology to support the learning process [16–20]. Presumably, how teachers activate these tools and the pedagogical need for such tools are essential factors in obtaining desirable benefits of technology in classrooms [21]. Therefore, the employment of technology enables learners to access a broad range of rich foreign language exposure, which may not be feasible in traditional classes. Abdel Latif emphasized that to overcome these challenges, there are several coping strategies to overcome these challenges such as "planning for online teaching, managing online classrooms, supporting students' mental health, enhancing students' ability to use/access technology, fostering active language learning engagement and motivation, and promoting teacher professional practices and wellbeing" [22].

2.3. Using Technology to Foster EFL Reading Comprehension

In traditional methods of language teaching, teachers normally depend on textbooks that are not stimulating enough for students to have an enjoyable and effective learning environment. Therefore, EFL learners tend to face difficulties in reading comprehension that stem from the inactivation of communicative reading strategies and a lack of motivation and authentic materials [23,24]. Thus, EFL teachers can employ different reading techniques as well as use technology to make the learning process easier and more enjoyable [25]. Improving reading skills using authentic materials is a required strategy to enhance EFL literacy and critical literacy skills. The use of technology in reading comprehension classrooms has been proven to improve learners' skills [26–28] and the issue has attracted considerable attention in supporting EFL learners. Teachers should choose the appropriate kind of technology and help learners deal independently with authentic texts [29].

It has also been hypothesized that employing technology in teaching EFL reading comprehension offers learners additional vital benefits along with learning outcomes [30]. Learners can read authentic texts independently with the help and guidance of the teacher. Accordingly, students' motivation to read in English increases by reading authentic texts outside the physical boundaries of the classroom. Furthermore, the use of technology would create a special teacher–student relationship, which, in turn, would reverse the teacher and students' roles in the learning environment [26].

Studies investigating the impact of technology on developing EFL learners' reading comprehension skills started concurrently with the spread of technology. Researchers and educators are constantly discovering new technologies that may improve literacy and reading comprehension skills for EFL learners [31]. Previously, technology and reading studies dealt with improving critical reading skills in a computer-networked environment [32]. For example, Usó-Juan and Ruiz-Madrid examined the effect of e-reading on EFL students' skills by comparing traditional and online reading teaching methods [33]. An academic reading test and a reading strategy questionnaire were used in said study to develop the students' reading skills. The results revealed that online reading was a helpful means to improve students' reading comprehension skills. The results of other studies are similar to the aforementioned study on the effectiveness of incorporating technology into teaching [29].

In light of recent research in technology and improving EFL reading comprehension, the use of technology is not confined to the use and application of computers, but also extends to telecommunications and recent mobile applications [32]. Taj et others explored the effect of both computer-assisted language learning and mobile-assisted language learning on EFL reading comprehension [34]. Using a quasi-experimental design, they employed computer-based reading comprehension exercises. Vocabulary was taught through WhatsApp. Posttest results of the reading comprehension achievement test indicated that the experimental group outperformed the control group. Such a study, along with other recent studies, provided evidence for the advantages of using technology to teach EFL reading comprehension skills [26–34].

Although extensive research has been conducted on the effect of technology on improving reading comprehension in EFL learners, there are three main reasons for conducting this study. First, there have been no studies in the EFL field, in general, and in a Saudi context in particular, that address the topic and impact of employing graded-reading websites/applications on enhancing the reading comprehension of EFL male and female undergraduates. Second, in studies that have already been conducted, the comparisons were made between a control group exposed to traditional learning environments (i.e., on-campus classrooms) with an experimental group exposed to blended or online learning environments (i.e., on-campus classrooms and/or virtual classrooms) and not between control and experimental groups exposed to the same blended learning environment or online learning environment. Third, there is a need to obtain information on the type of effect that the previously mentioned technology will have on the reading comprehension of said participants, when and where it will take place, and for how long it will last. Consequently, the following research questions were posed:

1. Are there any significant general and gender-specific differences within the control groups for various tests?
2. Are there any significant general and gender-specific differences within the experimental group for various tests?
3. Are there any significant general and gender-specific differences between the control and experimental groups for various tests?

3. Method

3.1. Research Design and Settings

This research has longitudinal experimental quantitative research designs [35] and was conducted at a university in Saudi Arabia where English is taught as an EFL subject.

The data were collected by administering adopted reading tests before, during, and after applying a reading-related experiment to Saudi EFL participants over a period of three semesters (approximately 10 months), starting from the academic year 2019–2020 to 2020–2021.

To this end, a questionnaire was distributed among students to investigate the above research questions between genders (male and female) and groups (control and experimental) and treated statistically using the SPSS program. Descriptive statistics were employed using IBM Statistical Programme for Social Science (SPSS® ver. 24; IBM, Armonk, NY, USA) to analyze students' responses to the questionnaire items in terms of frequency, percentage, mean, and standard deviation. Cronbach's alpha reliability test was computed using the SPSS® software. The value of Cronbach's alpha was 0.971, which indicated a significantly adequate level of reliability. The following sections explain in detail the research design and instruments for the current study.

3.2. Participants

Participants were recruited in three phases. In the first phase, at the beginning of the semester, an email was sent to 369 Saudi male and female EFL students registered for the level one Reading Comprehension 1 (Eng109) course, with the cooperation of the Student Academic Affairs Office. The email included information about the research and required the voluntary candidates to answer demographic questions regarding their names, the number of times they had registered for the reading course, years of exposure to English, and scores on the Standardized Test of English Proficiency (STEP). In the second phase, candidates' responses were analyzed to exclude those who did not meet the selection criteria. The criteria were that participants should: (a) have registered for the reading course for the first time; (b) have been exposed to English for 10 to 13 years, depending on their first exposure to formal instruction in English (either in level one or four in primary school); (c) enjoy reading in English for at least thirty minutes weekly; (d) have scored no less than 64 and no more than 74 in STEP—which equals 4.5 to 5 in the International English Language Testing System (IELTS) and B1 in Common European Framework for Reference of Language (CEFR). The reason for such criteria was not only to recruit participants with common characteristics but also to prevent extraneous variables from distorting the results [36]. In the third phase, participants who met the criteria (42 male, 88 female) received another email asking them to choose their group type (i.e., control or experimental) and to complete a pretest (see Appendix A). Finally, the results of the initial analysis of the participants' pretest after being assigned to the group of their choice indicated that there were no significant differences within groups and genders and across groups and genders. Table 1 presents the results of the independent sample t-test for the difference between the groups and genders.

Table 1. Independent Sample t-test Results for the Differences between Groups and Genders in the Pretest.

Group		N	Mean	Standard Deviation	t	*p*-Value
Entire Sample	Control	75	10.85	5.816	0.215	0.830
	Experimental	55	10.64	5.472		
Male	Control	21	12.76	6.595	1.666	0.107
	Experimental	21	10.14	2.903		
Female	Control	54	10.11	5.368	−0.616	0.540
	Experimental	34	10.94	6.606		
Control	Male	21	12.76	6.595	1.642	0.111
	Female	54	10.11	5.368		
Experimental	Male	21	10.14	2.903	−0.615	0.541
	Female	34	10.94	6.606		

Figure 1 shows the results of the independent sample t-test for the difference between the groups and genders.

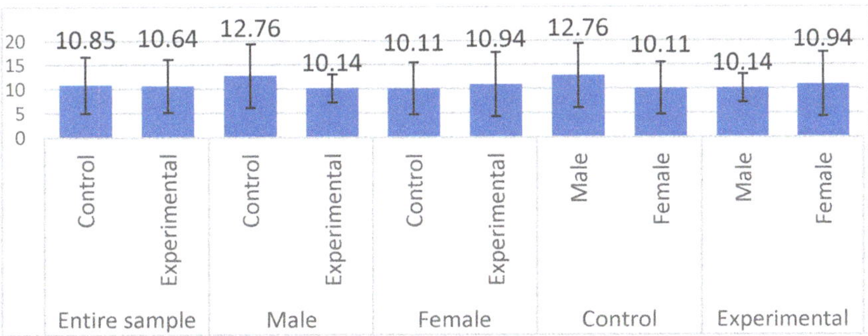

Figure 1. Mean Differences between Groups and Genders in the Pretest.

3.3. Teaching Approaches and Research Treatement

The teaching approaches incorporated for the reading courses in this research were executed through a collaboration between male and female colleagues and the researcher. Instructors who voluntarily agreed to participate in the research were debriefed about the research topic, requirements, teaching approaches, groups, websites/applications, and the duration of the experiment, as they would continue teaching the participants until they finished level two. Instructors were also assigned to their gender-based groups (i.e., control and experimental). Meanwhile, the researcher aided teaching and supervision of the experimental groups only.

To clarify, regardless of group type, all the learners underwent the same courses' textbook requirements, learning environments, tests, assignments, and activities, with one major difference being the incorporation of graded-reading websites/applications. For example, in the level one Reading Comprehension 1 (Eng109) course, all participants were required to undertake the first six chapters of the textbook *Reading Explorer 3* by Douglas and Bohlke [37], and, as they progressed academically to the second level, Reading Comprehension 2 (Eng115) course, they were required to study the last seven chapters from the same textbook. Although the specifications of these courses indicate that the assigned teaching environment was blended (i.e., traditional and virtual classrooms), the incorporation of said environment was only feasible in the first semester of the academic year. In the second semester, due to the COVID-19 pandemic, for all educational institutions in Saudi Arabia, classes were conducted online; students met their instructors virtually via electronic platforms, such as Blackboard.

The male and female control groups had a weekly two-hour reading class with their instructors, who relied heavily on the textbook(s) and could provide extra reading materials, websites, or applications that did not include graded-reading texts (i.e., one text presented at different levels of difficulty). Conversely, the male and female experimental groups had the same tasks as the control groups but had an additional weekly two-hour class with the researcher. In each of these additional classes, participants were first introduced to and subsequently instructed to use specific websites/applications that provided them with reading texts, each of which was presented at various levels of difficulty (i.e., a text could have three to seven versions of various difficulty levels). These websites/applications were News in Levels (available at https://www.newsinlevels.com/products/healthcare-workers-and-covid-19-vaccine-level-2/ (accessed on 5 February 2022)), Breaking News English (available at https://breakingnewsenglish.com/ (accessed on 27 February 2022)), and Tween Tribune (available at https://www.tweentribune.com/ (accessed on 5 March 2022)). In addition, the participants were required to use these websites/applications repetitively and retain a record of their reading progress for discussions conducted in a designated Telegram channel. The instructors raised reading questions in the Telegram group about a particular word, sentence, or idea in a specific text to encourage the students' participation. However, the additional classes, as well as the Telegram channel, were terminated at the end of the second semester, and the experimental groups were instructed to continue using the websites/applications until the end of the summer holidays.

Yet, as a monitoring tool and since the amount of exposure to written input is important for the successful treatment of reading approaches (Bamford & Day, 2004), all of these groups were required to fill in an online log of their readings. For control groups, the average amount of reading for the male participants outside class was 924 min (21 min × 44 weeks) and for the female participants it was 792 min (18 min × 44 weeks). For experimental groups, the male participants spent on average 1056 min (24 min × 44 weeks) whereas the female participants spent an average of 10,188 (27 min × 44 weeks) on reading outside class. All of these groups had an average of 1320 min (30 min × 44 weeks) in-class reading.

3.4. Instruments

The instruments used in this study were a pretest, posttest 1, posttest 2, and delayed posttest 3 (see Appendix A, for an example) all of which were adopted from the reading section in Cambridge's English proficiency test B1 Preliminary for Schools, previously known as PET: Cambridge English: Preliminary for Schools. The validity and reliability of this type of proficiency test has already been determined [38]. Even though this test was designed for school learners, rather than adults, it was chosen for the experiment for the following reasons. First, the reading section comprises various types of questions that Saudi EFL undergraduates are accustomed to. For example, the B1 reading section includes six parts: interpretation of signs, matching, multiple-choice, and filling in the blanks. Second, the level of the test is suitable for the participants' proficiency level because their equivalent score corresponds to the Common European Framework of Reference for Languages B1. Each of these tests was transferred into an electronic version using Google Forms and had to be completed within 45 min.

3.5. Data Collection

As all participants' email addresses had already been provided by the Student Academic Affairs Office, and we kept in touch with them regarding all educational aspects and, most importantly, regarding providing information about the time and links for the tests. As mentioned earlier, the pretest was conducted in the first week of the first semester. However, the other tests were conducted at different time points. For example, the first posttest was administered in the 16th week. The second posttest was administered in the 15th week of the second semester, while the third delayed posttest was administered in the first week of the third semester, which started after the summer holidays (i.e., after three months). To answer the research questions, various comparisons were made regarding each group and between genders based on the results of the four test groups. Two statistical tests (t-test and analysis of variance [ANOVA]) were used to analyze participants' test results and compare the results of the experimental and control groups. A repeated-measures ANOVA was performed five times to investigate the interaction effect between gender (male and female) and the groups (control and experimental). The validity of the yielded results was determined by a professional statistician.

4. Results

4.1. Descriptive Statistics

Table 2 presents the descriptive statistics of all participants (n = 130, 42 males [32.3%], 88 females [67.7%]). The control group consisted of 75 participants (57.7%) and the experimental group consisted of 55 participants (42.3%). Male participants scored the highest (M = 18.95, SD = 5.996) in the experimental group on posttest 3 and the lowest (M = 9.67, SD = 6.319) in the control group on posttest 1 (Table 2).

Figure 2 displays descriptive statistics for all groups.

4.2. Testing the Differences between Paired Tests within the Control Groups

Table 3 presents the results of the paired samples t-test of the differences between paired tests within the control group (n = 75).

4.2.1. Results for All Control Groups (n = 75)

There was a statistically significant difference ($p < 0.05$) between pretest and posttest 1, in favor of posttest 1, which had the highest mean score (M = 15.40, SD = 8.062). Additionally, there was a statistically significant difference ($p < 0.05$) between posttest 2 and posttest 3, in favor of posttest 2, which had the highest mean score (M = 11.81, SD = 5.306).

4.2.2. Results for Male Participants (n = 21)

There was a statistically significant difference ($p < 0.05$) between pretest and posttest 2, in favor of posttest 2, which had the highest mean score (M = 16.19, SD = 4.739). Further, there was a statistically significant difference ($p < 0.05$) between posttest 2 and posttest 3, in favor of posttest 2, which had the highest mean score (M = 16.19, *SD* = 4.739).

4.2.3. Results for Female Participants ($n = 54$)

There was a statistically significant difference ($p < 0.05$) between pretest and posttest 1, in favor of posttest 1, which had the highest mean score (M = 17.63, *SD* = 7.589).

Otherwise, there were no statistically significant differences in mean scores between other paired tests ($p > 0.05$) within the control group (Table 3).

Table 2. Descriptive Statistics for all Groups (N = 130).

Test	Group	Gender	N	Mean	Standard Deviation
Pretest	Control	Male	21	12.76	6.595
		Female	54	10.11	5.368
		Total	75	10.85	5.816
	Experimental	Male	21	10.14	2.903
		Female	34	10.38	5.954
		Total	55	10.29	4.980
Posttest 1	Control	Male	21	9.67	6.319
		Female	54	17.63	7.589
		Total	75	15.40	8.062
	Experimental	Male	21	10.86	4.016
		Female	34	13.91	5.600
		Total	55	12.75	5.232
Posttest 2	Control	Male	21	16.19	4.739
		Female	54	10.11	4.504
		Total	75	11.81	5.306
	Experimental	Male	21	15.43	5.390
		Female	34	13.32	5.639
		Total	55	14.13	5.591
Posttest 3	Control	Male	21	10.90	5.674
		Female	54	10.33	4.514
		Total	75	10.49	4.833
	Experimental	Male	21	18.95	5.996
		Female	34	12.38	3.516
		Total	55	14.89	5.590

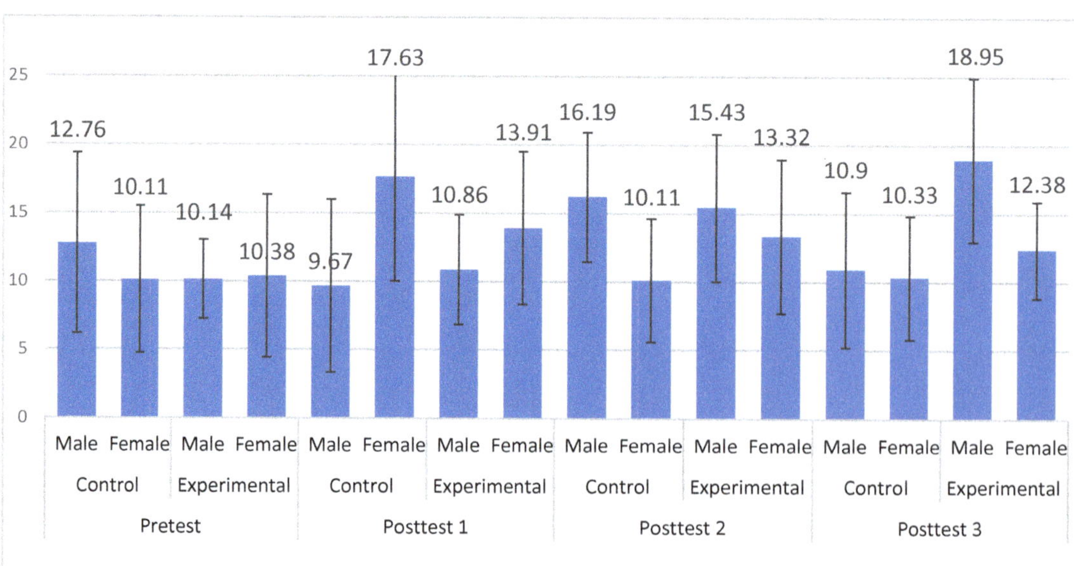

Figure 2. Descriptive Statistics for all Groups.

Table 3. Paired Samples t-test Results of the Differences between Paired Tests within the Control Groups.

	Paired Tests		Mean	Standard Deviation	t	p-Value
All control groups (n = 75)	Pair 1	Posttest 1 Pretest	15.40 10.85	8.062 5.816	3.874	0.000 **
	Pair 2	Posttest 2 Pretest	11.81 10.85	5.306 5.816	1.165	0.248
	Pair 3	Posttest 3 Pretest	10.49 10.85	4.833 5.816	−0.436	0.664
	Pair 4	Posttest 3 Posttest 2	10.49 11.81	4.833 5.306	−2.281	0.025 *
Male (n = 21)	Pair 1	Posttest 1 Pretest	9.67 12.76	6.319 6.595	−1.611	0.123
	Pair 2	Posttest 2 Pretest	16.19 12.76	4.739 6.595	2.095	0.049 *
	Pair 3	Posttest 3 Pretest	10.90 12.76	5.674 6.595	−1.190	0.248
	Pair 4	Posttest 3 Posttest 2	10.90 16.19	5.674 4.739	−3.177	0.005 **
Female (n = 54)	Pair 1	Posttest 1 Pretest	17.63 10.11	7.589 5.368	6.070	0.000 **
	Pair 2	Posttest 2 Pretest	10.11 10.11	4.504 5.368	0.000	1.000
	Pair 3	Posttest 3 Pretest	10.33 10.11	4.514 5.368	0.229	0.820
	Pair 4	Posttest 3 Posttest 2	10.33 10.11	4.514 4.504	0.772	0.444

Note: * Significant at 0.05; ** Significant at 0.01.

Figure 3 shows the mean differences between paired tests within the control groups.

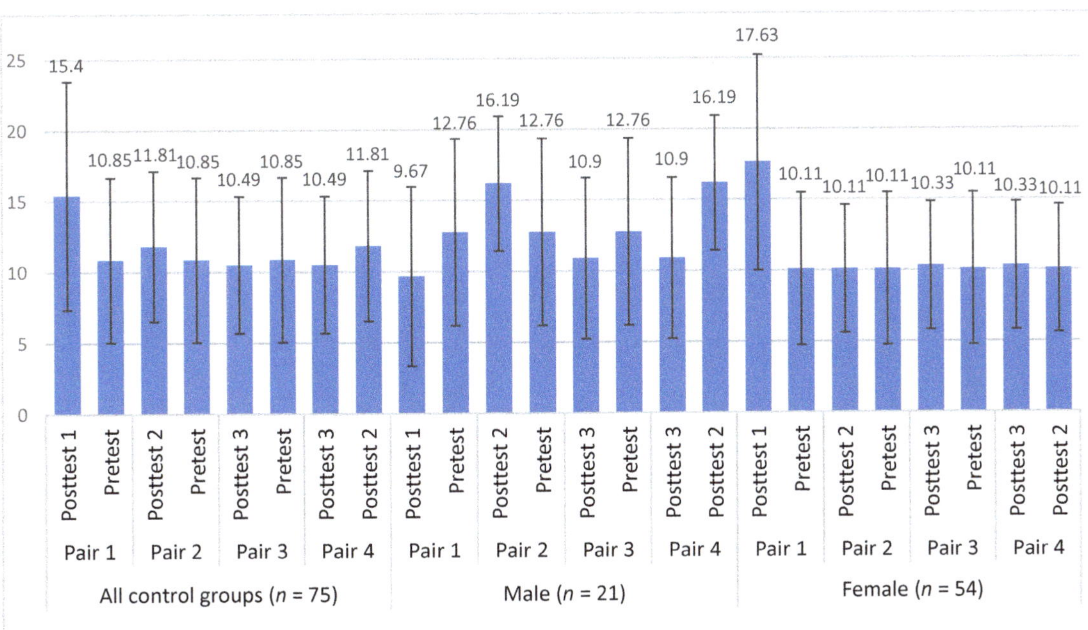

Figure 3. Mean Differences between Paired Tests within the Control Groups.

4.3. *Testing the Differences between Paired Tests within the Experimental Group*

Table 4 presents the results of the paired samples t-test for the differences between paired tests within the experimental group (n = 55).

Table 4. Paired Samples t-test Results for the Differences between Paired Tests within the Experimental Groups.

Group	Paired Tests		Mean	Standard Deviation	t	p-Value
All experimental groups (n = 55)	Pair 1	Posttest 1 Pretest	12.75 10.29	5.232 4.980	2.804	0.007 **
	Pair 2	Posttest 2 Pretest	14.13 10.29	5.591 4.980	4.848	0.000 **
	Pair 3	Posttest 3 Pretest	14.89 10.29	5.590 4.980	4.648	0.000 **
	Pair 4	Posttest 3 Posttest 2	14.89 14.13	5.590 5.591	0.885	0.380
Experimental groups (Male) (n = 21)	Pair 1	Posttest 1 Pretest	10.86 10.14	4.016 2.903	0.944	0.356
	Pair 2	Posttest 2 Pretest	15.43 10.14	5.390 2.903	4.165	0.000 **
	Pair 3	Posttest 3 Pretest	18.95 10.14	5.996 2.903	5.729	0.000 **
	Pair 4	Posttest 3 Posttest 2	18.95 15.43	5.996 5.390	2.580	0.018 *
Experimental groups (Female) (n = 34)	Pair 1	Posttest 1 Pretest	13.91 10.38	5.600 5.954	2.689	0.011 *
	Pair 2	Posttest 2 Pretest	13.32 10.38	5.639 5.954	2.955	0.006 **
	Pair 3	Posttest 3 Pretest	12.38 10.38	3.516 5.954	1.848	0.074
	Pair 4	Posttest 3 Posttest 2	12.38 13.32	3.516 5.639	−0.923	0.363

Note: * Significant at 0.05; ** Significant at 0.01.

4.3.1. For All Experimental Groups (n = 55)

There was a statistically significant difference ($p < 0.05$) between the pretest and posttest 1, in favor of posttest 1, which had the highest mean score (M = 12.75, *SD* = 5.232). Additionally, there was a statistically significant difference ($p < 0.05$) between the pretest and posttest 2, in favor of posttest 2, which had the highest mean score (M = 14.13, *SD* = 5.591). There was also a statistically significant difference ($p < 0.05$) between the pretest and posttest 3, in favor of posttest 3, which had the highest mean score (M = 14.89, *SD* = 5.590). There was no statistically significant difference ($p > 0.05$) between posttest 2 and posttest 3.

4.3.2. For Male Participants (n = 21)

There was no statistically significant difference ($p > 0.05$) between the pretest and posttest 1. There was a statistically significant difference ($p < 0.05$) between the pretest and posttest 2, in favor of posttest 2, which had the highest mean score (M = 15.43, *SD* = 0.390). Further, there was a statistically significant difference ($p < 0.05$) between the pretest and posttest 3, in favor of posttest 3, which had the highest mean score (M = 18.95, *SD* = 5.996). There was also a statistically significant difference ($p < 0.05$) between posttest 2 and posttest 3, in favor of posttest 3, which had the highest mean score (M = 18.95, *SD* = 5.996).

4.3.3. For Female Participants (n = 34)

There was a statistically significant difference ($p < 0.05$) between the pretest and posttest 1, in favor of posttest 1, which had the highest mean score (M = 13.91, *SD* = 5.600). Additionally, there was a statistically significant difference ($p < 0.05$) between the pretest and posttest 2, in favor of posttest 2, which had the highest mean score (M = 13.32, *SD* = 5.639). There was no statistically significant difference ($p > 0.05$) between the pretest and posttest 3.

There was also no statistically significant difference ($p > 0.05$) between the pretest, posttest 2, and posttest 3.

There was no significant difference in the mean score between some paired tests within the experimental group ($p > 0.05$) because of small differences (Table 4).

Figure 4 displays the mean difference between paired tests within the experimental groups.

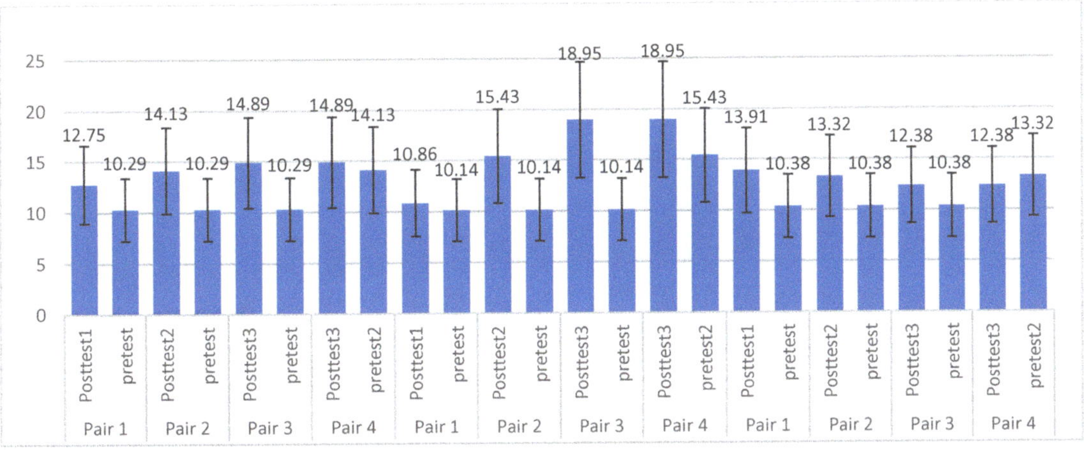

Figure 4. Mean Differences between Paired Tests within the Experimental Groups.

4.4. One-Way ANOVA

To investigate the difference in scores between groups according to the test (pretest, posttest 1, posttest 2, and posttest 3), one-way ANOVA tests were performed five times (all groups, male control group, female control group, male experimental group, and female experimental group). The results indicated that there were statistically significant differences between all groups ($p < 0.05$) (Table 5).

Table 5. ANOVA Results for the Differences between all Tests within all Groups.

	Test	N	Mean	Standard Deviation	F	*p*-Value
All groups ($n = 130$)	Pretest	130	10.62	5.465	8.311	0.000 **
	Posttest 1	130	14.28	7.105		
	Posttest 2	130	12.79	5.528		
	Posttest 3	130	12.35	5.589		
Male control group ($n = 21$)	Pretest	21	12.76	6.595	4.908	0.004 **
	Posttest 1	21	9.67	6.319		
	Posttest 2	21	16.19	4.739		
	Posttest 3	21	10.90	5.674		
Female control group ($n = 21$)	Pretest	54	10.11	5.368	23.568	0.000 **
	Posttest 1	54	17.63	7.589		
	Posttest 2	54	10.11	4.504		
	Posttest 3	54	10.33	4.514		
Male experimental group ($n = 21$)	Pretest	21	10.14	2.903	16.015	0.000 **
	Posttest 1	21	10.86	4.016		
	Posttest 2	21	15.43	5.390		
	Posttest 3	21	18.95	5.996		
Female experimental group ($n = 21$)	Pretest	34	10.38	5.954	2.929	0.036 *
	Posttest 1	34	13.91	5.600		
	Posttest 2	34	13.32	5.639		
	Posttest 3	34	12.38	3.516		

Note: * Significant at 0.05; ** Significant at 0.01.

Figure 5 shows the mean difference between all tests within all groups.

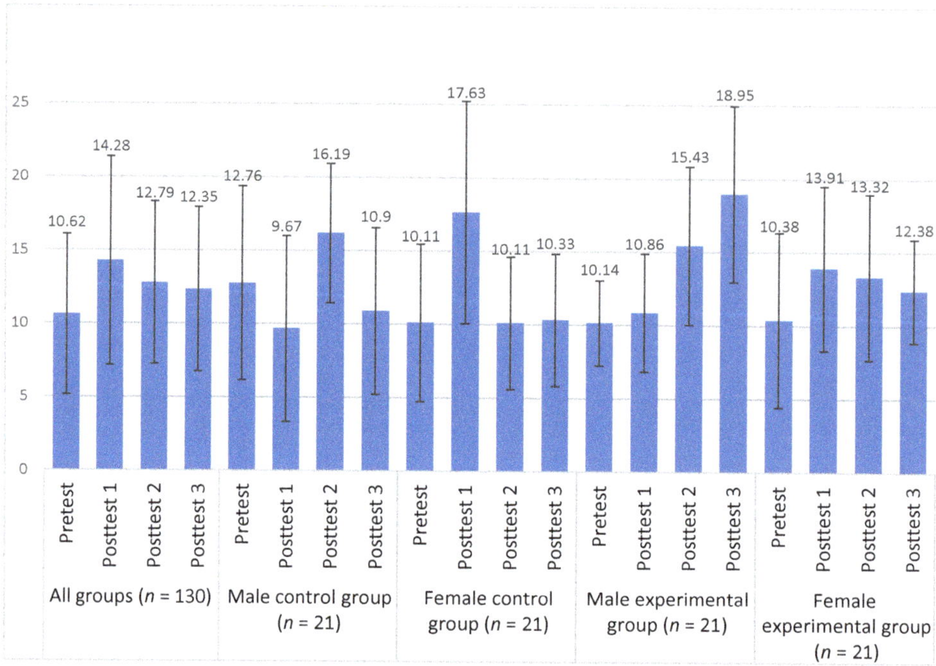

Figure 5. Mean Differences between all Tests within all Groups.

4.5. *Repeated Measures ANOVA*

To investigate the interaction effect between gender (male and female) and groups (control and experimental), repeated-measure ANOVA was performed five times. The results indicated that between the four tests (pretest, posttest 1, posttest 2, and posttest 3), there was no interaction effect between gender and groups ($p = 0.375$). However, for three tests (pretest, posttest 1, and posttest 3), there was a significant interaction effect between gender and groups ($p = 0.029$) (Table 6).

Table 6. Results of Inter-Subject Effects from Repeated-Measures ANOVA.

Test	Source	Type III Sum of Squares	df	Mean Square	F	Significance	Partial Eta Squared
Pre-Post1-Post2-Post3	Intercept	72,022.759	1	72,022.759	1998.172	0.000	0.941
	Group	102.779	1	102.779	2.851	0.094	0.022
	Gender	78.845	1	78.845	2.187	0.142	0.017
	Group * Gender	28.534	1	28.534	0.792	0.375	0.006
	Error	4541.585	126	36.044			
Post1-Post2-Post3	Intercept	59,374.018	1	59,374.018	1947.305	0.000	0.939
	Group	233.747	1	233.747	7.666	0.006 **	0.057
	Gender	43.215	1	43.215	1.417	0.236	0.011
	Group * Gender	111.899	1	111.899	3.670	0.058	0.028
	Error	3841.784	126	30.490			
Pre-Post1-Post2	Intercept	52,747.688	1	52,747.688	1460.819	0.000	0.921
	Group	13.688	1	13.688	0.379	0.539	0.003
	Gender	0.414	1	0.414	0.011	0.915	0.000
	Group * Gender	8.910	1	8.910	0.247	0.620	0.002
	Error	4549.646	126	36.108			
Pre-Post2-Post3	Intercept	53,104.075	1	53,104.075	1387.058	0.000	0.917
	Group	242.201	1	242.201	6.326	0.013 *	0.048
	Gender	732.487	1	732.487	19.132	0.000 **	0.132
	Group * Gender	1.746	1	1.746	0.046	0.831	0.000
	Error	4823.959	126	38.285			
Pre-Post1-Post3	Intercept	51,023.218	1	51,023.218	1657.400	0.000	0.929
	Group	63.476	1	63.476	2.062	0.153	0.016
	Gender	4.996	1	4.996	0.162	0.688	0.001
	Group * Gender	149.630	1	149.630	4.860	0.029 *	0.037
	Error	3878.921	126	30.785			

Note: * Significant at 0.05; ** Significant at 0.01.

Figure 6 displays partial eta squared for effects of the interaction between group and gender.

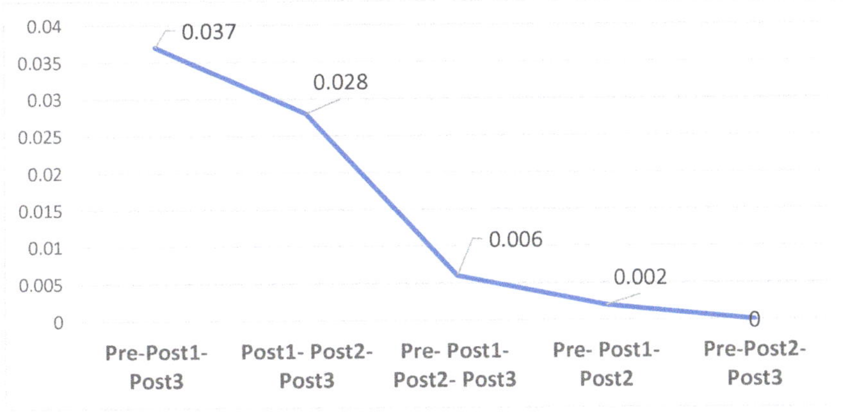

Figure 6. Partial Eta Squared for Effects of the Interaction between Group and Gender.

5. Discussion

Regarding the first research question, which was to determine whether there were any significant differences within the control group participants' tests (i.e., pretest, first posttest 1, posttest 2, and posttest 3), and within the tests for male and female participants, the results indicated the following. There was a significant difference between the control group's pretest and the first posttest, in favor of the first posttest, and there was a significant difference between the second posttest and delayed posttest, in favor of the second posttest. These results imply that although the reading comprehension of the control group participants had temporarily increased significantly, such an increase was not gradual. Moreover, their reading comprehension decreased significantly in the long term. In addition, the results of the male control group indicated similar results with the exception that the significant increase was in the second posttest compared to the pretest, which occurred after the application of the online learning environment. In contrast, the results of the female control group revealed that there was a significant increase in the first posttest in comparison to the pretest, which occurred at the end of the application of the blended learning environment. The results of the male and female control groups imply that each of the learning environments could lead to temporary positive effects on their reading comprehension depending on the gender, and such an effect could become negative in the long term, as for the male control group. The results are aligned with other researchers who advocate that discovering new technologies may improve literacy and reading comprehension skills for EFL learners [31].

Concerning the second research question, which questions whether there were any significant differences within all experimental group participants' tests (i.e., pretest, first posttest, second posttest, and delayed posttest) and the tests of each gender, the results indicated the following. There were significant differences between the experimental group's pretest and first posttest, second posttest, and delayed posttest, which were in favor of the latter three posttests. Thus, the reading comprehension of the experimental group participants had gradually and longitudinally increased significantly, which could be a result of the incorporation of the graded-reading text website/applications. These results were largely similar to those of the male experimental group, in that they showed gradual and longitudinal significant increments starting from the second posttest, which was after the application of the online learning environment, and to the delayed posttest, which was after relying solely on the graded-reading text websites/applications. The reason for the male experimental group not scoring significantly higher in their first posttest and after the application of the blended learning could be that they required more time to adjust as they might always have used technology for gaming only. Thus, it might have taken them time to adjust to utilize their potential capabilities—especially those used for learning purposes—to increase their reading comprehension abilities. In addition, the results of the male experimental group could imply that relying solely on the graded-text websites/applications after using it within the imposed learning environments aided such participants in maintaining their gradual skill increase in the long term. In contrast, the results of the female experimental group indicated that they had gradually significant increases starting from the first posttest, which was after the application of the blended learning environment, and subsequently the second posttest, which was after the application of the online learning environment. However, no significant differences were identified between the delayed posttest, which was after relying solely on graded-reading websites/applications, and the pretest or the second posttest. The justification for such results is that the female experimental group required formal instructional environments such as blended learning and online learning to benefit from the graded-reading websites/applications. The female experimental group demonstrated a lack of commitment during the summer. This was clear from the researchers' follow-ups with the students through the Telegram group. When the researchers contacted the participants regarding using the graded-reading websites/applications, the female experimental group did not attend all the sessions designed to meet with the researchers during the summer; therefore, their results could have been affected.

The third research question was aimed at determining which group outperformed the other, in general, and based on gender. The results indicated that although the control group outperformed the experimental group in the first posttest, the latter outperformed the former in all the other posttests (i.e., second posttest and delayed posttest). Likewise, the male control group outperformed the male experimental group only in the second posttest, while the latter performed better than the former in the first and delayed posttests. In addition, the female control group outperformed the female experimental group in the first posttest, whereas the latter performed better than the former in the second and delayed posttest.

Based on the results of this study, the effect of technology on the reading comprehension of Saudi male and female undergraduates is bound by the type of specialized technology (i.e., reading websites/applications) and the applied learning environments (i.e., blended and online). The use of the blended learning environment for all the participants of the control groups was the only environment that yielded a temporary significant increase in their reading comprehension; once such an environment was no longer available with the online learning facilities, the reading abilities decreased in the long term. The short-term significant increase in the male control group's reading comprehension could be a result of the online environment, whereas the long-term significant decrease in their reading ability could be attributed to the cessation of the learning environment. The reading comprehension of the female control group significantly increased in the short term as a possible consequence of the blended learning environment, but their reading abilities remained unchanged in the short-term while being exposed to the online learning environment and in the long term after the learning environment had stopped. Conversely, the reading comprehension of the participants of the experimental group continuously increased significantly (i.e., from the short term to the long term), even after the two imposed learning environments (i.e., blended or online) had ended, which could be the result of combining the learning environments (i.e., either the blended or the online) with the graded-reading websites/applications. These results remained relatively true for the male experimental group, in which participants' reading comprehension increased significantly in the short term after applying the online learning environment with the graded-reading websites/applications, and it continued to increase in the long term when relying on only the latter type of reading-related technology. Meanwhile, the female experimental group's reading comprehension gradually increased significantly in the short term after combining either the blended learning environment or the online one with the graded-reading websites/applications. However, their reading remained unchanged in the long term after relying only on the graded-reading websites/applications.

In other words, the fact that the four gender-based groups started the experiment with the same reading comprehension level and were exposed to the same kind of learning environment (blended and online) provides solid evidence for attributing the results of the experimental group to the use of the graded-reading websites/applications. This view is supported by many studies that advocate the role of using skill-related technology in developing reading comprehension skills, stressing the importance of technology as a powerful intervention that promises to deliver valuable help to EFL learners) [39–44].

Digital scaffolding, as used in this study, provides proof that exposing students to texts with different levels of difficulty using technology is a legitimate source of language input. The improvement in the experimental group in the general posttest scores provides conclusive evidence for its role in improving students' reading comprehension abilities. This is in line with the studies conducted to examine the effectiveness of using technology for improving language for EFL learners [1–4,15–19]. This has revealed a powerful positive effect of using technology to support students which is supported by earlier research [40–42]. The technology used gave students in the experimental group the boost they needed to score better on their comprehension assessments. By doing so, students built their background knowledge of the texts by having access to the Internet and reading various texts of different levels. Singer and Alexander [45] confirmed that adult students preferred to read digital texts and performed better on exams involving these texts.

Finally, this study raised different issues regarding the development of EFL reading comprehension abilities. First, using reading-specialized websites/applications provided students with different ways of understanding texts with different levels of difficulty. Hence, each student was influenced by many factors (such as using the Internet) in English reading comprehension learning. Gradual progress in reading skills with intensive exposure to various texts has proven to be an effective way of improving reading skills [32]. Second, if students are willing to learn, and if they are motivated using technology, they are given optimal opportunities for language learning to enhance their reading comprehension competencies. Thus, a learning environment accompanied by technology has created abundant resources and opportunities for language learners. Third, although many studies have been conducted on the role and importance of technology in improving reading comprehension, few studies have provided empirical data on the effective implementation of technology in the EFL field. Fourth, using the appropriate technology type, suitable teaching and reading strategies, and effective teaching methods for EFL, learners are variables that contribute to developing reading comprehension skills [2–4]. The results confirm that previous research conducted during the COVID-19 pandemic and the impact of the use of technology in education for better learning [39–45].

6. Conclusions

The purpose of this study was to explore the extent to which learning environments (i.e., blended and online) affect the reading comprehension of Saudi male and female EFL undergraduates with or without reading skill-related technology (i.e., websites/applications of graded-reading texts). All four control and experimental gender-based groups were exposed to the same learning environments; however, only the experimental groups (i.e., male and female) constantly used graded-reading websites/applications. The results of this study suggest that the learning environments alone had a limited positive effect on the reading comprehension of the male and female control groups, whereas the learning environments with the graded-reading websites/application led to gradual increases in the reading comprehension abilities of male and female experimental groups that lasted for the male group even when relying only on the said technology.

The results of this study suggest that the learning environments alone had a limited positive effect on the reading comprehension of the male and female control groups, whereas the learning environments with the graded-reading websites/application led to gradual increases in the reading comprehension abilities of male and female experimental groups that lasted for the male group even when relying only on the said technology. Specifically, the results showed that for all control groups, blended learning environment led to a significant increase in their reading comprehension as indicated by the comparison between their pretest (M = 10.85) and first posttest (M = 15.40; $p < 0.000$). However, no effect was found when applying the online environment. In a matter of fact, the control groups' reading comprehension decreased significantly after not receiving any instruction in any learning environment: second posttest (M = 11.81) and delayed posttest (M = 10.49; $p < 0.025$). Likewise, the male control group's reading comprehension increased significantly only when applying the online learning environment: pretest (M = 12.76) and second posttest (M = 16.19; $p < 0.049$). On the other hand, their reading comprehension decreased significantly in the long term after receiving no instruction in any learning environment as shown in the comparison between the second posttest (M = 16.19) and the delayed posttest (M = 10.90; $p < 0.005$). On the contrary, the reading comprehension of the female control group only increased significantly as a result of applying the blended learning environment: pretest (M = 10.11) and the first posttest (M = 17.63; $p < 0.000$). For all experimental groups, the combination of the learning environments (i.e., blended and online) with the graded-reading websites/application led to continuous gradual significant increases in their reading comprehension as revealed by the comparisons drawn between their pretest (M = 10.29), first posttest (M = 12.75), second posttest (M = 14.13), and delayed posttest (M = 14.89). The p values were: <0.007, 0.000, and 0.000, respectively. However, the reading comprehension of the male experimental group increased significantly after only applying the online environment with the graded-reading websites/application or only the latter: comparisons between the pretest (M = 10.14) second posttest (M = 15.43) and delayed posttest (M = 18.95) in favor of posttests ($p < 0.000$ and 0.000, respectively), and between second posttest (M = 15.43) and delayed posttest (M = 18.95), in favor of the delayed posttest ($p < 0.018$). In addition, the reading comprehension of the female experimental group increased significantly only when applying the blended and online learning environments in combination with the graded-reading websites/applications: pretest (M = 10.38), first posttest (13.91), and second posttest (M = 13.32), in favor of the posttests ($p < 0.011$ and 0.006, respectively). Finally, there were significant differences between all control groups and experimental groups across all tests ($p < 0.000$). However, the experimental male group outperformed their male counterparts across all posttests, except for the second posttest: the experimental male group mean was 15.43, whereas it was 16.19 for the control male group. Similarly, the female control group outperformed the experimental female group in only the first posttest (Means = 17.63 and 13.91, respectively).

Based on the results of this study, the effect of technology on the reading comprehension of Saudi male and female undergraduates is bound by the type of specialized technology (i.e., reading websites/applications) and the applied learning environments (i.e., blended and online). The application of the blended learning environment on all the participants of the control groups was the only environment that yielded a temporary significant increase in their reading comprehension. The study contributes to ongoing research on the use of reading website in EFL context, especially in Saudi context where English is used as a foreign language. In turn, this study in websites/application usage systems in Saudi universities can offer some effective solutions for online/distance learning during the COVID-19 pandemic. Finally, the study indicated that there is a need to investigate other important factors related to technology used in Saudi institutes, as well as their effects on students' learning process in ongoing changes in education sector in Saudi Arabia.

7. Limitations and Recommendations for Future Research

Although this is the first study to investigate this topic and the related variables in general and in Saudi Arabia in particular, it is susceptible to some limitations. These limitations are somehow connected: the number of participants, their proficiency level, learning environments, test type, graded-reading application/websites, and duration of the application. To verify the results, future researchers are recommended to longitudinally (over years) address the above-mentioned results by tracing the reading skill abilities of many low, intermediate, and advanced proficiency Saudi EFL undergraduates. These participants should ideally represent different universities and be exposed to different learning environments (traditional, blended, and online), in which various reading-focused websites/applications (such Oxford graded literature) are applied, using various reading-related tests. By generating accumulative research, it would be possible not only to generalize the results of this study but also to know what, where, when, how long, and for whom a particular type of reading-related technology should be used.

8. Pedagogical Implications

Despite incorporating two learning environments (i.e., blended and online), in which technology is the essence of their execution, the comparisons drawn between the groups exposed to the same unconventional environments have indicated that purposeful technological-learning websites/applications could lead to better performance and possibly a gradual development in reading skills. In other words, the only difference between the instructions of the groups in each learning environment relies on demanding the consistent use of graded-reading texts for one of the groups, which resulted in a significant improvement in the participants' reading abilities. However, there are other implications as well. The reading abilities of participants in the male experimental group started to significantly develop at the end of the second semester, in which the online environment was used, and continued to do so months after the learning environment was terminated. This could imply that participants who are accustomed to playing video games required more time to adjust their understanding of the possible useful effect of technology on their reading abilities. Meanwhile, the female experimental group started to develop reading abilities in the first and second semesters (blended and online learning environment), which ceased during the summer holidays even though they continued to use the graded-reading text websites/applications. This could imply that such technological reading aids would help female participants only if they are accompanied by institutional learning environments.

Author Contributions: Conceptualization, T.M.A.; methodology, T.M.A. and T.E.; software, T.M.A.; validation, T.M.A. and T.E; formal analysis, T.M.A.; investigation, T.M.A.; resources, T.M.A. and T.E; data curation, T.M.A.; writing—original draft preparation, T.M.A. and T.E.; writing—review and editing, T.M.A. and T.E.; visualization, T.M.A. supervision, T.M.A.; project administration, T.M.A. All authors have read and agreed to the published version of the manuscript.

Funding: This research received no external funding.

Informed Consent Statement: Informed consent was obtained from all subjects involved in the study. Official approval was obtained from the college dean's office. Participants consented to participation and were given the option to withdraw at any time.

Conflicts of Interest: The authors declare no conflict of interest.

Appendix A Reading Skill Pretest

Name:
Group:
Time: 45 min
Instructions to Participants
Do not open this question paper until you are told to do so.
Write your name and Group on this first paper sheet.
Read the instructions for each part of the paper carefully.
Answer all the questions here and not on a separate paper. Use a pen.
You **must** complete your answers within the time limit.
INFORMATION FOR CANDIDATES

Questions **1–32** carry one mark.

Part 1

Questions 1–5.

For each question, choose the correct answer.

1. 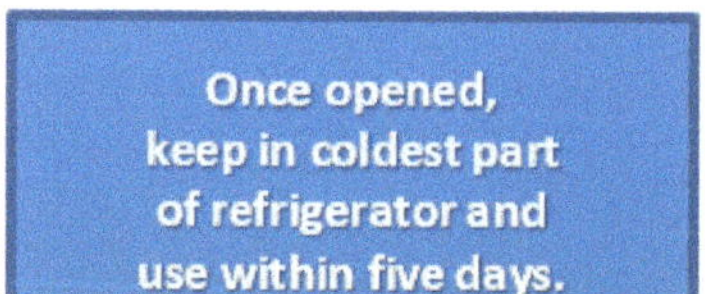

(A) The product can last five days after opening.

(B) The product should be stored in a refrigerator immediately.

(C) The product doesn't have to be kept cool.

2. 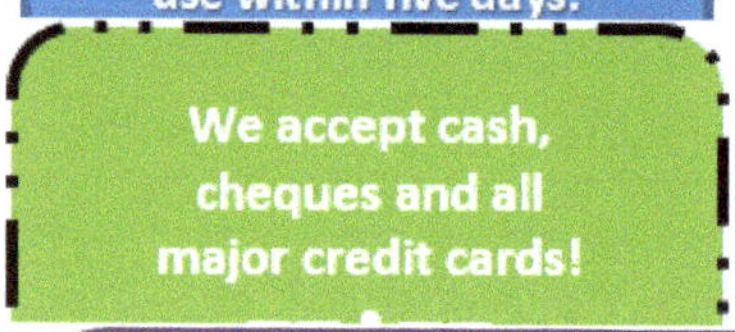

(A) We prefer you to pay with a credit card or in cash.

(B) You must have a credit card and a cheque as well as cash.

(C) You may pay with a credit card, cheque or in cash.

3. 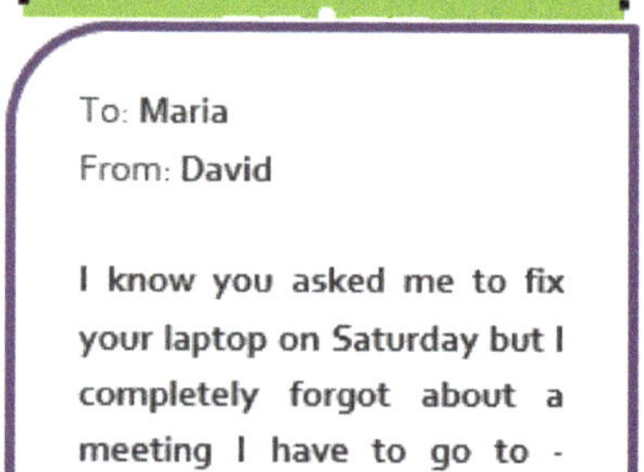

Why is David apologising to Maria?

(A) He cannot do the favour he had promised.

(B) He hasn't had time to fix her laptop.

(C) He won't be able to go to the meeting with her.

4. 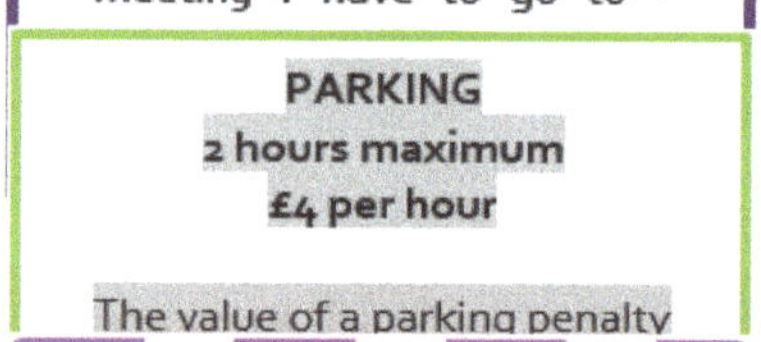

(A) It costs at least £50 to park here for two hours.

(B) There are no parking charges after 2.00 pm.

(C) You have to pay a fine if you park here for three hours.

5. 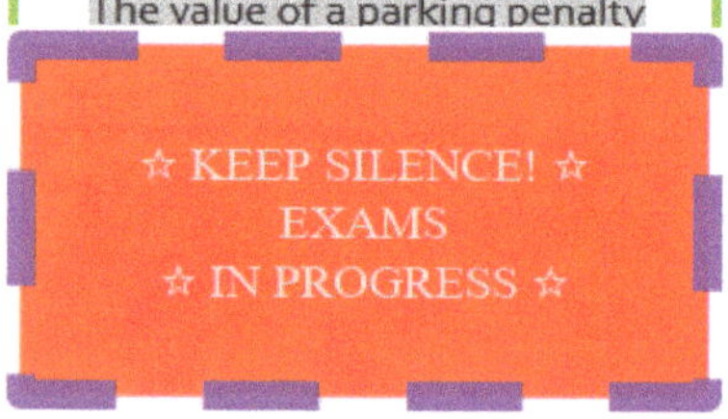

(A) You mustn't speak during the examination.

(B) Please respect others and remain quiet during the examination.

(C) Do not talk to the teacher.

Part 2

Questions 6–10.

For each question, choose the correct answer.

The young people below all want to do a cycling course during their school holidays.

On the opposite page there are descriptions of eight cycling courses.

Decide which course would be the most suitable for the people below.

6 Nancy is fourteen and cycles quite well. She needs to learn how to cycle safely from her home to school on busy city roads. She's only free at the weekends.

7 Markus is an excellent cyclist and he wants the excitement of riding on countryside and woodland tracks. He'd also like to learn more about looking after his bke. He can't attend a morning course.

8 Ellie is nine and knows how to ride her bike, but isn't confident about starting and stopping. She'd love to meet other cyclists with a similar ability and have fun with them.

9 Leo can't cycle yet, and wants to learn on his own with the teacher. He'd prefer a course with sessions twice a week. He'd also like some practical information about cycling clothes and equipment.

10 Josh is eleven and a skilled cyclist. He's keen to learn to do exciting cycling tricks in a safe environment. He'd like to be with people of a similar age.

Cycling Courses

(A) Two Wheels Good!
Mountains! Rivers! Forests!
Our 'off-road' course offers you the chance to get out of the city. You'll need very good cycling skills and confidence. You will be with others of the same ability. Expert advice on keeping your bike in good condition also included.
Mondays 2.00–6.00 p.m. or Fridays 3.00–7.00 p.m.

(B) On Your Bike!
Can't ride a bike yet, but really want to? Don't worry. Our beginners-only group (4–10 pupils per group) is just what you're looking for. Excellent teaching in safe surroundings. Makes learning to cycle fun, exciting and easy.
Mondays 9.00–11.00 a.m. and Thursdays 2.00–4.00 p.m.

(C) Fun and Games
Do you want some adventure? Find out how to do 'wheelies' (riding on one wheel), 'rampers' (cycling off low walls), 'spins' and much more... We offer a secure practice ground, excellent trainers and loads of fun equipment. Wear suitable clothes. Only for advanced cyclists.
(Age 11–12) Saturdays 1.00–4.00 p.m.

(D) Pedal Power
A course for able cyclists. We specialise in teaching riders of all ages how to manage difficult situations in heavy traffic in towns and cities. We guarantee that by the end of the course, no roundabout or crossroads will worry you!
Saturdays 2.00–4.00 p.m.

(E) Cycling 4U
Not a beginner, but need plenty of practice? This course offers practical help with the basics of balancing and using your brakes safely. You'll be in a group of pupils of the same level. Improve your cycling skills and enjoy yourself at the same time! Open to all children up to the age of ten.
Sundays 10.00 a.m.–12.00 p.m.

(F) Bike Doctors
Have you been doing too many tricks on your bike? Taken it up mountains and through rivers? Then it probably needs some tender loving care. Bike Doctors teach you to maintain and repair your bike. (Some basic equipment required.) Ages 11–19
Tuesdays 9.00 a.m.–12.00 p.m. or Wednesdays 3.00–6.00 p.m.

(G) Safety First
We teach cycling safety for the city centre and country lane biker. We'll teach you the skills you need to deal with all the vehicles using our busy roads. All ages welcome from 10+. Thursdays 9.00–11.00 a.m.

(H) Setting Out
A course for absolute beginners needing one-to-one instruction to get off to a perfect start. We also give advice on helmets, lights, what to wear and much more. A fantastic introduction to cycling! Mondays and Tuesdays 9.00–11.00 a.m.

Part 3

Questions 11–15.

For each question, choose the correct answer.

Jacques Cousteau: A Remarkable Man

Jacques-Yves Cousteau was an explorer, ecologist, filmmaker, inventor and conservationist. He was a man, who spent nearly his whole life underwater exploring the hidden depths of the ocean and who did more to educate the world about the mysteries of the deep sea than any other scientist before or since. He was born in June, 1910 in the village of Saint-André-de-Cubzac, in south western France. Jacques was a sickly boy and spent much of his time in bed, reading books and dreaming about a life at sea. In 1920, Jacques' family moved to New York and he was encouraged to start swimming to build up his strength. This was the beginning of his fascination with water and the more he learnt through his own experiences, the more passionate he became about "looking through nature's keyhole". Nevertheless, his career in underwater exploration came about by accident. After entering France's naval academy and travelling around the world, he was involved in an almost fatal car accident that left him seriously injured with two broken arms. He began swimming in the Mediterranean Sea to strengthen his arm muscles as part of his recovery process and rediscovered his love of the ocean. Cousteau developed a pair of underwater breathing apparatus to allow him to stay underwater for long periods of time. His experiments led to the development of the first Aqua-Lung which was a great commercial success. During World War II, he worked for the French Resistance and experimented with underwater photographic equipment. He helped to get rid of German mines and was awarded the Legion D'Honneur and the Croix de Guerre medals for his bravery. In 1942, he filmed his first underwater film Sixty Feet Down. It was 18 min long and was entered in the Cannes Film Festival.

(11) What is the writer trying to do in the text?

Top of Form

(A) teach readers how to make films
(B) explain how Jacques-Yves Cousteau has made a lot of money
(C) introduce readers to the filmmaker Jacques-Yves Cousteau
(D) describe particular films directed by Jacques CousteauBottom of Form

(12) Being a child, Cousteau had....

Top of Form

(A) strong will
(B) bright mind
(C) heart attacks

(D) delicate health

(13) In a car accident he...
Top of Form

(A) burnt both of his arm
(B) broke his extremities
(C) injured his leg
(D) hurt his eyesBottom of Form

(14) Cousteau developed underwater breathing equipment
Top of Form

(A) to extend his underwater investigations
(B) to gain fame
(C) to achieve commercial success
(D) having no certain goalsBottom of Form

(15) During World War II Cousteau collaborated with ...
Top of Form

(A) Polish resistance movement
(B) German antifascists
(C) American troops
(D) underground resistance fighters in France

Part 4

Questions 16–20

Five sentences have been removed from the text below.
For each question, choose the correct answer.
There are three extra sentences which you do not need to use.

Antarctica

Antarctica is the coldest, emptiest and driest place on Earth. Ninety-nine percent of Antarctica is covered by ice about 5 metres thick. The coldest temperature ever recorded on Earth was minus 89.2 degrees Celsius, registered on 21 July 1983, at Antarctica's Vostok station. Antarctica's climate is also very dry and windy. **[16:**] There is an area called Dry Valleys that has not had rain for more than a million years!
The existence of Antarctica was completely unknown until the continent was first discovered in 1820. Antarctica doesn't have a government and belongs to no country. **[17:**] There are 30 various countries that operate 80 research stations located around the continent. In summer, more than 4000 scientists from all over the world work in research stations. Tourists arrive here, too. **[18:**].
Antarctica has no trees or bushes. The only plants that can live in a place that cold are algae, moss and fungi. **[19:**] They live close together in large colonies and build their nests on the ice. In the ocean around the continent you can see seals, whales and orcas but there are no big and large native land animals on the continent. **[20:**].

(A) More than 56,000 people travelled to Antarctica during the 2018–2019 season.

(B) Also hiding under the Antarctic ice is an entire lake called Lake Vostok.

(C) But there are a lot of penguins.

(D) Winds in some places of the continent can reach 320 km/h.

(E) But Antarctica hasn't always been an icy land.

(F) It's just too cold!

(G) The Antarctic is land surrounded by ocean.

(H) It is the only region in the world which is not ruled by any nation.

Part 5

Questions 21–26

For each question, choose the correct answer.

London Parks

London is famous **(21)** .... its parks and gardens. Some of them belong to the Crown but they are all open to the public and the entrance is free of charge. In St James's Park you can watch and **(22)** .... swans, ducks, geese and other water birds. Hyde Park **(23)** .... to be a hunting ground and is still popular with horse riders.
Those who like a good argument should go to the Speakers' Corner to listen to individuals **(24)** .... their speeches on various subjects. Regent's Park now houses London Zoo and open-air theatre where Shakespeare's plays are staged in summer. Not **(25)** .... the parks are in the city centre. Greenwich and Richmond are located in the suburbs. All these areas of green give the city dwellers an excellent **(26)**............ to enjoy some peace and quiet away from traffic and crowded streets.

21	(A)	by	(B)	for	(C)	from	(D)	with
22	(A)	feed	(B)	eat	(C)	breed	(D)	lead
23	(A)	should	(B)	ought	(C)	used	(D)	have
24	(A)	doing	(B)	giving	(C)	taking	(D)	talking
25	(A)	each	(B)	whole	(C)	every	(D)	all
26	(A)	chance	(B)	knowledge	(C)	account	(D)	source

Part 6

Questions 27–32

For each question, write the correct answer. Write one word for each gap.

Our Holiday in Spain

Our trip to Spain was wonderful! First, we flew to Valencia, one of the **(27)** .... beautiful cities in Spain. It's a nice and elegant port city. We stayed at Hampton by Hilton there for three nights. We went sightseeing and just relaxed at the swimming pool.

From Valencia, we flew to Ibiza, arriving **(28)** .... Saturday morning. We went to Las Salinas, **(29)** .... is one of the most popular beaches in Ibiza. The next day, we had a go **(30)** .... water skiing or parasailing. One night, we took a bus tour to a traditional Ibizan village and stayed for dinner and a Flamenco show. We heard Spanish songs for voice and guitar, and we saw traditional dances-it **(31)** .... a very special evening.

From Sant Jordi, we drove to San Rafael. We stayed there for two nights. The very next day, we drove back to Ibiza and flew back to Valencia. We plan to come back to Spain soon, **(32)** .... for now, we're on our way to Portugal!

References

1. Rahman, R.A.; Ahmad, S.; Hashim, U.R. A study on gamification for higher education students' engagement towards education 4.0. In *Intelligent and Interactive Computing*; Piuri, V., Balas, V.E., Borah, S., Syed Ahmad, S.S., Eds.; Springer: Singapore, 2019; pp. 491–502.
2. Oraif, I.; Elyas, T. Applicability of Collaborative Work in the COVID-19 Era: Use of Breakout Groups in Teaching L2 Translation. *Electronics* **2021**, *10*, 2846. [CrossRef]
3. Oraif, I.; Elyas, T. The Impact of COVID-19 on Learning: Investigating EFL Learners' Engagement in Online Courses in Saudi Arabia. *Educ. Sci.* **2021**, *11*, 99. [CrossRef]
4. Affouneh, S.; Salha, S.N.; Khlaif, Z. Designing Quality E-Learning Environments for Emergency Remote Teaching in Coronavirus Crisis. *Interdiscip. J. Virtual Learn. Med. Sci.* **2020**, *11*, 135–137.
5. Al-Bogami, B.; Elyas, T. Promoting Middle School Students' Engagement through Incorporating iPad Apps in EFL/ESL Classes. *SAGE Open* **2020**, *10*. [CrossRef]
6. Grabe, W.; Stoller, F.L. *Teaching and Researching Reading*, 2nd ed.; Routledge Press: Oxfordshire, UK, 2013.
7. Gilmore, A. Authentic materials and authenticity in foreign language learning. *Lang. Teach.* **2007**, *40*, 97–118. [CrossRef]
8. Ardeshiri, M. Technology-Based Intervention in L2 Reading Comprehension: Toward Digital Scaffolding. Ph.D. Thesis, Graduate Department of Curriculum, Teaching and Learning, University of Toronto, Toronto, ON, Canada, 2019.
9. Rydland, V.; Aukrust, V.G.; Fulland, H. How word decoding, vocabulary and prior topic knowledge predict reading comprehension: A study of language-minority students in Norwegian fifth grade classrooms. *Read. Writ.* **2012**, *25*, 465–482. [CrossRef]
10. Wood, D.; Bruner, J.S.; Ross, G. The role of tutoring in problem solving. *J. Child Psychol. Psychiatry Allied Discip.* **1976**, *17*, 89–100. [CrossRef]
11. Vygotsky, L.S. *Mind in Society: The Development of Higher Psychological Processes*; Harvard University Press: Cambridge, MA, USA, 1978.
12. Wood, D.; Wood, H. Vygotsky, tutoring and learning. *Oxford Rev. Educ.* **1996**, *22*, 5–16. [CrossRef]
13. Idrus, M.M.; Ismail, I.S. Learning language through wireless technology: An impact study. *Procedia—Soc. Behav. Sci.* **2013**, *90*, 381–387. [CrossRef]
14. Golonka, E.M.; Bowles, A.R.; Frank, V.M.; Richardson, D.L.; Freynik, S. Technologies for foreign language learning: A review of technology types and their effectiveness. *Comput. Assist. Lang. Learn.* **2012**, *27*, 70–105. [CrossRef]
15. Gilakjani, A.P.; Leong, L.-M. EFL teachers' attitudes toward using computer technology in English language teaching. *Theory Pract. Lang. Stud.* **2012**, *2*, 630–636. [CrossRef]
16. Bahrani, T. Exposure to authentic audiovisual programs in informal setting and language learning. *Turk. Online J. Des. Art Commun.* **2012**, *2*, 10–17. [CrossRef]
17. Khamis, I.K. Technological Devices in Teaching English Vocabulary in Tanzanian Primary Schools: The Case of Chake-Chake District. Master's Thesis, The University of Dodoma, Dodoma, Tanzania, 2015.
18. Mohamad, F. Internet-based grammar instruction in the ESL classroom. *Int. J. Pedagog. Learn.* **2009**, *5*, 34–48. [CrossRef]
19. Mwaamukange, G.H.P. Assessing the Impact of Videos in Promoting Learners' English Second Language Comprehension and Listening in Omusati Region. Ph.D. Thesis, Department of Linguistics, University of Namibia, Windhoek, Namibia, 2018. Unpublished doctoral dissertation.
20. Yunus, M.M.; Nordin, N.; Salehi, H.; Embi, M.A.; Mahamod, Z. Managing problems and planning activities involving ICT tools in teaching ESL reading and writing. *Asian Soc. Sci.* **2013**, *9*, 222–230. [CrossRef]
21. Balajthy, E. Technology and current reading/literacy assessment strategies. *Read. Teach.* **2007**, *61*, 240–247. [CrossRef]
22. Abdel Latif, M.M.M. Coping with COVID-19-related online English teaching challenges: Teacher educators' suggestions. *ELT J.* **2021**, *76*, 20–33. [CrossRef]
23. Kim, J. Extensive reading in EFL contexts: How do EFL learners perceive use of extensive reading in Korean university classrooms? *Engl. Lang. Teach.* **2016**, *28*, 47–68. [CrossRef]
24. Park, J. Integrating reading and writing through extensive reading. *ELT J.* **2015**, *70*, 287–295. [CrossRef]

25. Mohammadian, A.; Saed, A.; Shahi, Y. The effect of using video technology on improving reading comprehension of Iranian intermediate EFL learners. *Adv. Lang. Lit. Stud.* **2018**, *9*, 17–23. [CrossRef]
26. Dreyer, C.; Nel, C. Teaching reading strategies and reading comprehension within a technology-enhanced learning environment. *System* **2003**, *31*, 349–365. [CrossRef]
27. Vahdat, R. Computer assisted reading (CAR) versus traditional print format in EFL academic reading comprehension. *Uloom Trbiati* **2009**, *15*, 71–88.
28. Williams, H.S.; Williams, P.N. Integrating reading and computers: An approach to improve ESL students reading skills. *Read. Improv.* **2000**, *37*, 97.
29. Levine, A.; Ferenz, O.; Reves, T. EFL academic reading and modern technology: How can we turn our students into independent critical readers. *Electron. J. Engl. Second Lang.* **2000**, *4*, 1–9.
30. Sun, Y.; Wang, T.-H.; Wang, L.-F. Implementation of Web-Based Dynamic Assessments as Sustainable Educational Technique for Enhancing Reading Strategies in English Class during the COVID-19 Pandemic. *Sustainability* **2021**, *13*, 5842. [CrossRef]
31. Alfaleh, A.R. Using technology for developing ESL students reading comprehension and motivation. In Proceedings of the 2015 WEI International Academic Conference, Athens, Greece, 19–21 March 2015.
32. Khusniyah, N.L. Implementation online reading strategy on English reading comprehension skills. *Elite J.* **2019**, *1*, 87–94.
33. Usó-Juan, E.; Ruiz-Madrid, M. Reading printed versus online texts: A study of EFL learners' strategic reading behavior. *Int. J. Engl. Stud.* **2009**, *9*, 59–170.
34. Taj, I.H.; Ali, F.; Sipra, M.A.; Ahmad, W. Effect of technology enhanced language learning on EFL reading comprehension at tertiary level. *Arab. World Engl. J.* **2017**, *8*, 108–129. [CrossRef]
35. Creswell, J.W. *Research Design: Qualitative, Quantitative, and Mixed Methods Approaches*, 3rd ed.; Sage Publication, Inc.: Thousand Oaks, CA, USA, 2009.
36. Al-Nujaidi, A. The Relationship between Vocabulary Size, Reading Strategies, and Reading Comprehension of EFL Learners in Saudi Arabia. Ph.D. Thesis, Oklahoma State University, Stillwater, OK, USA, 2003. Unpublished doctoral dissertation.
37. Douglas, N.; Bohlke, D. *Reading Explorer 3*, 3rd ed.; Cengage Learning: Boston, MA, USA, 2020.
38. Docherty, C.; Corkill, D. Test construction: The Cambridge English approach. *Camb. Engl. Res. Notes* **2015**, *59*, 10–14.
39. Sofianidis, A.; Meletiou-Mavrotheris, M.; Konstantinou, P.; Stylianidou, N.; Katzis, K. Let students talk about emergency remote teaching experience: Secondary students' perceptions on their experience during the COVID-19 pandemic. *Educ. Sci.* **2021**, *11*, 268. [CrossRef]
40. Habib, M.K.; Nagata, F.; Watanabe, K. Mechatronics: Experiential learning and the stimulation of thinking skills. *Educ. Sci.* **2021**, *11*, 46. [CrossRef]
41. Alshehri, Y.A.; Mordhah, N.; Alsibiani, S.; Alsobhi, S.; Alnazzawi, N. How the Regular Teaching Converted to Fully Online Teaching in Saudi Arabia during the Coronavirus COVID-19. *Creat. Educ.* **2020**, *11*, 985–996. [CrossRef]
42. Vollbrecht, P.J.; Porter-Stransky, K.A.; Lackey-Cornelison, W.L. Lessons learned while creating an effective emergency remote learning environment for students during the COVID-19 pandemic. *Adv. Physiol. Educ.* **2020**, *44*, 722–725. [CrossRef] [PubMed]
43. Almaiah, M.A.; Ayouni, S.; Hajjej, F.; Lutfi, A.; Almomani, O.; Awad, A.B. Smart Mobile Learning Success Model for Higher Educational Institutions in the Context of the COVID-19 Pandemic. *Electronics* **2022**, *11*, 1278. [CrossRef]
44. Alam, G.M.; Parvin, M. Can online higher education be an active agent for change?—Comparison of academic success and job-readiness before and during COVID-19. *Technol. Forecast. Soc. Change* **2021**, *172*, 121008. [CrossRef]
45. Singer, L.M.; Alexander, P.A. Reading on paper and digitally: What the past decades of empirical research reveal. *Rev. Educ. Res.* **2017**, *87*, 1007–1041. [CrossRef]

electronics

Article

Will–Skill–Tool Components as Key Factors for Digital Media Implementation in Education: Austrian Teachers' Experiences with Digital Forms of Instruction during the COVID-19 Pandemic

Flora Woltran [1,*], Katharina-Theresa Lindner [2], Tanja Dzojic [2] and Susanne Schwab [1,2,3]

1 Centre for Teacher Education, University of Vienna, Porzellangasse 4, 1090 Vienna, Austria; susanne.schwab@univie.ac.at
2 Department of Education, University of Vienna, Sensengasse 3A, 1090 Vienna, Austria; katharina-theresa.lindner@univie.ac.at (K.-T.L.); a01403951@unet.univie.ac.at (T.D.)
3 Optentia Research Focus Area, North-West University Vanderbijlpark, 1174 Hendrick Van Eck Boulevard, Vanderbijlpark 1900, South Africa
* Correspondence: flora.woltran@univie.ac.at

Citation: Woltran, F.; Lindner, K.-T.; Dzojic, T.; Schwab, S. Will–Skill–Tool Components as Key Factors for Digital Media Implementation in Education: Austrian Teachers' Experiences with Digital Forms of Instruction during the COVID-19 Pandemic. *Electronics* **2022**, *11*, 1805. https://doi.org/10.3390/electronics11121805

Academic Editors: Boni García, Carlos Alario-Hoyos, Mar Pérez-Sanagustín, Miguel Morales and Oscar Jerez

Received: 29 April 2022
Accepted: 4 June 2022
Published: 7 June 2022

Abstract: Although comprehensive digitalization (e.g., the provision of skills and resources) had already been placed on Austria's education policy agenda prior to the emergence of COVID-19, there is evidence that educators had some difficulty ensuring digital learning opportunities for their students when schools closed in early 2020. Against this backdrop, the present study, which drew on qualitative data from the large-scale INCL-LEA (Inclusive Home Learning) study, aimed to determine whether secondary school teachers (n = 17) from Viennese schools met the prerequisites for successfully implementing digital instruction, formulated in the Will–Skill–Tool model developed by Christensen and Kzenek (2008). Findings reveal that teachers primarily associated their sufficient digital skills with three factors: (1) basic interest and competence, (2) recently attended training, and/or (3) a positive attitude toward changing teaching practices. Interestingly, some educators recognized that digitization offers great potential for implementing individualized teaching approaches. However, the findings point to the didactic necessity of digital socialization in terms of social communication and inclusion when establishing emergency digital education.

Keywords: COVID-19; spring 2020; school closures; Austria; secondary school teachers; will–skill–tool model; digital media implementation

1. Introduction

The COVID-19 crisis is considered an unprecedented exceptional situation for all involved in education and has led to numerous profound challenges for students, teachers, principals, and parents or guardians. In March 2020, 107 countries introduced national school closures as the first immediate response to the sharp increase in new infections [1]. As a result, educators around the world were forced to temporarily abandon traditional teaching methods. Furthermore, teachers had to quickly adapt to this unusual disruption in education by moving to online teaching within a few days [2,3].

As digital instruction became necessary to promoting educational continuity, access to and competencies in using technological devices and digital instructional formats, as well as the affective-motivational conditions (e.g., attitudes) of teachers, would have been required for successful implementation in the classroom [3,4]. Against this backdrop, the current study draws on Christensen and Knezek's theoretical Will–Skill–Tool model [5] to examine various factors that influenced teachers' use of digital technologies during the period of emergency distance education due to the materialization of COVID-19 in spring 2020. The current study is of great importance as it contributes to building and sharing sound knowledge about the preparation and use of digital tools by teachers. In addition, the results provide insights that may promote further development of individualized didactic approaches using information and communication technologies.

1.1. COVID-19 and School in Austria: A Chronology of Events during the 2019/20, 2020/21, and 2021/22 School Years

As in many other countries [6], the 2019/20, 2020/21, and 2021/22 school years in Austria have not been spared from the effects of the COVID-19 crisis. School closures were initiated in March 2020. This rapid and unpredictable change in regular school operations led to a transition to synchronous and asynchronous digital teaching and learning methods at all levels of education. However, schools remained open to those students whose parents or guardians were engaged in employment away from home or who were unable to allow students to attend classes from home [7]. After two months of distance teaching and learning, Austrian schools finally reopened in May 2020 [8].

Although the 2020/21 school year started with regular on-site teaching, education policymakers decided to suspend regular classes in November 2020 until further notice for the secondary school level and in December for all other school levels. Again, schools remained open for care and educational support [9]. At the end of the semester break, all students returned to school and attended classes in shifts, beginning in February 2021. Accordingly, classes were divided into two groups, with one group attending school on Mondays and Tuesdays and the other group attending on Wednesdays and Thursdays. Distance education was maintained on the other day of the week [10]. In April 2021, the Federal Ministry published a "phased plan for re-activating the schools" [8], providing the return of all students to schools. However, classes continued to be taught to two separate groups of students until the end of the school year [8].

In view of the rising vaccination rate among teachers, the Federal Ministry envisaged the reintroduction of continuous in-person classroom teaching for the 2021/22 school year. However, the rapid spread of seasonal COVID-19 prompted the implementation of a three-week safety period at the beginning of the semester (which included COVID-19 testing three times a week and wearing face masks), a basic safety plan for all schools, and specific measures depending on the number of new infections in each region [11]. Due to the rapidly increasing number of infections nationwide, the Federal Ministry issued a decree in November 2021 allowing absences from class without justification or teacher approval [12]. These specific measures were not removed until the end of February 2022, i.e., at the start of the new semester. Further, students living with at-risk individuals were allowed to remain in distance learning until the end of the school year. In summary, teachers in Austria were faced with the task of constantly adapting to changing education policies and flexibly responding to challenging situations.

1.2. The Role of Digitalization in Austria's Education System

Until spring 2020, cooperative learning and face-to-face encounters between teachers and students characterized regular classroom situations in Austria. However, because of the COVID-19 crisis, about 128,000 teachers and about 1.1 million students [13] in Austria were thrust into unfamiliar digital teaching and learning environments. Given the inevitable need to urgently adapt to the pandemic-induced governmental and social requirements in early 2020, educators felt compelled to maintain a minimum level of communication with students to subsequently support and monitor their learning processes [3]. As a result, teachers had to decide overnight how to teach virtually to prevent the spread of the COVID-19 virus.

While digital forms of teaching and learning have played rather minor roles in the history of the Austrian education system [14], teachers' digital literacy and digital literacy training opportunities crucially impacted the adaptation to online teaching during COVID-19 [3]. In addition, the results of a study by Dincher and Wagner [15] indicate that teachers' affinity (i.e., acceptance, skills, and technophobia) for technology and perceived learning effectiveness with regard to technology were strongly associated with the use of web-based teaching and learning methods during pandemic-related distance education. However, several study findings suggest that a large number of teachers in Austria, as in other countries, felt inadequately prepared to deliver instruction in digital form with little notice [16,17]. Furthermore, a literature review conducted by Helm et al. [18], considering studies from Austria, Germany, and Switzerland, indicates that about 10–30% of teachers rated both their own skills and those of their colleagues as insufficient for providing high-quality digital learning opportunities. In addition, several studies report large differences in technical equipment and Internet connections between school locations and within schools [16,19]. Overall, the unexpected global emergence of COVID-19 led to the discovery of previously unnoticed grievances and deficiencies in schools across the country.

This fact is all the more astonishing given that efforts to foster digitalization in Austrian schools date back to 2000 when the Federal Ministry of Education sought to ensure the availability and use of technical devices (e.g., laptops and computers) in schools [14]. Even today, Austria's education policymakers are committed to further advancing digitalization in schools. Accordingly, projects such as the "master plan for digital education" [20] and the "8-point plan for digital teaching" [21] were recently initiated. The 'Master Plan for Digital Education,' published by the Ministry of Education in 2018, provides for curricula revisions, teacher training, and the expansion of technological infrastructure [20]. Furthermore, the "8-Point Plan for Digital Teaching" contains even more ambitious targets, such as ensuring nationwide IT infrastructure by 2023 and embedding digital learning in all Austrian schools by 2024 [21].

Although these developments continue against the backdrop of the inexorable growth of digitalization, Fraillon et al. [22] point out that many schools in European countries lag behind the expected progress toward transforming information and communication technologies. Thus, the 2018 International Computer and Information Literacy Study (ICILS) considering twelve countries (i.e., Chile, Denmark, Finland, France, Germany, Italy, Kazakhstan, the Republic of Korea, Luxembourg, Portugal, the United States, and Uruguay) revealed interesting results. Across countries, the vast majority of participating teachers indicated that they already had experience with various information and communication technologies but still experienced some barriers to their use in the classroom [22]. Accordingly, it appears that while a large majority of educators recognized the benefits of using information and communication technologies (ICT) in the classroom, those teachers who reported using ICT less frequently were concerned about possible negative effects. Thus, teachers fear that students may copy material from the internet or deteriorate in their ability to express themselves in writing. In summary, several findings indicate that while there are significant differences between countries in ICT availability and use, teacher collaboration, and conditions of professional learning, there is a positive correlation between the frequency of ICT use and teachers' attitudes toward ICT [22].

1.3. Pedagogical Digital Competencies

Before detailing the theoretical model that underlies the current study, a brief look at the European Framework for the Digital Competence of Educators (DigCompEdu), which was developed by the European Commission's Joint Research Center to promote the digital literacy of education staff and drive digital innovation in education in all member states, is provided [23].

Given the rapid pace of technological advancement, it is increasingly necessary for educational institutions to be equipped with high-quality didactic resources to prepare youth for the demands of the industrial world. For example, as described in the concept of learning factories, learners are introduced to innovative teaching and learning methods at an early stage of their education, where they can test and consolidate their theoretically acquired knowledge in real, technology-based learning spaces [24]. However, in order for students to develop digital and technical skills and be prepared to act safely and productively in a technology-infused environment, teachers must have a wide range of competencies.

The European Framework for the Digital Competence of Educators (DigCompEdu) provides a general reference frame for developers of teacher digital literacy models [23]. DigCompEdu proposes 22 core competencies organized in six areas: (1) professional engagement, (2) digital resources, (3) teaching and learning, (4) assessment, (5) empowering learners, and (6) facilitating learners' digital competence. Accordingly, educators need to be able to (1) use digital technologies for their professional development and for the common good and (2) select digital resources that best fit their teaching and learning objectives and create new digital educational resources. In addition, teachers must be capable of (3) planning, designing, and implementing digital learning strategies and (4) using digital technologies to enable innovative assessment approaches. Finally, the DigCompEdu framework formulates the need for teachers (5) to use digital technologies to enhance learners' active engagement in the learning process and (6) to facilitate students' digital competencies. Overall, the DigCompEdu framework, with its six defined basic competencies, provides the skills that twenty-first-century teachers need to navigate an educational environment characterized by technology-driven change [25].

1.4. *The Will–Skill–Tool Model (WST)*

The model for technology integration proposed in this paper was developed for use in educational contexts at the beginning of the twenty-first century. This much-noted model for explaining technology integration in classrooms is comprised of three core constructs that predict 90% of the variance in digital media usage by classroom teachers. These constructs aim to examine the relationships between teachers' attitudes toward computers (*Will*), which can be described in contemporary terms as attitudes toward digital technologies, teachers' technology competence (*Skill*), and available digital resources (*Tool*). According to Christensen and Knezek [5], the importance and originality of this model lie in these three constructs, which are essential ingredients for effectively integrating information technology (digital technology) into classroom practice. There are needs for positive attitudes, corresponding competencies, and, finally, sufficient technical infrastructures (Tool).

1.4.1. The Will Key Factor

The confidence and positive attitude with which a teacher uses digital technologies can greatly influence the effective implementation of technology methods in the classroom. The *Will* key factor relates to the emotional and perceptual level by surveying feelings: lack of anxiety [26], lack of teacher confidence [27], attitude and beliefs toward technology and its usefulness for teaching, level of IT curricular expectations, computer confidence, and openness to change [28].

1.4.2. The Skill Key Factor

According to Knezek and Christensen, the *Skill* key factor is the ability to use and experience digital technologies. This includes, above all, self-perceived confidence and a willingness to use digital technology. Knezek and Christensen describe this aspect as a *"self-assessment process"* [28] (p. 311), in that teachers' positive attitudes and beliefs, acceptance of the need for digital technologies, and competencies in technology use have significantly positive relationships. The indicators that can be categorized under *Skill* are (1) digital technology educational qualification, (2) digital teaching experience, (3) dealing with different software, hardware, and apps, (4) technological and pedagogical teaching competencies, and (5) media didactics.

1.4.3. The Tool Key Factor

The *Tool* key factor is the measure of access to digital technologies. For example, the use and deployment of certain tools, such as Microsoft Teams, Moodle, SchoolFox, learning videos, and Skype, among others. To capture the *Tool* construct, the elements that were used show the number and usage behavior regarding digital devices, which means how often the devices and tools are used for teaching [29].

The *WST* model has been empirically tested several times. In 1999, Knezek et al. tested their model in Texas [30]. They surveyed 39 teachers regarding the three predictors (*Will–Skill–Tool*) and were able to explain an 84 percent variance in ICT use in the classroom using regression analysis. The model was tested in countries such as the United States [31], Mexico [32], Switzerland [33], Ghana [26], the Philippines [34], and Ireland [35]. Petko [33] used the *WST* model with self-selected measures to determine the influence of a constructivist teaching style on digital media use in Swiss classrooms. He confirmed a variance resolution of 39 percent of the model's validity, which is not nearly as high as the results of Christensen and Knezek's studies. Nevertheless, Petko considers the predictors crucial to a high level of digital technology integration (cf. ibid., 33–34). The results of the study in the Philippines (a total of 325 teachers) indicated that the *WST* model examines significant (up to 54%) variances in ICT integration in science and mathematics education. The results show, furthermore, that science teachers had higher scores in the *WST* key factors and ICT integration than mathematics teachers [34].

In 2005, Knezek and Christensen presented an extended version of the model (*WSTP* model) by adding a fourth construct: pedagogy [28] (p. 314). The *WST* model was extended with the *Pedagogy* construct to capture different teaching styles (constructivist and behaviorist), which are typically understood as teaching approaches or instructional strategies. The fourth factor is also related to technology integration and explores the level of confidence teachers demonstrate in using digital instructional strategies to enhance student learning.

1.5. The Current Study

Against the backdrop of previous studies dealing with teachers' digital competencies, the Will–Skill–Tool model (*WST*) offers a multifaceted approach to capturing teachers' perceptions of professional demands, which, until before the pandemic, were not considered mandatory and therefore tended to be implemented according to individual preference. Thus, the objective of the paper is to reveal teachers' experiences and perceptions of digitalized teaching on three levels: the will level (RQ1), skill level (RQ2), and tool level (RQ3). The study, therefore, addresses three research questions, as follows:

RQ1: What are secondary school teachers' attitudes toward digital teaching during virtual teaching due to the COVID-19 pandemic?

RQ2: How did secondary school teachers perceive their digital skills during virtual teaching due to the COVID-19 pandemic?

RQ3: How did secondary school teachers implement digital tools in their teaching during virtual teaching due to the COVID-19 pandemic?

2. Materials and Methods

2.1. Procedure

The analyzed data were derived from the longitudinal Inclusive Home Learning (INCL-LEA) research project [36]. As a quick response to government education mandates altering the delivery methods of education, the research project was initiated to scientifically accompany various participants in the Austrian education system during this unusual and unpredictable situation. To shed light on teachers' initial reactions to the transition from classroom teaching to digital teaching with regard to their own competencies and professional opportunities, this study refers to data from interviews conducted during the beginning of the pandemic.

2.2. Interviews

The interview guideline was developed by a research group at the University of Trier (SCHELLE study) [17]. The guideline included different topics addressing teachers' perceptions about their professional and personal experiences during the beginning of the COVID-19 pandemic. The following focal points were addressed: (1) perceived well-being during remote teaching, (2) perceived opportunities for a digitalized educational future, (3) differentiated teaching methods, (4) support of students' learning processes, (5) performance testing and error correction, and (6) evaluation and assessment.

In the present study, qualitative data from secondary school teachers are used, which were collected through semi-structured interviews using an interview guideline. Interviews with 56 Austrian teachers were conducted and audio-/video-recorded via Zoom (average duration of 50 min).

2.3. Sample

For the present study, the material for analysis was selected from interviews with lower secondary teachers from Viennese schools. This focus stems from the mandatory school subject "basic digital education" at the lower secondary level [33] (BMBWF, 2018a), which is assumed to lead to a greater affinity for and professional habits in dealing with digital teaching methods among teachers at the lower secondary level than, for example, primary school teachers. The sample of secondary school teachers consisted of 17 participants with a mean age of 44.76 years (SD = 13.11) and an average teaching experience of 18.88 years (SD = 14.80) (for further information, see Table 1).

Table 1. Sample description.

	Gender	Age	Teaching Experience
Teacher 1	Female	25	2
Teacher 2	Female	28	4
Teacher 3	Male	30	4
Teacher 4	Female	32	1
Teacher 5	Female	33	7
Teacher 6	Female	33	7
Teacher 7	Female	41	19
Teacher 8	Female	43	18
Teacher 9	Female	44	5
Teacher 10	Female	45	12
Teacher 11	Male	49	28
Teacher 12	Male	54	29
Teacher 13	Female	58	38
Teacher 14	Female	60	38
Teacher 15	Female	61	41
Teacher 16	Male	62	40
Teacher 17	Female	63	38

2.4. Data Analysis

After successful completion of the data collection, the structuring qualitative content analysis, according to Mayring, was used for subsequent analysis [37]. This evaluation method brings the advantage of a theory-based, rule-guided reduction of large amounts of text to its manifest components. Prior to the analysis of the interviews, a system of deductive categories was formed on the basis of in-depth theoretical considerations. Following Mayring [37], a guideline for category formation was created that contains definitions of individual categories, examples (text excerpts) that fall under respective categories, and coding rules for correctly assigning the text contents.

Based on the underlying theoretical framework regarding the *WST* model and the results of the qualitative data, the following three categories (Table 2)—displaying not only the model but also the previously presented research questions—were elaborated upon:

Table 2. Categories created during data analysis.

<table>
<tr><td colspan="3">Category 1
Teachers' Attitudes toward Digital Teaching</td></tr>
<tr><td>1.1 Attitudes with positive connotations</td><td colspan="2">1.2 Attitudes with negative connotations</td></tr>
<tr><td colspan="3">Category 1 addresses teachers' attitudes toward digital instruction during the COVID-19 pandemic. This focus, related to the theoretical basis of this study (namely the will–skill–tool model), is about teachers' experiences at the first level: the dimension of teachers' will. The theoretical model focuses only on attitudes with positive connotations (SOURCE). As a counterpart, one additional category dealing with negative attitudes is also necessary. This addition seems important because the transition to digital instruction was not a personal choice made by teachers according to their individually intended adaptations but forced teachers' adaptations to sudden changes in basic working conditions.</td></tr>
<tr><td colspan="3">Category 2
Teachers' digital competencies in the context of education</td></tr>
<tr><td>2.1 Positive self-perception</td><td>2.2 Negative self-perception</td><td>2.3 Acquisition of skills</td></tr>
<tr><td colspan="3">Category 2 refers to teachers' media pedagogical competence (skill), characterized by two opposing perspectives. These two perspectives are characterized by the teachers' "positive self-perception" or "negative self-perception" of their competencies. Depending on how the teachers assessed their media pedagogical competencies, further conclusions could be drawn regarding the integration of digital media.</td></tr>
<tr><td colspan="3">Category 3
Teachers' implementation of digital tools</td></tr>
<tr><td colspan="3">3.1 Use of digital tools</td></tr>
<tr><td colspan="3">Category 3 looks at digital tools used by teachers during school closures. Accordingly, this category provides information about the types of digital devices (e.g., computer, iPad, and laptop) and digital platforms (e.g., Microsoft Teams, LMS, Moodle, and Skype) that teachers accessed during distance learning. In addition, the results assigned to this category provide information about how educators used these applications.</td></tr>
</table>

3. Results

The following section provides in-depth insight into the results of the data analysis. To offer a deeper understanding, the findings are supported by several selected participant statements.

3.1. *Teachers' Attitudes toward Digital Teaching*

To capture the *will* component of the will–skill–tool model, which serves as the theoretical foundation of this study, teachers' attitudes were considered a construct with two diametrically opposed perspectives throughout the data analysis. Thus, the results reveal whether participants reported fundamentally positive or negative attitudes toward digital teaching methods.

3.1.1. Attitudes with Positive Connotations

In the course of the analysis, the overarching category "Positive attitudes" was broken down into the following seven sub-categories: (1) "Previous experiences with digital technologies," (2) "Change of perspective," (3) "New experiences," (4) "Innovative action," (5) "Differentiation and individualization," (6) "Expansion of media-pedagogical competence," and (7) "New possibilities of regular instruction." During the school closures, participants perceived the aspect of "Previous experiences with digital technologies" as particularly enriching. For instance, one teacher stated that she had already participated in numerous projects and programs promoting digitalization in schools and had therefore already acquired some knowledge about digital technologies and their educational use. Furthermore, a first sign of rethinking digital instruction can be seen in the category "Change of perspectives," as illustrated by the following interview extract:

> "I notice that, at least at my school, a change in thinking is taking place because we now realize that this digitization actually only brings benefits, and I definitely see a lot of progress, especially in the attitude of my colleagues, but also in my

own. There is actually only one [type of] person who is kind of resisting a little bit. Otherwise, everyone is very active, regardless of the generation of teachers. They all hold Skype classes, and I think most of them are very happy with that as well." **(Teacher 4)**

In some interviews, the increase in "New experiences" with digital media was also seen as an opportunity to redesign instruction. Accordingly, several participants repeatedly emphasized that the new and unfamiliar classroom situation opened up opportunities for "Innovative action":

> "Actually, much more attention should be paid to the progress of digitalization because, especially in the area of differentiation, online platforms offer much more variability. I am able to respond to students much more individually. I think that we should work much more in this regard because there are also other platforms, such as apps for cell phones you can use. I think that school, as we know it at the moment, is partly outdated and not up to date anymore. And that we should invest in digitalization and technical progress as far as differentiation is concerned." **(Teacher 4)**

With regard to the sub-category "Expansion of media-pedagogical competence," those teachers who had already gained experience with digital technologies and media were more confident in implementing inclusive distance instruction than educators who had not previously used digital approaches in their classrooms. In this context, it did not matter whether teachers had acquired digital knowledge and skills from in-service training or were self-taught. Since educators indicated that individualization and differentiation measures could now be implemented in more diverse ways, two of the teachers interviewed found the use of digital tools helpful in guiding students to learn independently. Finally, teachers reported that students were now gaining experience with digital media and tools that they may need for their current and future life situations, including their professional practice. Regarding "New possibilities of regular instruction," educators expressed their wish to use digital resources more often in regular classes in the future. Teachers justified this by saying that using digital teaching methods opens up a form of teaching that leaves more room for individualization and open work:

> "Yes, definitely, so I think that we will maintain a lot of what we are doing now, or at least I am sure that I will maintain some of it. We have now dealt with systems and tools that we would perhaps not have dealt with for a long time, which are also totally usable in regular teaching." **(Teacher 5)**

3.1.2. Attitudes with Negative Connotations

In the course of the interviews, teachers expressed both positive and negative attitudes toward digital instruction. Based on educators' statements, the following seven sub-categories could be identified: (1) "Lack of previous experiences with digital technologies," (2) "Lack of media education competencies," (3) "Structure and time management," (4) "Lack of social contact," (5) "Lack of resources," (6) "Social and educational inequality," and (7) "Skeptical attitude." The category "Lack of previous experiences with digital technologies" includes both educators' acquired knowledge about digital teaching approaches and the extent to which these digital teaching approaches were previously applied in class. Some participants did not feel well prepared for this new and unforeseen situation, as the following quotation suggests:

> "Oh no, not at all ((laughs)), not at all. We've had in-school training, one or two that might have been quite helpful, but no, not at the university and also not as part of my internship." **(Teacher 5)**

In addition, teachers often expressed feeling that they were facing excessive demands at work due to their current "Lack of media education competencies." Accordingly, teachers indicated that their lack of ability to use media meaningfully resulted in significant additional work and other burdens. The excessive demands perceived by teachers are also reflected in the category "Structure and time management." Several educators reported that they did not receive any information from school administrators about the structure or time management of digital instruction during the early 2020 school closures. Furthermore, in contrast to the positive experiences of teachers presented earlier, several participants indicated that they found it easier to implement inclusive teaching strategies, such as differentiation, during regular face-to-face instruction than during distance teaching.

Teachers also cited a "Lack of social contact" with their students and unavailable technological equipment ("Lack of resources") in students' homes as important stressors:

> "I'm a person who likes to be around people, and I think that a lot of teachers feel that way, and it's just a shame that our current work doesn't give back as much as regular work." **(Teacher 5)**

It appears that teachers were more comfortable working face-to-face with students in the classroom than in distance education, as the teachers considered the social aspect as one of the most important components of the teaching profession. With regard to a "Lack of resources," some teachers reported that it was difficult for them to stay in touch with students who did not have access to necessary technical equipment (e.g., laptop, computer, and internet connection) at home. In fact, one participant stated that students' restricted accessibility due to a lack of resources proved to be the biggest obstacle during distance teaching. The teachers interviewed cited an increase in "Social and educational inequality" as the most significant negative consequence of these differences in access to technical resources. Accordingly, several educators indicated that the achievement gap between students from socioeconomically disadvantaged backgrounds, on the one hand, and students from socioeconomically privileged backgrounds, on the other hand, may be further exacerbated by the conditions of distance learning.

Finally, some of the teachers were rather critical of the use of digital media in the classroom. One reason for this "Skeptical attitude," according to many educators, is that using digital media in the classroom essentially depends on first addressing current barriers, such as a lack of digital knowledge and skills and a lack of resources in schools and among students.

3.2. Teachers' Digital Competence in the Context of Education

The second important superordinate category ("Teachers' digital competence in the context of education") refers to teachers' subjectively perceived skills in handling digital media. In this regard, the categorized content elements are based on educators' self-assessments and were collected using the following questions: (1) How well do you feel your teaching degree/traineeship/in-service training prepared you for digital teaching? (2) Have you noticed any changes in your competency since you first used digital media in the classroom? (3) Have you had any experience with digital elements in the classroom? If yes, which ones?

As with responses on attitudes (*will* component), educators' responses on self-assessments of their competencies (*skill* component) could each be assigned to one of two sub-categories: positive and negative self-perceptions.

3.2.1. Positive Self-Perception

Teachers' self-perception of media pedagogical competencies is based on various situations, procedures, and experiences. In the course of the interview, a 32-year-old teacher showed a positive attitude toward digital instruction and expressed the following in terms of subjectively perceived competencies:

> "It's easier for me to switch to e-learning via Skype and co. because I usually sit in front of a computer a lot and I don't ever have technical problems. I've always been an IT expert." **(Teacher 4)**

It is also striking that teachers who have only recently completed their teacher training evaluated their digital competencies better than teachers who have been in their profession for a longer period of time. Thus, the age of those teachers who reported positive self-perception ranged from 25 to 45 years:

> "So I was lucky that I just recently graduated from university (laughs), so I know some online materials and I am also good with computers." **(Teacher 1)**

In general, many of the teachers said they were confident in their own digital skills. However, it is interesting to note that participants often expressed this positive self-perception in relation to their enforced exposure to digital media, triggered by distance teaching. Accordingly, educators indicated that the development of their skills has led to an increase in the potential scope for implementing digital instruction:

> "Yes, my media competence has certainly become greater than it was before, and I will benefit from this. Yes, definitely. My media competence has definitely improved." **(Teacher 10)**

3.2.2. Negative Self-Perception

Some teachers made negative comments about their self-perceptions of their own digital competencies. However, these specific teachers often expressed general dissatisfaction with digital instruction. Accordingly, educators found this new and unfamiliar way of teaching overwhelming and time-consuming. Teachers with plenty of work experience, in particular, indicated that the analog teaching and learning materials they had been using for years could only be used to a limited extent during distance instruction and that they perceived the digital preparation of analog worksheets as burdensome:

> "I finished my teacher training 40 years ago. There were no computers then. I worked with a typewriter, but I do not feel like an expert. Most of the time I don't even know what to ask, but I do not feel ready." **(Teacher 14)**

Other teachers said they had no choice but to use digital media and prepare online lessons under psychological and time pressures, even if they lacked the necessary skills, as they saw it as their responsibility to continue to provide learning opportunities for students.

3.2.3. Acquisition of Skills

Based on the data, it is apparent that participants showed different approaches to using digital technologies. On the one hand, several educators said they had no choice but to face the task of acquiring media education skills to provide instruction to their students from a distance. As a result, teachers had to acquire the competence to use certain digital media effectively in the classroom, either in collaboration with colleagues or on their own initiative:

> "Then I turned to my colleagues and asked them how they do it and what they do, and there was a young colleague who told me that he had been using a certain online program for a long time and that it worked very well, and as a result I decided to use it too." **(Teacher 16)**

On the other hand, ten out of seventeen teachers stated that they had already obtained media education knowledge and skills as part of their studies, seminars, or in-service training and were, therefore, able to implement their previously acquired competencies in the course of distance instruction. Nonetheless, all these teachers indicated that they felt compelled to upgrade their skills in spring 2020 because designing and delivering instruction through videoconferencing, correcting work assignments online, or creating instructional videos required a higher-than-average skill level.

3.3. Teachers' Implementation of Digital Tools

This category provides insight into teachers' use of various digital devices, such as computers, laptops, iPads, and various digital tools and platforms, during distance instruction. Teachers' experiences using digital resources in the classroom were surveyed with the following questions: (1) How did you implement digital instruction? (2) What technical tools did you use during distance teaching and learning?

Use of Digital Tools

The use of digital tools was broad and diverse, as the survey revealed, since teachers' experiences were characterized by various difficulties. According to the participants, challenges primarily arose from data security issues and overloading the learning platforms. In the survey, teachers related a major problem in the use of digital tools to students' lack of technical resources. Thus, without comprehensive and uniform equipment, participants could only partially require the use of digital tools for all students:

> "So at least that every student is equipped with the same technical resources, that is, that every child has access to the Internet, a laptop or a tablet, and a printer. Yes, so, of course, it would be a relief to know that everyone has the same requirements. This would also save a lot of phone calls." **(Teacher 10)**

Other difficulties for teachers included the inconsistent use of digital tools at all levels and the related problem of providing students with access to these tools. Teachers complained of an overabundance of options regarding digital devices and platforms used for distance learning. Among other things, this has caused students and faculty to feel overwhelmed by the variety and resulted in the difficulty of providing children with access to all these digital instruments.

> "Some of the kids send me the completed work tasks via WhatsApp, which I don't think is great at all, but I have to accept that, especially with those I can't reach otherwise. Some have responded on Google Drive. Since we don't have a platform for everyone, Google Teams is under construction, but that won't be ready until the fall. So we have to rely on different platforms right now, and that means a big loss of communication with a lot of follow-up calls." **(Teacher 4)**

To avoid this situation, one participant's school site had already selected a specific platform to use during distance learning before schools closed, set it up with students, and clarified how to work with it:

> "When we heard that schools were closing, there was an internal training on how to use Microsoft Teams. After that, I showed my class. again in more detail how they can work with it." **(Teacher 8)**

Tools, such as SchoolFox and Schoolupdate, were primarily used to communicate with students and parents or guardians. In addition, participants reported using email, WhatsApp, Zoom, Skype, and Microsoft Teams (Teacher 4; Teacher 10; Teacher 11; Teacher 13; Teacher 16).

However, the use of digital tools was not only characterized by the difficulties mentioned above but also opened up new ways of teaching by allowing educators to use creative and diverse strategies. Table 3 provides an overview of open-source platforms, apps, and other tools used by the participants during distance teaching.

Table 3. Digital tools during Inclusive Home-Learning/-Teaching.

Communication and Learning Platform	Management, Organization, and Task Processing (Apps, etc.)	Other
www.LMS.at (learning platform)	Microsoft Teams (platform for meetings, notes and attachments, uploading and downloading files)	PowerPoint (presentation software for digital illustrations and lesson design)
SchoolFox (digital software for internal communication and LMS)	Digi4School (digital bookshelf)	Voice Over (audio recording for PowerPoint)

Table 3. *Cont.*

Communication and Learning Platform	Management, Organization, and Task Processing (Apps, etc.)	Other
Google Meet (for video conferences)	Google Classroom (digital classroom management and deployment)	YouTube (learning videos)
Edhu (teaching and learning app)	One Note (digital notebook software)	Podcasts (learning videos and audios)
Anton App (learning app for German, German as a second language, mathematics, music education, subject lessons for secondary school I)	ÖBV—Österreichischer Bundesverlag (digital teaching materials and books for all school levels)	
HELBING e-zone (learning platform)	One Drive (digital storage location for children's completed work tasks)	
Kinderzeitmaschine (game-based learning platform)		
Kahoot (game-based learning platform)		

4. Discussion

In this study, we took the opportunity to explore teachers' experiences and perceptions of digitalized teaching on three levels, prompted by the radical transformation from regular in-person instruction to digital teaching and learning caused by the COVID-19 pandemic. Building on the theoretical foundation of Knezek and Christensen's *WST* model [28], we used qualitative interviews with a total of 17 teachers to investigate the relationships between teachers' media-related attitudes *(will)*, abilities to use digital technologies *(skill)*, and available technical resources *(tool)* that should lead to media integration in the classroom. In the course of performing an in-depth analysis of participants' statements, awareness was raised concerning a variety of factors that can influence teachers' use of digital teaching approaches. In line with other studies in the context of digital forms of instruction [28] and broader contexts such as inclusive education [38], the results of this study underpinned the important role of teachers' attitudes in implementing high quality (digitalized) teaching. At first glance, the results indicated that both positive and negative attitudes toward digitalized teaching could be found in teachers' responses. Two aspects were especially linked with positive attitudes: being "up-to-date" with teaching methods and addressing students' individual needs. Clearly, teachers recognize that previous teaching has not been connected with digital innovations outside of school. However, the use of digital tools tends to support existing teaching methods rather than developing new teaching methods [38,39].

It is interesting to note the causes to which the surveyed teachers attributed their perceived adequate or inadequate competencies regarding digital teaching during the pandemic. Focusing on the positive perceptions of their own digital teaching skills, teachers mainly linked their sufficient skills to three factors: (1) basic interest and competence, (2) recently attended training, and/or (3) a positive attitude toward a change in teaching methods. It is encouraging to compare these findings with the work of König et al. [3], noting that the extent to which teachers previously attended training dealing with digital literacy has had a crucial impact on how in-service teachers were able to adapt to online teaching during school closures.

In line with this, teachers identified major concerns with regard to their digital skills that were inevitably expected due to changing educational circumstances. The biggest challenge, especially for teachers with years of teaching experience, was that established teaching methods and strategies suddenly became obsolete and could not be implemented in the context of school education in times of pandemic. The results regarding self-perceived skills in digital teaching accord with our earlier research, indicating that digital forms of teaching and learning in Austrian schools played little to no role before the pandemic [14].

For this reason, teachers' professional reliance on well-used school pedagogies and didactic methodologies seems inadequate for rapid transformation according to social characteristics and needs. This may be related to changing educational circumstances during the pandemic. However, as we have already seen previously, it may also be due to other societal trends and the unpredictable needs of certain groups of students (e.g., students with a refugee history [40].

Interestingly, some teachers linked the use of digitalization with another shift in education: inclusion in the sense of addressing the individual needs of each student. Individualized instruction is a well-known requirement of inclusive education [36], and teachers have recognized that there is a huge potential to implement individualization through digitalization [3]. From this perspective, future studies could gain insight into students' perceptions to determine whether they perceived a higher level of individualization during the COVID-19 phase.

As in the context of inclusive education, a positive correlation was found between teachers' attitudes and prior experience in digitization, which is directly related to teachers' competencies and willingness to implement digital learning and teaching formats [41]. In addition to a lack of experience, organizational barriers (e.g., structure and time management) were also raised as factors hindering digitization. In addition, two critical aspects of digitization, in particular, were raised among teachers: "Lack of social contact" and "Social and educational inequality." Research in the context of COVID-19 and education [2,42] has already shown that teachers and students suffer from limited social contact due to digitalization. However, this could be explained by the fact that teachers were less likely to use synchronous digital instruction (e.g., videoconferencing) at the beginning of distance education [18] and, as a result, often did not consider the social-emotional aspects of learning. Against this background, the sudden transition from face-to-face to digital-only teaching highlights the didactic necessity of digital socialization in terms of social communication and inclusion. Especially when considering that the pandemic was accompanied by far-reaching negative psychological consequences for individuals, which made it difficult to even operate successfully in environments that were already familiar and explored, finding one's way into a new facet of the digital world—namely, online school—seems like a comprehensible challenge for all involved [43]. This assumption underlines the fact that it is not digital communication and the associated social inclusion as such that poses a problem, but rather the sudden conversion of the traditionally proven face-to-face communication in the school system to digital communication. Therefore, the perceived challenges may be more related to the non-established digital communication structures in the compulsory school system than digital interaction per se [21]. This can potentially be countered by implementing the government's digitization plan [20]. In addition, introducing innovative digital tools that enable synchronous and adaptive delivery of instructional content and are easy to use, such as ASYMPTOTE [44], may serve as a solution to the perceived lack of face-to-face contact associated with digital teaching and learning.

Following the *WST* model, it would be necessary for all these components to be fully represented in the student sample. For instance, while teachers spoke positively about the potential for digital tools to support digitalization, they also pointed to the potential for reinforcing educational inequity. Thus, students who are at risk for academic development (e.g., students with special educational needs, students from low socioeconomic backgrounds, and students with linguistic barriers in the language of instruction) tend to have limited access to digital devices and, therefore, often have lower digital literacy skills. In this sense, digitization also has the potential to increase educational inequality, as was the case for Austrian students during the COVID-19 period [45,46].

5. Conclusions

Even though previous education policy efforts pushed for digitization in schools, little progress was made before the emergence of COVID-19. This raises the question of whether a change in the ideas and implementation of education and pedagogy requires a radical cut to be implemented. One explanation might be that the major organizational aspects of digitalization processes [47] have been underestimated. However, the use of digital tools and technologies does not automatically lead to new and improved educational practices [39]. Accordingly, recent studies indicate that several factors other than teachers' attitudes *(will)*, competencies *(skill)*, and resources *(tool)* are important to ensure high-quality online instruction, such as increased interaction between teachers and parents or guardians, teachers' ability to guide students in digital instruction, and teachers' support and motivation [48]. Further, improving education to provide equity for students, regardless of their characteristics and needs, is perceived as a complex issue. In this context, several aspects have to be taken into account from an intersectional perspective (awareness of the cumulation of systematic discrimination due to the intersection of diverse differential categories [49,50], e.g., pedagogical, organizational (tools), but also the personal aspects (will and skills) of all actors involved (school policy makers, principals, teachers, students, etc.).

The results of the current study provide insights into teachers' self-perception regarding their attitudes toward digital teaching and competencies in implementing and creating digital teaching methods and learning situations for their students during the beginning of the COVID-19 pandemic. In addition, the empirical findings in this study provide a new understanding and contextualization of educational demands, particularly the need to focus on newly arisen challenges that teachers face regarding their pedagogical goal to ensure quality and equity in education.

Author Contributions: Conceptualization, T.D. and F.W.; methodology, T.D. and K.-T.L.; data analysis, T.D.; writing—original draft preparation, F.W., K.-T.L., T.D. and S.S.; writing—review and editing, K.-T.L. and S.S.; visualization, K.-T.L.; project administration, S.S. and K.-T.L. All authors have read and agreed to the published version of the manuscript.

Funding: Open Access Funding by the University of Vienna.

Informed Consent Statement: Informed consent was obtained from all subjects involved in the study.

Data Availability Statement: The data presented in this study are available on request from the corresponding author. The data are not publicly available due to sensitive information from which conclusions about the individual participants could be drawn.

Conflicts of Interest: The authors declare no conflict of interest.

References

1. Viner, R.M.; Russel, S.J.; Packer, J.; Ward, J.; Stansfield, C.; Mytton, O.; Bonell, C.; Booy, R. School closure and management practices during coronavirus outbreaks including COVID-19: A rapid systematic review. *Lancet Child Adolesc. Health* **2020**, *4*, 397–404. [CrossRef]
2. Kim, L.E.; Asbury, K. 'Like a Rug Had Been Pulled from Under You': The Impact of COVID-19 on Teachers in England during the First Six Weeks of the UK Lockdown. *Br. J. Educ. Psychol.* **2020**, *90*, 1062–1083. [CrossRef] [PubMed]
3. König, J.; Jäger-Biela, D.J.; Glutsch, N. Adapting to online teaching during COVID-19 school closure: Teacher education and teacher competence effects among early career teachers in Germany. *Eur. J. Teach. Educ.* **2020**, *43*, 608–622. [CrossRef]
4. Tas, Y.; Eminoglu, S.; Atila, G.; Yildiz, Y.; Bozkurt, U. Teachers self-efficacy beliefs and opinions about distance education during the COVID-19 pandemic. *Turk. Online J. Distance Educ.* **2021**, *22*, 229–253. [CrossRef]
5. Christensen, R.; Knezek, G. Self-report measures and findings for information technology attitudes and competencies. In *International Handbook of Information Technology in Primary and Secondary Education*; Voogt, J., Knezek, G., Eds.; Springer: New York, NY, USA, 2008; pp. 349–366. ISBN 978-0-387-73314-2.
6. UNESCO. Education: From Disruption to Recovery. Available online: https://en.unesco.org/covid19/educationresponse (accessed on 27 May 2022).
7. BMBWF. Maßnahmen in Schule und Unterricht. Available online: https://www.bmbwf.gv.at/Themen/schule/beratung/corona/corona_info.html (accessed on 10 March 2022).
8. BMBWF. Etappenplan zur Aktivierung des Schulsystems. Available online: https://www.bmbwf.gv.at/Themen/schule/beratung/corona/corona_info/corona_etappenplan.html (accessed on 10 March 2022).
9. BMBWF. Oberstufen im Distance-Learning—Das Weitere Vorgehen an den Schulen. Available online: https://www.bmbwf.gv.at/Ministerium/Presse/20201102.html (accessed on 10 March 2022).
10. BMBWF Umsetzung des Etappenplans für Schulen. Richtlinien für die Unterrichtsorganisation und die Pädagogische Gestaltung. 2020. Available online: https://evw.schule.wien.at/fileadmin/s/evw/system/user_upload/corona_etappenplan_info_20200507.pdf (accessed on 10 March 2022).

11. BMBWF. Sichere Schule—Schulbetrieb im Schuljahr 2021/22. 2021. Available online: https://www.bmbwf.gv.at/Themen/schule/beratung/corona/sichereschule.html (accessed on 11 March 2022).
12. BMBWF. Maßnahmen für den Schulbetrieb in ganz Österreich ab dem 22 November 2021—Die Schulen Bleiben Offen. 2021. Available online: https://www.bmbwf.gv.at/Themen/schule/beratung/corona/sichereschule.html (accessed on 6 June 2022).
13. Statistik Austria. Bildung. Available online: https://statistik.at/web_de/statistiken/menschen_und_gesellschaft/bildung/index.html (accessed on 17 March 2022).
14. Brandhofer, G.; Micheuz, P. Digitale Bildung für die österreichische Lehrerschaft. *Beiträge Zur Lehr.* **2011**, *29*, 185–198. [CrossRef]
15. Dincher, M.; Wagner, V. Teaching in times of COVID-19: Determinants of teachers' educational technology use. *Education* **2021**, *29*, 461–470. [CrossRef]
16. Huber, S.G.; Günther, P.S.; Schneider, N.; Helm, C.; Schwander, M.; Schneider, J.A.; Pruitt, J. *COVID-19 und Aktuelle Herausforderung in Schule und Bildung. Erste Befunde des Schul-Barometers in Deutschland, Österreich und der Schweiz*; Waxmann: New York, NY, USA, 2020; ISBN 978-3-830-94216-0.
17. Letzel, V.; Pozas, M.; Schneider, C. Energetic students, stressed parents, and nervous teachers: A comprehensive exploration of inclusive homeschooling during the COVID-19 crisis. *Open Educ. Stud.* **2020**, *2*, 159–170. [CrossRef]
18. Helm, C.; Huber, S.; Loisinger, T. Was wissen wir über schulische Lehr-Lern-Prozesse im Distanzunterricht während der Corona-Pandemie?—Evidenz aus Deutschland, Österreich und der Schweiz. *ZfE* **2021**, *24*, 237–311. [CrossRef]
19. Schober, B.; Lüftenegger, M.; Spiel, C. Lernen unter COVID-19 Bedingungen. Erste Ergebnisse einer Studie mit Lehrer*innen. Available online: https://lernencovid19.univie.ac.at/fileadmin/user_upload/p_lernencovid19/Erstergebnisse_Lehrpersonen.pdf (accessed on 14 March 2022).
20. BMBWF. Masterplan für die Digitalisierung im Bildungswesen. Available online: https://www.bmbwf.gv.at/Themen/schule/zrp/dibi/mp.html (accessed on 10 March 2022).
21. BMBWF. Digitale Schule. Der 8-Punkte-Plan für den Digitalen Unterricht. Available online: https://digitaleslernen.oead.at/fileadmin/Dokumente/digitaleslernen.oead.at/Dokumente_fuer_News/201015-4_Folder_Digitale_Schule_DINlang_A4_BF.pdf (accessed on 10 March 2022).
22. Fraillon, J.; Ainley, J.; Schulz, W.; Friedmann, T.; Duckworth, D. *Preparing for Life in a Digital World. IEA International Computer and Information Literacy Study 2018 International Report*; Springer International Publishing AG: Cham, Germany, 2020; ISBN 978-3-030-38781-5.
23. Redecker, C. European Framework for the Digital Competence of Educators: DigCompEdu. Available online: https://joint-research-centre.ec.europa.eu/digcompedu_en (accessed on 6 June 2022).
24. Stravropoulos, P.; Pikas, H.; Mourtzis, D. Collaborative Machine Tool design: The Teaching Factory paradigm. *Procedia Manuf.* **2018**, *23*, 123–128. [CrossRef]
25. Caena, F.; Redecker, C. Aligning teacher competence frameworks to 21st century challenges: The case for the European Digital Competence Framework for Educators. *Eur. J. Educ.* **2019**, *54*, 356–369. [CrossRef]
26. Agyei, D.D.; Voogt, J.M. Exploring the potential of the will, skill, tool model in Ghana: Predicting prospective and practicing teachers' use of technology. *Comput. Educ.* **2011**, *56*, 91–100. [CrossRef]
27. Buabeng-Andoh, C. Factors influencing teachers' adoption and integration of information and communication technology into teaching: A review of the literature. *Int. J. Educ.* **2012**, *8*, 136.
28. Knezek, G.; Christensen, R. Extending the will, skill, tool model of technology integration: Adding pedagogy as a new model construct. *J. Comput. High. Educ.* **2016**, *28*, 307–325. [CrossRef]
29. Knezek, G.; Christensen, R.; Miyashita, K.; Ropp, M. Instruments for Assessing Educator Progress in Technology Integration. Institute for the Integration of Technology into Teaching and Learning. Available online: https://iittl.unt.edu/sites/default/files/Instruments/InstrumentsforAssessingEducatorProgressinTechnologyIntegration_0.pdf (accessed on 15 March 2022).
30. Knezek, G.; Christensen, R.; Fluke, R. Testing a Will, Skill, Tool Model of Technology Integration. In Proceedings of the Annual Meeting of the American Educational Research Association, Chicago, IL, USA, 21–25 April 2003.
31. Grant, D. Predicting Web 2.0 use among US teens—Expanding the power of the skill, will and tool model. *Br. J. Educ. Technol.* **2019**, *50*, 3405–3419. [CrossRef]
32. Velazquez, C.M. Testing Predictive Models of Technology Integration in Mexico and the United States. *Comput. Sch.* **2007**, *24*, 153–173. [CrossRef]
33. Petko, D. Teachers' pedagogical beliefs and their use of digital media in classrooms: Sharpening the focus of the 'will, skill, tool' model and integrating teachers' constructivist orientations. *Comput. Educ.* **2012**, *58*, 1351–1359. [CrossRef]
34. Sasota, R.S.; Cristobal, R.R.; Sario, I.S.; Biyo, J.T.; Magadia, J.C. Will–skill–tool (WST) model of technology integration in teaching science and mathematics in the Philippines. *J. Comput. Educ.* **2021**, *8*, 443–464. [CrossRef]
35. Scully, D.; Lehane, P.; Scully, C. 'It is no longer scary': Digital learning before and during the COVID-19 pandemic in Irish secondary schools. *Technol. Pedagog. Inf.* **2021**, *30*, 159–181. [CrossRef]
36. Schwab, S.; Lindner, K.-T. Auswirkungen von Schulschließungen und Homeschooling während des ersten österreichweiten Lockdowns auf Bildungsungleichheit. *WISO* **2020**, *43*, 49–63.
37. Mayring, P. *Qualitative Inhaltsanalyse. Grundlagen und Techniken*, 12th ed.; Beltz: Weinheim, Germany, 2015; ISBN 978-3-407-29393-0.
38. Yada, A.; Leskinen, M.; Savolainen, H.; Schwab, S. Meta-analysis of the relationships between teachers' self-efficacy and attitudes toward inclusive education. *Teach. Teach. Educ.* **2022**, *109*, 103521. [CrossRef]

39. Islam, S.; Grönlund, Å. An international literature review of 1:1 computing in schools. *J. Educ. Chang.* **2016**, *17*, 191–222. [CrossRef]
40. Kast, J.; Lindner, K.-T.; Gutschik, A.; Schwab, S. Austrian teachers' self-efficacy beliefs regarding at-risk students during home learning due to COVID-19. *Eur. J. Spec. Needs Educ.* **2021**, *36*, 11–126. [CrossRef]
41. Pozas, M.; Letzel, V. "Do You Think You Have What it Takes?"—Exploring Predictors of Pre-Service Teachers' Prospective ICT Use. *Technol. Knowl. Learn.* **2021**, 1–19. [CrossRef]
42. Pelikan, E.R.; Lüftenegger, M.; Holzer, J.; Korlat, S.; Spiel, C.; Schober, B. Learning during COVID-19: The role of self-regulated learning, motivation, and procrastination for perceived competence. *ZfE* **2021**, *24*, 393–418. [CrossRef] [PubMed]
43. Misirli, O.; Ergulec, F. Emergency remote teaching during the COVID-19 pandemic: Parents experiences and perspectives. *Educ. Inf. Technol.* **2021**, *26*, 6699–6718. [CrossRef] [PubMed]
44. Barlovits, S.; Caldeira, A.; Fesakis, G.; Jablonski, S.; Koutsomanoli Filippaki, D.; Lázaro, C.; Ludwig, M.; Mammana, M.F.; Moura, A.; Oehler, D.-X.K.; et al. Adaptive, Synchronous, and Mobile Online Education: Developing the ASYMPTOTE Learning Environment. *Mathematics* **2022**, *10*, 1628. [CrossRef]
45. Holtgrewe, U.; Lindorfer, M.; Siller, C.; Vana, I.; Šalamon, N. "Plötzlich bin ich in der Situation eines Youtubers"—Lehrer_innen und ihre Erfahrung mit Schüler_innen im Lockdown. Available online: https://zenodo.org/record/4550640#.YmBQgi-21QI (accessed on 20 April 2022).
46. Steiner, M.; Köppnig, A.; Leitner, A.; Pessl, G.; Lassnig, L. Lehren und Lernen unter Pandemiebedingungen Was tun, damit aus der Gesundheits-nicht auch eine Bildungskrise Wird? Available online: https://irihs.ihs.ac.at/id/eprint/5873/24/ihs-report-2021-steiner-koepping-leitner-pessl-lassnigg-lehren-und-lernen-unter-pandemiebedingungen.pdf (accessed on 20 April 2022).
47. Pettersson, F. Digitally Competent School Organizations—Developing Supportive Organizational Infrastructures. *Int. J. Lifelong Educ.* **2018**, *14*, 132–143. [CrossRef]
48. Duraku, Z.H.; Hoxha, L. The impact of COVID-19 on education and on the well-being of teachers, parents, and students: Challenges related to remote (online) learning and opportunities for advancing the quality of education. In *Impact of the COVID-19 Pandemic on Education and Wellbeing. Implications for Practice and Lessons for the Future*; Duraku, Z.H., Ed.; University of Prishtina: Prishtina, Albania, 2021; pp. 17–45.
49. Zheng, Y.; Walsham, G. Inequality of what? An intersectional approach to digital inequality under COVID-19. *Inf. Organ.* **2021**, *31*, 100341. [CrossRef]
50. Gandolfi, E.; Ferdig, R.E.; Kratcoski, A. A new educational normal an intersectionality-led exploration of education, learning technologies, and diversity during COVID-19. *Bull. Sci. Technol. Soc.* **2021**, *66*, 101637. [CrossRef]

electronics

Article

Factors Influencing Students' Continuous Intentions for Using Micro-Lectures in the Post-COVID-19 Period: A Modification of the UTAUT-2 Approach

Tommy Tanu Wijaya [1] and Robert Weinhandl [2,*]

1 School of Mathematical Sciences, Beijing Normal University, Beijing 100875, China; 202139130001@mail.bnu.edu.cn
2 Linz School of Education, STEM Education, Johannes Kepler University Linz, 4040 Linz, Austria
* Correspondence: robert.weinhandl@jku.at

Abstract: Micro-lectures, i.e., short learning videos on a specific aspect of a topic, have become one of the most effective technology-based learning media approaches and were widely used during the COVID-19 pandemic. However, in the post-pandemic era starting from early 2022, as K-12 students have been allowed to resume going to school, it is necessary to evaluate students' intentions to continuously use micro-lectures for learning mathematics. Therefore, this study aims to explore attitudes and continuous intentions of students towards the utilization of micro-lectures. To investigate students' intentions of using micro-lectures, we utilized the unified theory of acceptance and use of technology (UTAUT-2). Data were collected from 321 junior high school students (14–17 years old) in Bandung, Indonesia, who used online classes and micro-lectures to learn mathematics during the pandemic. A structural equation model was also used to analyze the independent (performance expectancy, effort expectancy, social influence, facilitating condition, hedonic motivation, and habit) and dependent (attitude and continuous intention) variables. Furthermore, online questionnaires were used to obtain data on students' attitudes and continuous intention to utilize micro-lectures in the post-COVID-19 era. The results suggested that effort expectancy (EE) and hedonic motivation (HM) had a significant effect on attitudes, whose correlation with habit also influenced the continuous intention during this post-pandemic period. Despite these results, the habit variable was found to be the factor most influencing continuous intention. These results provide information to teachers, schools, and the government to continuously increase the use of micro-lectures based on improving student learning performances in the post-pandemic era.

Keywords: post-pandemic; learning media; behavior intention; mathematics education

Citation: Wijaya, T.T.; Tang, J.; Cao, Y.; Weinhandl, R.; Lavicza, Z. Factors Influencing Students' Continuous Intentions for Using Micro-Lectures in the Post-COVID-19 Period: A Modification of the UTAUT-2 Approach. *Electronics* **2022**, *11*, 1924. https://doi.org/10.3390/electronics11131924

Academic Editors: Boni García, Carlos Alario-Hoyos, Mar Pérez-Sanagustín, Miguel Morales and Oscar Jerez

Received: 8 May 2022
Accepted: 26 May 2022
Published: 21 June 2022

1. Introduction

The COVID-19 pandemic has reportedly caused significant changes in many sectors, including education [1], as people are advised to develop and rely on technologies towards carrying out activities while avoiding direct contact [2]. This has led to the rapidly increasing demand for the use of appropriate technologies [3]. Due to the increasing number of positive Indonesian COVID-19 patients in March 2020 [4,5], the government declared a lockdown where students were temporarily unable to attend school towards carrying out face-to-face teaching and learning activities, leading to the commencement of online education [6,7]. To ensure the effectiveness of these online activities, various learning approaches were applied to all educational levels [8–10]. Many Indonesian institutions utilized video conferencing tools such as Zoom and Google Meet as learning media [11,12], due to their having a similar traditional style to schools where the teacher explains and the students listen. Based on these conditions, the pandemic situation is a challenge for institutions and teachers in this country, where circumstances forced the transformation of teaching and learning in classrooms to online learning. During the lockdown period, micro-lectures were observed as one of the most widely used learning media to help students understand mathematical materials [13,14]. In 2020, the pandemic condition became an opportunity for this medium to develop and support online learning [10,15,16]. Additionally, the initial aim of the learning medium was to focus on important understandable points towards improving students' comprehensive ability when learning new concepts and knowledge [17].

Despite these advantages of online learning, some previous studies have argued that the pandemic shifted from a face-to-face teaching method to online learning to deliver new knowledge to students [18,19]. According to Kramarski [20], the use of the formative assessment concept was very important to continue teaching and learning activities through online educational media. In online education, the model for delivering materials is divided into two categories, namely synchronous and asynchronous [21]. The combination of several digital platforms based on increasing communication and interaction in teaching and learning activities is observed in the synchronous approach [22], while the asynchronous method uses technology-based media to enrich educational processes, as well as help students, by repeating the subject matter taught when taking virtual lessons [23]. This confirms that one of the asynchronous-based learning media is the utilization of micro-lectures. Before the pandemic, the development of this learning medium was not widely used as a supplement to improve students' understanding of new theories [24,25]. However, teachers were increasingly using micro-lectures in learning activities after the emergence of COVID-19, although video conferencing and other media had weaknesses within the country, such as a lack of virtual facilities or expensive internet prices [26–28]. This proves that micro-lectures are likely to essentially increase the usefulness of their effects and uses due to the pandemic situation, which forced the transformation of the learning approaches [29].

Previous studies are observed to be very limited in the analysis of students' intention towards micro-lectures. There is only one study from K. Wang et al. [25] based on the use of micro-lectures learning media before the pandemic (2016–2017), through the technology acceptance model (TAM). However, the other reports focused more on experimental pretest–posttest analyses to evaluate the effect of micro-lectures on student outcomes [30–32]. This shows that no experiment has been carried out on the factors influencing students to continuously utilize micro-lectures after the COVID-19-related school shutdowns. By June 2022, Indonesia is expected to return to the "new normal" era, where all business and trade sectors are to be carried out as previously performed [2,33]. In the education sector, students are also expected to start attending school for face-to-face learning for the new academic year (2021–2022). Therefore, this study aims to determine the continuous intention of students towards using micro-lectures as a supplement for mathematics learning in the post-pandemic period. This is expected to provide several contributions towards appropriately understanding the factors influencing students' continuous intention to use micro-lectures when returning to the traditional direct learning method. The results are also expected to help in appropriately understanding the use and benefits of the learning media, especially in mathematics, to improve academic performance in the post-pandemic period. In addition, it contributes to the provision of significant suggestions to micro-lecture developers, school principals, and policymakers, to relevantly consider the factors influencing students' continuous intention during this period.

2. Literature Review

2.1. Micro-Lectures in Mathematics Education

Micro-lectures have been widely used and proven to improve students' mathematical abilities, as studies have continuously promoted in the educational sector since the early 2000s [24,34], and were also increasingly used during the COVID-19 pandemic [29]. This explains that the learning media aims to actively promote students to learn according to their needs, through the combination with online education [25]. Before and during the pandemic, micro-lectures were often used to review lessons [24,35], defined as a short, less than 10 min, video focusing on new knowledge and concepts [36]. When students do not understand the teacher's in-class explanation, this learning media helped them in reviewing important points [31,34]. Many platforms were also used before and during the COVID-19 pandemic as the sources of micro-lectures, such as Moodle [37,38], YouTube [39, 40], and others. Due to the convenience of this media, teachers created their educational sources according to the learning circumstances and needs. In this decade, micro-lectures are often used in schools to support the flipped learning approach and review lessons capable of meeting the needs of students as individualized education (Zhang and Xu, 2015). Meanwhile, there is no empirical study analyzing students' continuous intention towards the utilization of micro-lectures after a post-pandemic period.

In such learning environments, a significant change was observed before and after the pandemic. This explained that some studies emphasized specific influential factors and analyzed the effectiveness of these learning tools before and during the pandemic, respectively [29,41,42]. Before this condition, many reports focused on increasing the use of micro-lectures as learning media for students [24,25,43]. Meanwhile, the online learning approach became more advanced, with the educational tools being increasingly used as complementary media after the 2-year existence of the pandemic [8,44]. This indicates the need for subsequent studies on the factors influencing the continuous intention to use micro-lectures during a post-pandemic period when more strong evidence is observed on the effect of the learning media on students' abilities.

2.2. *Unified Theory of Acceptance and Use of Technology (UTAUT)*

Based on many technological models, UTAUT has been widely used as a comprehensive tool for measuring the acceptance and use of online learning media [45–47]. Venkatesh et al. (2012) believe that the original UTAUT model can still be developed to predict better user behavior intention and usage behavior towards new technologies. This led to the development of the UTAUT-2 by Venkatesh et al. [48] from the previous 2003 UTAUT model. The UTAUT model is also a combination of the theories of reasoned action and planned behavior (TRA and TPB), as well as the technology acceptance model (TAM) [2,49].

Based on Figure 1, the UTAUT2 model was formed from 7 elements, namely performance and effort expectancies, social influence, facilitating conditions, hedonic motivation, price value, and habit [50]. Compared to the initial version, the proposed UTAUT-2 extension produces a substantial increase to explain the variance (R^2) in intention behavior from 56% to 74% and usage behaviors from 40% to 52%, respectively [51]. Moreover, the model is used to explain students' behavioral intentions and teachers' attitudes towards online learning media [38,51,52]. Despite these conditions, some studies criticized that UTAUT-2 was still rarely used in various countries and contexts [51]. This led to the selection of the proposed model, which is expected to increase the variance explained in the factors influencing students' continuous intention to use micro-lectures after the pandemic. Figure 2 shows a proposed model illustrating the initial hypothesis to explain the assumed influential factors of the learning media's continuous intention. This proposed model was derived from UTAUT-2 (Venkatesh et al. [48]) and modified through the addition of attitude, as a variable affecting students' continuous intention based on previous reports. The hypotheses of this study are as follows.

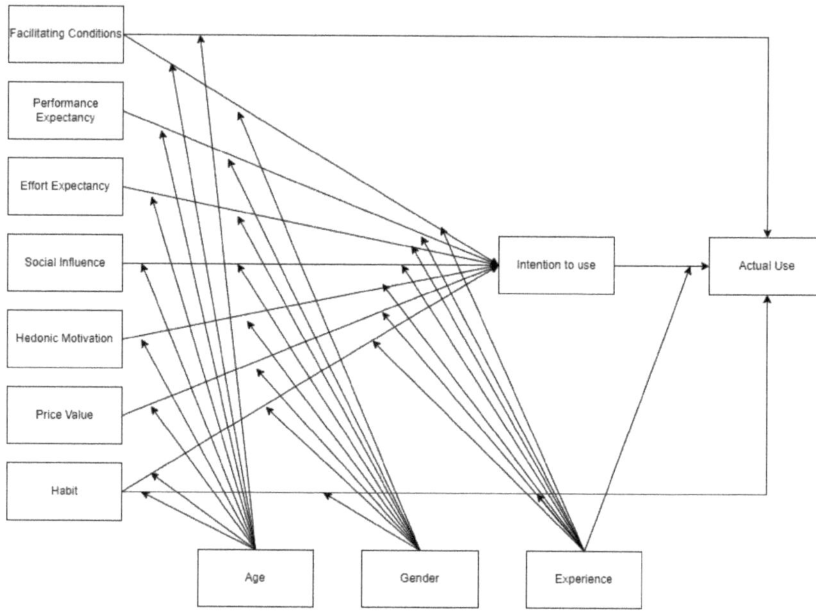

Figure 1. The original UTAUT-2 model with moderator variable.

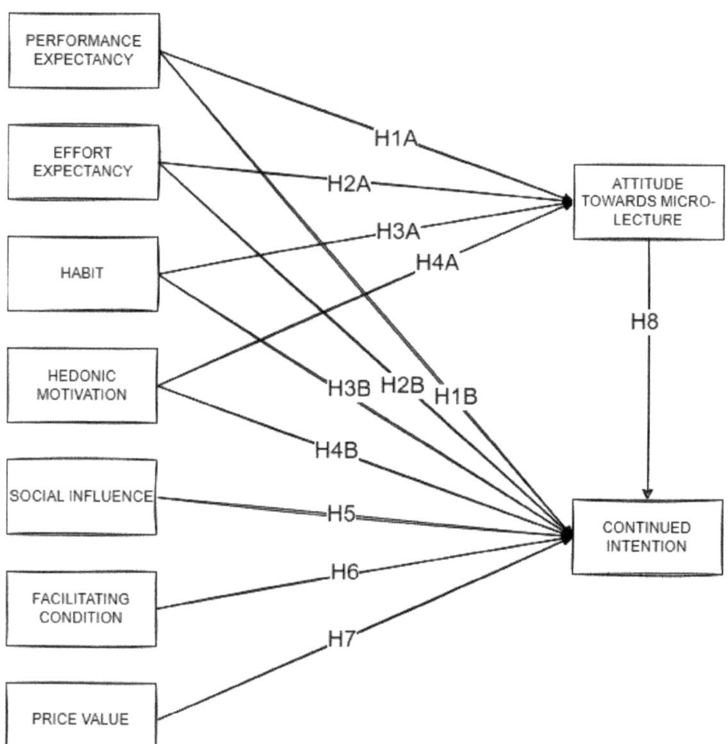

Figure 2. The proposed model and hypothesis.

2.2.1. Performance Expectancy (PE)

Performance expectancy is the level to which a person believes that technology improves efficiency and effectiveness [48,53,54]. It is also the degree to which students believe that using micro-lectures enhances their mathematical abilities and scores in the post-pandemic period. This shows that students need to have a positive attitude and believe that the learning media has mathematical improvement capabilities. In this condition, intentions are mostly predicted based on the confidence level of people in the assistance system or tool used in their daily lives. This shows that performance expectancy is significant when students believe that micro-lectures have the capability to improve their mathematical abilities and learning outcomes in the post-pandemic period. For example, students believed that these tools often helped in understanding mathematical materials when used at the end of the pandemic. In this case, their continuous intention to keep using the micro-lectures after the COVID-19 condition is very possible. Thus, in the present study the following is hypothesized:

Hypothesis 1 (H1). *There is a significant positive effect of PE on students' attitudes towards micro-lectures in the post-pandemic period.*

Hypothesis 2 (H2). *There is a significant positive effect of PE on the continuous intention to use micro-lectures in the post-pandemic period.*

2.2.2. Effort Expectancy (EE)

Effort expectancy is the degree to which a person does not have a problem with the use of the technology (costly and slow) [53]. This often occurs when a person has previous experience with the utilized technologies [55]. Furthermore, UTAUT has been widely used to analyze the influential level of EE on a person, as well as students and teachers, towards the acceptance of online learning media in the educational sector [56,57]. Effort expectancy is also confirmed to have a strong influence on the intention to use technology-based media, as some previous reports explained that it was an essential factor affecting the utilization of online learning in the educational sector [58–60]. In this present study, EE was effortlessly used to explore students' continuous intention to use micro-lectures in the post-pandemic period. Based on the mathematics teachers, previous reports stated that EE affected the use of the learning media in China [35]. Despite these results, it was still not significant for technology-based learning media according to a few reviews [59], [61]. Thus, in the present study the following is hypothesized:

Hypothesis 3 (H3). *There is a significant positive effect of EE on students' attitudes towards micro-lectures in the post-COVID-19 period.*

Hypothesis 4 (H4). *Effort expectancy has a significant effect on students' continuous intention to use micro-lecture in the post-pandemic period.*

2.2.3. Habit (HB)

Habit is one of the interesting determinants introduced by Venkatesh in the development of UTAUT-2 [48]. In some previous reports, habit was observed as a determinant with a greatly significant effect on a person towards the utilization of online learning media [51,62]. Meanwhile, it was the main factor influencing the use of new technologies according to other reviews [51,52]. This confirms that the intention of people towards the use of habit increases when they utilize new technologies [51]. These observations are reinforced by the study of de Guinea and Markus [63], where intention behavior is transformed into a habit. This explained that when the use of technologies became a person's habit, the utilization intention was no longer needed. These results were due to the definition of habit as a mechanism for the repetition of unrealized behaviors, indicating its ineffectiveness on a person's attitude. Thus, in the present study the following is hypothesized:

Hypothesis 5 (H5). *Habit does not have a significant effect on students' attitude to use micro-lectures in the post-pandemic period.*

Hypothesis 6 (H6). *Habit has a significant effect on students' continuous intention to use micro-lectures in the post-pandemic period.*

2.2.4. Hedonic Motivation (HM)

The pleasure obtained when using a new technology affects the intention to utilize digital media [40,64]. This is because some previous studies confirmed that hedonic motivation (HM) predicted behavioral intention (BI) towards using online learning media [40,65,66]. It is also observed to have influential abilities on teachers and students regarding the utilization of mobile learning technology [62,67]. Furthermore, various reports have reportedly used HM with ATT (attitude toward technology) in UTAUT-2, with a positive effect being observed on attitude towards the use of interactive social media [68]. Another result also indicated that the enjoyment felt by students significantly influenced the use of online media [69], leading to the hypothesis that HM affects the continuous intention to use micro-lectures in a post-pandemic period. The pleasure obtained from using this media to learn mathematics also positively affects students' continuous intention to use micro-lectures even when the pandemic is over. Thus, in the present study the following is hypothesized:

Hypothesis 7 (H7). *Hedonic motivation affects students' attitudes towards micro-lectures in the post-pandemic period.*

Hypothesis 8 (H8). *Hedonic motivation influences students' continuous intention to use micro-lectures in the post-pandemic period.*

2.2.5. Social Influence (SI)

Social influence is the degree to which a person believes that the environment leads to the use of technologies in their daily activities [53]. This has a big impact, especially at the beginning of technological utilization [70]. It also affects the intention to use new technologies when an individual obtains societal support for utilization [38,71,72]. Moreover, social influence affects the utilization of online learning media even when unwanted [71,73]. Using UTAUT-2, Venkatesh (2012) found that it was the biggest factor influencing an individual towards the use of technological equipment under mandatory conditions. These effects nevertheless become insignificant when observed under voluntary conditions. Based on various reports, SI was found to have a positive and significant effect on a person's and students' intentions to ordinarily and educationally use technology-based media, respectively [53,74,75]. This was in line with other previous studies, where SI affected students towards using video-based learning media [35]. Thus, in the present study the following is hypothesized:

Hypothesis 9 (H9). *Social Influence has a significant effect on students' continuous intention to use micro-lectures in the post-pandemic period.*

2.2.6. Facilitating Conditions (FC)

The intention to use new technology is influenced by resources and institution support, which are often observed as facilitating conditions [38,69,76]. This indicates that a supportive team and knowledge are very important in innovative technological utilization [37]. According to some previous reports, FC had a significant impact on BI [35], with a direct influence observed on the intention of an individual to use online media [69]. However, others found that FC did not have a significant impact on BI and UB [46,77]. This is then analyzed as a factor influencing the continuous intention of junior high school students in using micro-lectures during the post-pandemic period. Thus, in the present study the following is hypothesized:

Hypothesis 10 (H10). *Facilitating conditions have a significant effect on students' continuous intention to use micro-lectures in the post-COVID-19 period.*

2.2.7. Price Value (PV)

Price value is widely used to predict teacher intentions to use online learning media [62], with a study confirming a positive relationship between PV and mobile internet utilization [52]. These conditions revealed that the micro-lectures used by students required an internet quota to download and replay learning activities, with price value determining its significance level on students' continuous intention to use micro-lectures after the pandemic. In addition, sensible and low-cost mobile internet access influences the intention to use the learning media during this period [69]. Thus, in the present study the following is hypothesized:

Hypothesis 11 (H11). *Price value has a significant effect on students' continuous intention to use micro-lectures in the post-pandemic period.*

2.2.8. Attitudes towards Micro-Lectures (ATT)

According to a previous review, the attitude towards behavior measured a person's interest and use of new technologies [78]. This confirmed that the factor defined students' feelings when using micro-lectures to help them learn mathematics during the post-pandemic period. Meanwhile, other reports proved that attitude was influenced by several factors, namely performance and effort expectancies, social influence, facilitating conditions, hedonic motivation, price value, and habit [51,79]. A positive correlation was also found between the attitude towards technology and the intention to use new technology [80,81]. This verified that the factor was important in influencing students' continuous intention to use micro-lectures after the pandemic. Thus, in the present study the following is hypothesized:

Hypothesis 12 (H12). *Attitude has a significant effect on students' continuous intention to use micro-lectures post-COVID-19.*

3. Methods

This study used a quantitative approach: the researcher made an initial hypothesis about what factors influence the students' continuous intention to use micro-lectures in the post-COVID-19 period and finally tested the initial hypothesis using the SEM technique to find the answer. The UTAUT-2 model (Venkatesh et al., 2012) was developed to analyze the determinants with a positive influence on students' intention to use micro-lectures after the pandemic. The adopted questionnaire used a 5-point Likert scale [82], where 1, 2, 3, 4, and 5 = SD, D, *n*, A, and SA (strongly disagree, disagree, neutral, agree, and strongly agree), respectively [83]. Subsequently, each question item was developed based on the existing literature and instruments in previous studies, as shown in Appendix A.

3.1. Study Instrument

The original English questionnaire was translated and validated by two professors and doctoral students, containing nine determinants with a total of 23 items. Subsequently, the link was transferred to the mathematics teacher for distribution to the students experienced in the utilization of micro-lectures during the pandemic. From January to March 2022, 336 data were obtained, with 321 analytically used due to the incomplete nature of 15. Based on the average, each respondent filled out the questionnaire more than 8 min before submission, with the complete instrument containing various demographic data, such as gender, class, and micro-lectures experience (Table 1).

Table 1. The demographic data.

Factor		*n*	%
Gender	Male	194	60.44%
	Female	127	39.56%
Use micro-lectures to learn math	More than 4 times a week	201	62.62%
	1–3 times a week	91	28.35%
	Once a week	29	9.03%
Class	7	116	36.14%
	8	111	34.58%
	9	94	29.28%
The device used to open the micro-lectures	Smartphone	297	92.5%
	Tablet	241	75.08%
	Laptop/computer	201	62.61%

According to Table 1, 60.44% and 39.56% were female and male students, with 29.28, 34.58, and 36.14% being 9th, 8th, and 7th graders, respectively. Based on the device of study, more than 60% of students had smartphones, tablets, and laptops regarding the utilization of micro-lectures during the pandemic. Approximately 62.62, 28.35, and 9.03% also used micro-lectures to learn mathematics more than four times, once–thrice (1–3), and once a week, respectively.

3.2. Data Collection Technique

A purposive sampling technique [65] was used at one public junior high school in Bandung, where the target samples were students from grades 7–9 (14–17 years) who used micro-lectures as a supplement to learning mathematics during the pandemic. This selection process was conducted based on experience and familiarity with micro-lectures, as participation was voluntary without any form of duress. The samples were also provided with a gift in the form of learning tools as a sign of gratitude for seriously participating in the exercise. In addition, informed consent was obtained from all the study participants.

3.3. Ethical Considerations

Although the participants were provided with rewards, compliance with all study ethics was still very important during coding, analysis, and data collection. All responses also remained anonymous for the prevention of bias [8].

3.4. Data Analysis

A structural equation modelling (SEM) was used to test the predictive models and analyze the relationship among variables. This approach has been widely used in the social science fields to analyze and evaluate the suitability of the theoretical model through empirical data [60,84,85]. SEM is also appropriately considered to test the assumptions in student or teacher acceptance of online learning media [86,87], such as micro-lectures. Meanwhile, the literature on SEM has demonstrated the need to adopt different approaches, based on the type of study, as well as sample normality and size [88]. In the present report, covariance-based SEM was used according to the referential recommendations from Hair et al. [89], where the objectives were to test, confirm, and compare alternative theories. This indicated that the first and second reflective measures were used according to the techniques frequently used in UTAUT-2, the theoretical approach, and the modelling. Based on a strong literature review, UTAUT-2 was modified by adding an attitude being influenced by PE, EE, HB, and HM, with SEM analyzing and interpreting the reliability of measurement and structural models, which specifically incorporated a linear specification reflecting the dependencies between the latent and the measured variable. In this present condition, all the analyses were performed with AMOS software, with the reliability test being conducted through the CFA-based Cronbach alpha and CR (composite reliability) [90,91]. Additionally, subsequent analysis was performed on each latent variable, *t*-value, as well as the item loading and AVE for each question and construct, respectively [84,92].

4. Results

Before conducting a path analysis and developing conclusions on the influential factors, the data normality, as well as assessment of the measurement and structural models, were initially evaluated. In this condition, the reliability, as well as convergent and discriminant validity (CV and DV) tests, were used to evaluate the measurement model based on suggestions [93,94]. Furthermore, the structural models were often used to measure the strength and direction of each construct.

4.1. Normality Analysis

The data normality test (Table 2) was analyzed with the skewness and kurtosis values on each variable, which were between −0.795 and −0.112 and between −0.978 and 4.156. With both values between ±3 and ±10 for skewness and kurtosis, the data were observed to be normally distributed [50].

Table 2. Normality testing and descriptive statistics.

Variable	Min	Max	Skewness	c.r.	Kurtosis	c.r.
HM1	1.000	5.000	−0.332	−2.430	0.544	1.989
HM2	1.000	5.000	−0.216	−1.579	0.479	1.753
HM3	1.000	5.000	−0.395	−2.892	0.557	2.036
PE1	1.000	5.000	−0.475	−3.473	0.242	0.887
PE2	1.000	5.000	−0.387	−2.830	0.002	0.006
CI2	1.000	5.000	−0.136	−0.997	0.463	1.692
CI1	1.000	5.000	−0.138	−1.009	0.838	3.063
ATT1	1.000	5.000	−0.112	−0.818	0.488	1.783
ATT2	1.000	5.000	−0.418	−3.057	0.539	1.972
ATT3	1.000	5.000	−0.169	−1.236	0.443	1.621
PV1	1.000	5.000	−0.318	−2.324	−0.267	−0.978
PV2	1.000	5.000	−0.382	−2.797	−0.067	−0.244
PV3	1.000	5.000	−0.427	−3.127	0.062	0.225
FC1	1.000	5.000	−0.795	−5.815	0.965	3.530
FC2	1.000	5.000	−0.512	−3.741	0.111	0.404
SI1	1.000	5.000	−0.474	−3.469	0.514	1.879
SI2	1.000	5.000	−0.262	−1.915	1.137	4.156
HB1	1.000	5.000	−0.192	−1.406	0.463	1.693
HB2	1.000	5.000	−0.182	−1.334	0.031	0.114
HB3	1.000	5.000	−0.268	−1.959	0.469	1.716
EE1	1.000	5.000	−0.343	−2.508	0.304	1.111
EE2	1.000	5.000	−0.233	−1.707	0.077	0.283
EE3	1.000	5.000	−0.197	−1.442	0.041	0.149
Multivariate					196.099	51.802

4.2. Measurement Model for Measuring Reliability and Validity

The reliability test is often found to reflect the internal stability and consistent level of each measurement questionnaire [95]. This indicated that a good reliability questionnaire was obtained when the Cronbach $\alpha > 0.7$ [89]. In this report, the Cronbach's alpha value on each construct ranged between 0.74–0.95, confirming the existence of high questionnaire reliability and internal consistency between latent variables. It also analyzed the loading factor, composite reliability, and average extraction variation to assess convergence validity. In Table 3, the AVE, loading factor, and CR values are found to be more than 0.5, 0.7, and 0.5, respectively [96], indicating a very good convergent validity for this model.

Table 3. The measurement construct validity ($n = 321$).

Construct	Items	Mean	SD	Factor Loading	Cronbach Alpha	CR	AVE
Performance expectancy	PE2	3.56	0.90	0.81	0.927	0.81	0.68
	PE3	3.42	0.94	0.80			
Effort expectancy	EE1	3.55	0.83	0.56	0.742	0.77	0.53
	EE2	3.38	0.86	0.65			
	EE3	3.27	0.89	0.78			
Habit	HB1	3.36	0.82	0.92	0.892	0.93	0.81
	HB2	3.43	0.87	0.86			
	HB3	3.35	0.83	0.81			
Hedonic motivation	HM1	3.34	0.83	0.93	0.955	0.96	0.90
	HM2	3.34	0.88	0.93			
	HM3	3.33	0.87	0.93			
Social influence	SI1	3.45	0.90	0.82	0.76	0.83	0.71
	SI2	3.31	0.75	0.76			
Facilitating conditions	FC1	3.78	0.89	0.75	0.78	0.71	0.55
	FC2	3.43	0.93	0.66			
Price value	PV1	3.42	1.03	0.89	0.93	0.93	0.81
	PV2	3.31	0.97	0.90			
	PV3	3.29	0.95	0.89			
Attitudes towards micro-lectures	ATT1	3.52	0.86	0.90	0.94	0.96	0.89
	ATT2	3.42	0.78	0.94			
	ATT3	3.42	0.83	0.92			
Continuous intention	CI1	3.31	0.75	0.94	0.91	0.95	0.90
	CI2	3.38	0.81	0.91			

4.3. Discriminant Validity

To analyze the discriminant validity (DV) for each construct, the square root of the AVE values was calculated and compared with the correlation coefficient. The bold diagonal values are the square root of AVE, which needs to be greater than the inter-construct correlations below to meet discriminant validity. This shows that DV is accepted when the results are greater than the correlations for each construct. In this case, the square root of the AVE values exceeded the correlation coefficients for all constructs [97], as shown in Table 4.

Table 4. The discriminant validity test.

	PV	FC	SI	HM	HB	EE	PE	ATT	CI
PV	**0.770**								
FC	0.370	**0.375**							
SI	0.400	0.309	**0.325**						
HM	0.543	0.343	0.383	**0.673**					
HB	0.475	0.366	0.369	0.507	**0.554**				
EE	0.409	0.293	0.311	0.384	0.360	**0.322**			
PE	0.275	0.233	0.254	0.322	0.316	0.249	**0.530**		
ATT	0.481	0.332	0.356	0.509	0.449	0.362	0.282	**0.587**	
CI	0.425	0.313	0.338	0.425	0.441	0.331	0.286	0.416	**0.494**

4.4. Confirmatory Factor Analysis

Based on the recommendation from Hair et al. [98], the structural model should be measured using the highest likelihood estimation process, through the goodness-of-fit analysis. Table 5 shows the goodness-of-fit values for the assessment model. This revealed that the absolute (CMIN/df = 2.83; GFI = 0.87, RMR = 0.02; RMSEA = 0.07), incremental (AGFI = 0.82; NFI = 0.92; CFI = 0.95), and parsimonious (PNFI = 0.72; PGFI = 0.62) SEM measurements of the proposed model had a good fit. According to Doll [99] and Hair [100], the GFI and NFI greater than 0.8 showed acceptable fits.

Table 5. Measurement for model fit *.

Measurement	Fit Indication	Obtained Value	Recommended Criteria
Absolute fit measures	CMIN/df	2.83	1< x < 3
	GFI	0.87	>0.8
	RMSEA	0.07	<0.08
	RMR	0.02	<0.08
Incremental fit measures	NFI	0.92	>0.8
	CFI	0.95	>0.8
	AGFI	0.822	>0.8
Parsimonious fit measures	PNFI	0.72	>0.05
	PGFI	0.62	>0.05

* Source: Browne and Cudeck [101].

For endogenous variables, the R^2 data were considered the most important criterion when testing a structural model, although no agreed standard was recently observed [60]. Based on this condition, a Cohen's standard [102] was used to interpret the R^2 value, where a large and small explanatory power was found above 0.26 and 0.02, respectively. Furthermore, the R^2 values for students' attitude and continuous intention towards the use of micro-lectures in the post-pandemic period were 0.745 (74.5%) and 0.781 (78.1%), respectively, indicating that the model had a large explanatory power. This was then compared to the UTAUT-2 model (54% explanatory power) to explain usage behavior for new technology.

4.5. Structural Model and Hypothesis Test

An assessment of the structural coefficient was presented to analyze the proposed model, which aimed to predict students' continuous intention to use micro-lectures in mathematics lessons (Figure 3). The results confirmed that this model was quite suitable for behavioral explanations, due to the goodness-of-fit being observed in acceptable ranges.

Based on Figure 3 and Table 6, students' attitude towards mathematical micro-lectures was positively influenced by EE ($\beta = 0.620$, $p < 0.001$) and HM ($\beta = 0.313$, $p < 0.001$). Meanwhile, the continuous intention to use micro-lectures during the post-pandemic period was positively influenced by ATT ($\beta = 0.175$, $p < 0.05$) and HB ($\beta = 0.445$, $p < 0.001$). This states that two constructs had significant positive and direct impacts on students' attitudes and continuous intention.

Table 6. Hypothetical analysis. Note: *** $p < 0.001$.

Hypothesis	Relationship			Estimate	S.E.	C.R.	*p*-Value	Interpretation	
								0.05	0.01
H1A	Performance expectancy	→	Attitude	−0.031	0.058	−0.537	0.591	Rejected	Rejected
H1B	Performance expectancy	→	Continuous intention	0.028	0.060	0.462	0.644	Rejected	Rejected
H2A	Effort expectancy	→	Attitude	0.620	0.162	3.834	***	Accepted	Accepted
H2B	Effort expectancy	→	Continuous intention	0.063	0.620	0.102	0.919	Rejected	Rejected
H3A	Habit	→	Attitude	0.138	0.106	1.301	0.193	Rejected	Rejected
H3B	Habit	→	Continuous intention	0.445	0.114	3.892	***	Accepted	Accepted
H4A	Hedonic motivation	→	Attitude	0.313	0.079	3.974	***	Accepted	Accepted
H4B	Hedonic motivation	→	Continuous intention	−0.125	0.087	−1.431	0.152	Rejected	Rejected
H5	Social Influence	→	Continuous intention	0.624	0.758	0.824	0.410	Rejected	Rejected
H6	Facilitating conditions	→	Continuous intention	−0.214	0.244	−0.873	0.382	Rejected	Rejected
H7	Price value	→	Continuous intention	−0.009	0.072	−0.127	0.899	Rejected	Rejected
H8	Attitude	→	Continuous intention	0.175	0.076	2.288	0.022	Accepted	Rejected

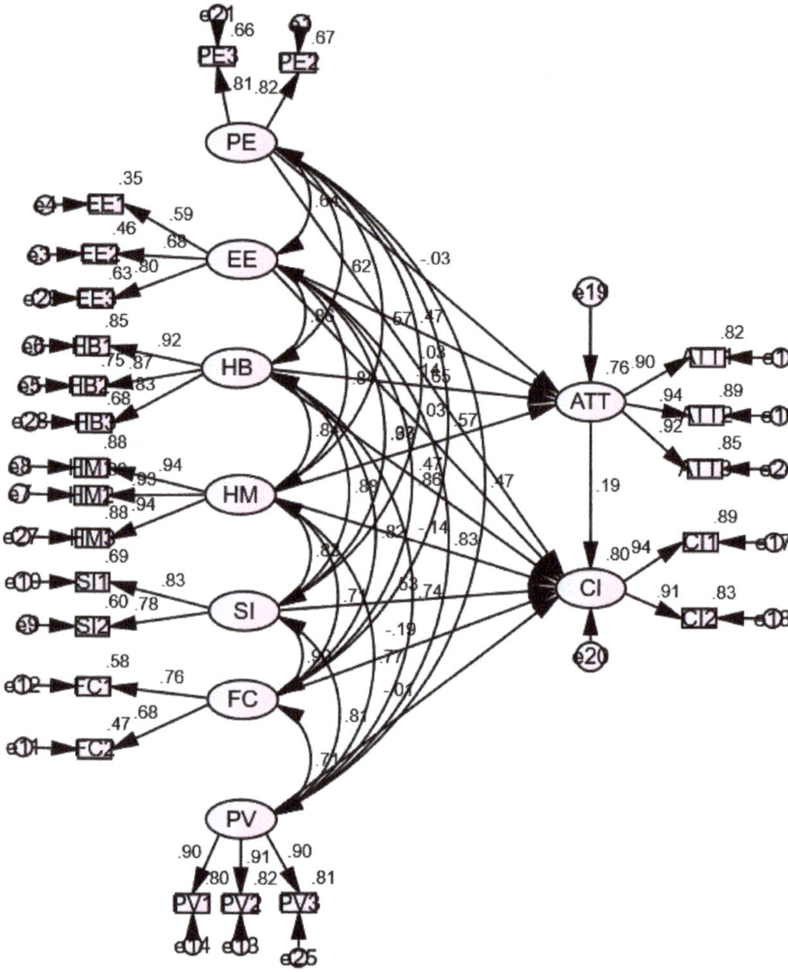

Figure 3. The SEM for the proposed model. Note: PV = price value, FC = facilitating conditions, SI = social influence, HM = hedonic motivation, HB = habit, EE = effort expectancy, PE = performance expectancy, ATT = attitude, and CI = continuous intention.

The most significant positive factor that directly affects students' attitudes towards micro-lectures was found to be effort expectation. The results show that micro-lectures were easy to use during the post-pandemic period, so students' attitudes towards micro-lectures also improved. The factor that had a positive direct effect on attitude was hedonic motivation. The enjoyable learning experience felt by students has a relationship with students' attitudes towards micro-lectures. Furthermore, the most significant factor that had a direct effect on continuous intention was habit. The results of this study suggest that when the use of micro-lectures has become a habit, the continuous intention of students will also increase. Finally, students' attitudes towards micro-lectures have a relationship with the continuous intention to use micro-lectures post-COVID-19. If students' attitudes towards micro-lectures are high, then their continuous intention to use micro-lectures is also high.

5. Discussion

The proposed model was developed by extending the UTAUT-2 to examine the factors affecting students' attitudes and continuous intention to use micro-lectures in mathematics lessons. This experiment succeeded in theoretically and empirically supporting the applicability of UTAUT-2 as a useful model for appropriately understanding CI (continuous intention). Using questionnaires and SPSS/AMOS software, data were also obtained from junior high school students and analyzed through the SEM technique. The influence results showed that effort expectancy affected the attitude towards micro-lectures. This proved that if students find that reviewing mathematics lessons with the help of micro-lectures did not require a big effort, then students will have a good attitude towards mathematics learning. Moreover, attitude also had an indirect effect on continuous intention. This was in line with a previous study, which showed that effort expectancy affected attitude and intention towards using new technologies [79,103,104]. Another report also verified that people often utilize new technology that is easily understood [105]. Furthermore, hedonic motivation was found to positively affect the attitude towards micro-lectures, indicating that students observed the media to be fun and exciting when learning mathematics. Based on the results, the HM-based micro-lecture had the ability to transform students' perceptions of the difficulties and boringness of the subject. Hedonic motivation also had a large indirect effect on students' continuous intention to use micro-lectures after the pandemic. This was in line with previous studies, where HM had a positive effect on the intention to use new technology. Hedonic motivation is the strongest factor and has the most significant effect on students using animation-based media because of the entertainment effect that exists in animation-based media [106]. In addition, the HM factor also has a significant positive effect that affects the possibility of teachers increasing their use of the mobile internet for teaching because they feel that using mobile internet is an enjoyable experience [62].

Attitude also had a direct effect on students' continuous intention to use micro-lectures after the pandemic. This was in line with previous reports, where attitude was the main factor influencing the intention to use new technologies [77,107,108]. However, it was not the determinant with the biggest effect on students' CI in the present analysis. This suggests that subsequent experiments should conduct additional analyses to appropriately understand the effect of attitude on the intentions to use new technologies. According to the results, no significant effect was observed for facilitating conditions due to its inability to influence students' CI to use micro-lectures during the post-pandemic period, as consistently observed in other reports [109]. Moreover, price value did not have a significant effect on mathematical micro-lectures continuous intention, as shown in Martins et al. [68], where PV did not influence the students using digital textbooks. This was not in line with Choi [110], where the factor affected a person's use of MOOC. It was also supported by a previous report, where PV was more significant in the business and economic fields than in the educational sector [48]. Habit was observed to have the most significant effect on students' CI to use micro-lectures after the pandemic. This was because the utilization of the media to learn mathematics had become a habit during the COVID-19 period, for approximately 2 years. In this condition, students were found to often use micro-lectures to prepare and review lessons before and after classes. The results also stated that they liked watching micro-lectures while performing their homework. Therefore, both students and teachers are found to continuously use new technology when it has become a habit. This was in line with a previous review, where HB affected the intention and behavior to use new technology.

6. Theoretical and Practical Implications

Theoretical and practical implications are provided for the maintenance of using micro-lectures during the post-pandemic period. The increased use of this media is also expected to improve students' learning outcomes and mathematical abilities.

This is the first study to extend UTAUT-2 in explaining the continuous intention of using micro-lectures after the COVID-19 pandemic. The results suggest that the extended model provided empirical evidence of CI in the mathematical learning media. This theoretically provides valuable insight into the attitude and continuous intention of this educational field. It also provides additional knowledge on the CI theory of using micro-lectures. Therefore, the development of the UTAUT-2 model theoretically provides possible prospects for future reports. This is useful for micro-lecture developers and schools to identify the influential factors that should be considered when developing several strategies. It also confirms that habit is a significantly important predictor in influencing students' continuous intention to use micro-lectures after the pandemic.

Firstly, principals and teachers should continuously familiarize students with the internal and external use of micro-lectures as additional learning media, based on the influence of ATT, EE, and HB on CI. Secondly, mathematics teachers need to create many interesting and attractive educational media. Thirdly, the resolution of micro-lectures should be optimal and appropriate to meet the needs of students, as problems are often encountered with internet quotas. Fourthly, the ease of using this media for learning must be considered, with teachers also explaining the methods of effectively using the educational tools for mathematical knowledge. Fifthly, the duration of the micro-lectures is also another factor to be considered, as a long watch often breaks concentration and causes fatigue. When developing this learning media, teachers and schools should abide by the duration suggested by previous studies, which is not more than 10 min. This enables students to stay focused, as well as to not become bored and frustrated when using micro-lectures to learn mathematics.

After the pandemic, mathematics teachers are expected to be busy again with various activities, such as teaching preparation, administration, homework, and developing practice questions, which allows less time to create micro-lectures as a supplement for students. This explains that the key to using the learning media remains with the teacher, as the school and creation team are expected to support the educator's development needs. When supported with adequate infrastructures such as hardware, software, and the internet, teachers are more likely to continuously develop micro-lectures for students. This confirms that the habit of using the media for learning is influenced by many aspects, including teachers and schools. Therefore, familiarizing students with the use of micro-lectures as an additional mathematical learning supplement is one of the methods of improving and increasing educational abilities and effectiveness during the post-pandemic period.

7. Conclusions

This study was conducted to explain and predict the factors influencing students' continuous intention to use micro-lectures to learn mathematics after the pandemic based on the UTAUT-2 model, with SEM being utilized to validate and develop conclusions. The empirical results showed that ATT and HB were the strongest predictors of factors influencing students' continuous intention to use micro-lectures after the pandemic, with HM and EE also having a significant effect on learning attitude. Moreover, habit was the most significant factor influencing the CI to use this educational media. This report contributed to the increase in using micro-lectures after the COVID-19 period, which proved to improve learning outcomes. It also contributed to the awareness of developing countries with various factors in mathematical fields, such as attitudes towards online learning media.

8. Limitations

Firstly, the study is limited to the junior high school level in Bandung, Indonesia, with the sole focus on the continuous intention to use mathematical micro-lectures after the COVID-19 pandemic. The results are also not generalized to all educational levels, indicating a lack of studies at elementary, senior high, and tertiary institutions. Secondly, other reports commonly considered the extension of the TAM-2 model, due to it being widely used to analyze behavioral intention towards the use of new technology. Thirdly, it is undeniable that the use of micro-lectures is also influenced by teachers and schools, with many studies showing that students often used online learning media when instructed. Therefore, a subsequent study is suggested to examine the factors influencing the intention of teachers to teach mathematical micro-lectures. Fourthly, the moderating effects of gender and educational level were not assessed on the continuous intention to use this learning media. This indicates that subsequent studies need to examine this relationship. Fifthly, there is a need for future research to support a deeper understanding through a mixed-method approach when investigating the factors affecting junior high school students to use mathematical micro-lectures after the pandemic.

Author Contributions: Conceptualization, T.T.W.; methodology, T.T.W.; validation, R.W.; formal analysis, T.T.W. and R.W.; investigation, R.W.; project administration, T.T.W.; writing—original draft, T.T.W. and R.W.; writing—review and editing, T.T.W. and R.W. All authors have read and agreed to the published version of the manuscript.

Funding: This work was supported by Guangxi Zhuang Autonomous region postgraduate Education Project (2020JGA129) and improve teaching ability of elementary and secondary school teaching project (JSSKT2020012).

Data Availability Statement: The data presented in this study are available upon request from the corresponding author. The data are not publicly available due to privacy.

Acknowledgments: We want to thank Professor Jianlan Tang and Professor Yiming Cao who generously offered their time, insights in our study. And we also wish to thank Professor Zsolt Lavicza for his valuable comments on an earlier version of this paper.

Conflicts of Interest: The authors declare no conflict of interest.

Appendix A

Item	Performance Expectancy	Source
PE1	I continue to use YouTube after the pandemic because it helps me understand math material	[53]
PE2	I continue to use YouTube after the pandemic because it improves my math scores	
	Effort expectancy	
EE1	I do not need much effort when using micro-lectures to learn math	[53]
EE2	I assume learning math through YouTube is easy	
EE3	Micro-lectures are easy to use	
	Social influence	
SI1	My teacher advised me to use YouTube to study maths after the pandemic	
SI2	My friends use YouTube to study maths after the pandemic	
	Facilitating condition	
FC1	I have a cellphone, laptop, or tablet to learn math through YouTube	[53]
FC2	People are helping me when I do not know how to use YouTube to study maths after the pandemic	
	Hedonic motivation	[48]
HM1	I continue to use YouTube to study maths after the pandemic because it is fun	
HM2	I continue to use YouTube to study maths after the pandemic because it is entertaining	
HM3	I continue to use YouTube to study maths after the pandemic because it is so much fun	
	Price value	[48]
PV1	I continue to use YouTube to study maths after the pandemic because it is cheap	
PV2	I continue to use YouTube to study maths after the pandemic because internet costs are affordable	
PV3	I continue to use YouTube to study maths after the pandemic because the internet price is acceptable	
	Continuous intention	[111,112]
CI1	I continue to use YouTube to learn math after the pandemic	
CI2	I recommend the micro-lectures to learn maths to my friends	
	Habits	
HB1	I continue to use YouTube to learn math after the pandemic because I am used to it	[48]
HB2	I continue to use YouTube to learn after the pandemic because I am used to repeating math lessons	
HB3	I continue to use YouTube to study maths after the pandemic because I am used to using it to do my homework	
	Attitude	
ATT1	Learning math using YouTube after the pandemic is a good idea	[92,113]
ATT2	Learning math using YouTube after the pandemic is very interesting for me	
ATT3	Learning math using YouTube after the pandemic is so much fun	

References

1. Al-Balas, M.; Al-Balas, H.I.; Jaber, H.M.; Obeidat, K.; Al-Balas, H.; Aborajooh, E.A.; Al-Taher, R.; Al-Balas, B. Correction to: Distance learning in clinical medical education amid COVID-19 pandemic in Jordan: Current situation, challenges, and perspectives. *BMC Med. Educ.* **2020**, *20*, 341. [CrossRef] [PubMed]
2. Prasetyanto, D.; Rizki, M.; Sunitiyoso, Y. Online Learning Participation Intention after COVID-19 Pandemic in Indonesia: Do Students Still Make Trips for Online Class? *Sustainability* **2022**, *14*, 1982. [CrossRef]
3. Rannastu-Avalos, M.; Siiman, L.A. *Challenges for Distance Learning and Online Collaboration in the Time of COVID-19: Interviews with Science Teachers*; Springer International Publishing: Berlin/Heidelberg, Germany, 2020; Volume 12324 LNCS. [CrossRef]
4. Nuere, S.; de Miguel, L. The Digital/Technological Connection with COVID-19: An Unprecedented Challenge in University Teaching. *Technol. Knowl. Learn.* **2020**, *26*, 931–943. [CrossRef]
5. Nayak, B.; Bhattacharyya, S.S.; Goswami, S.; Thakre, S. Adoption of online education channel during the COVID-19 pandemic and associated economic lockdown: An empirical study from push–pull-mooring framework. *J. Comput. Educ.* **2022**, *9*, 1–23. [CrossRef]
6. Navarro, M.M.; Prasetyo, Y.T.; Young, M.N.; Nadlifatin, R.; Redi, A.A.N.P. The perceived satisfaction in utilizing learning management systems among engineering students during the COVID-19 pandemic: Integrating task technology fit and extended technology acceptance model. *Sustainability* **2021**, *13*, 10669. [CrossRef]
7. Uymaz, P.; Uymaz, A.O. Assessing acceptance of augmented reality in nursing education. *PLoS ONE* **2022**, *17*, e0263937. [CrossRef]
8. Mulyono, H.; Suryoputro, G.; Jamil, S.R. The application of WhatsApp to support online learning during the COVID-19 pandemic in Indonesia. *Heliyon* **2021**, *7*, e07853. [CrossRef]
9. Kusumadewi, R.F.; Islam, U.; Agung, S. Developing electronic teaching materials through comic mathematics media to increase student learning independence during the COVID-19 pandemic in Indonesia. *İlköğretim Online* **2021**, *20*, 708–716. [CrossRef]
10. Ramadhani, R.; Fitri, Y. EPUB3 Based Mathematical E-Modules Using the Sigil Application as A Solution in Teaching and Learning Process through COVID-19 Pandemic. *Form. J. Ilm. Pendidik. MIPA* **2021**, *11*, 35–48. [CrossRef]
11. Amalia, R.R.; Anggoro, S.; Eka, K.I. Identification of Teachers and Students' Readiness to E-Learning Implementation. *J. Teach. Learn. Elem. Educ.* **2021**, *4*, 170–179. [CrossRef]
12. Sarah, L.L. The implementation of discovery learning using personal site in physics online classroom. *J. Phys. Conf. Ser.* **2021**, *1957*, 012042. [CrossRef]
13. Aryanti; Wijaya, T.T. Elementary Students ' Attitudes towards an Online Visiting Lecture in COVID-19 School Closure Period. *J. Teach. Learn. Elem. Educ.* **2022**, *5*, 81–94.
14. Weinhandl, R.; Houghton, T.; Lindenbauer, E.; Mayerhofer, M.; Lavicza, Z. Integrating Technologies Into Teaching and Learning Mathematics at the Beginning of Secondary Education in Austria Integrating Technologies into Teaching and Learning Mathematics at the Beginning of Secondary Education in Austria. *EURASIA J. Math. Sci. Technol. Educ.* **2021**, *17*, em2057. [CrossRef]
15. Zhu, X.; Liu, J. Education in and after COVID-19: Immediate Responses and Long-Term Visions. *Postdigital Sci. Educ.* **2020**, *2*, 695–699. [CrossRef]
16. Maqableh, M.; Alia, M. Evaluation online learning of undergraduate students under lockdown amidst COVID-19 Pandemic: The online learning experience and students' satisfaction. *Child. Youth Serv. Rev.* **2021**, *128*, 106160. [CrossRef]
17. Jevsikova, T.; Stupuriene, G.; Stumbriene, D.; Juškevičiene, A.; Dagiene, V. Acceptance of Distance Learning Technologies by Teachers: Determining Factors and Emergency State Influence. *Informatica* **2021**, *32*, 517–542. [CrossRef]
18. Cao, Y.; Zhang, S.; Chan, M.C.E.; Kang, Y. Post-pandemic reflections: Lessons from Chinese mathematics teachers about online mathematics instruction. *Asia Pac. Educ. Rev.* **2021**, *22*, 157–168. [CrossRef]
19. Alabdulwahhab, K.M.; Kazmi, S.Y.; Sami, W.; Almujel, K.N.; Alanazi, M.H.; Alanazi, K.F.; Moyana, A.M.; Ahmad, M.S.; Alasbali, T.A.; Al Alwadani, F. Use of online resources by undergraduate medical students at College of Medicine, Majmaah University, Kingdom of Saudi Arabia. *PLoS ONE* **2021**, *16*, e0255635. [CrossRef]
20. Kramarski, B.; Liberman, A. *Using Electronic Mail Communication and Metacognitive Instruction to Improve Mathematical Problem Solving*; Springer: Boston, MA, USA, 2003. [CrossRef]
21. Padaguri, V.; Pasha, S.A. Investigating the Intentions to Adopt E-Learning using UTAUT-3 model: A Perspective of COVID-19. In Proceedings of the AUBH E-Learning Conference Innovative Learning and Teaching: Lessons from COVID-19 SYNCHRONOUS, Manama, Bahrain, 24–26 May 2021; pp. 1–9.
22. Fitzgerald, D.A.; Scott, K.M.; Ryan, M.S. Blended and e-learning in pediatric education: Harnessing lessons learned from the COVID-19 pandemic. *Eur. J. Pediatr.* **2021**, *181*, 447–452. [CrossRef]
23. Lapitan, L.D.; Tiangco, C.E.; Sumalinog, D.A.G.; Sabarillo, N.S.; Diaz, J.M. An effective blended online teaching and learning strategy during the COVID-19 pandemic. *Educ. Chem. Eng.* **2021**, *35*, 116–131. [CrossRef]
24. Cao, M. An Overview of Micro-Lecture in College English Teaching in Mainland China. *Int. J. Lang. Linguist.* **2017**, *5*, 34–38. [CrossRef]
25. Wang, K.; Zhu, C.; Tondeur, J.; Wang, K. Using micro-lectures in small private online courses: What do we learn from students ' behavioural intentions? *Technol. Pedagog. Educ.* **2021**, *30*, 427–442. [CrossRef]

26. Fauzi, I.; Hermawan, I.; Khusuma, S. Teachers' Elementary School in Online Learning of COVID-19 Pandemic Conditions. *J. Iqra' Kaji. Ilmu Pendidik.* **2020**, *5*, 58–70. [CrossRef]
27. Chick, R.C.; Clifton, G.T.; Peace, K.M.; Propper, B.W.; Hale, D.F.; Alseidi, A.A.; Vreeland, T.J. Using Technology to Maintain the Education of Residents During the COVID-19 Pandemic. *J. Surg. Educ.* **2020**, *77*, 729–732. [CrossRef] [PubMed]
28. Osorio-saez, E.M.; Eryilmaz, N.; Sandoval-hernandez, A. Parents' Acceptance of Educational Technology: Lessons from around the World. *Front. Psychol.* **2021**, *12*, 719430. [CrossRef]
29. Xie, Z.; Xiao, L.; Hou, M.; Liu, X.; Liu, J. Micro classes as a primary school–level mathematics education response to COVID-19 pandemic in China: Students' degree of approval and perception of digital equity. *Educ. Stud. Math.* **2021**, *108*, 65–85. [CrossRef] [PubMed]
30. Peng, Y. Application of Micro-lecture in Computer Teaching. *IOP Conf. Ser. Earth Environ. Sci.* **2019**, *234*, 012053. [CrossRef]
31. Lu, S. The Application of Micro Lecture in English Reading Teaching. In Proceedings of the 5th International Conference on Social Science, Education and Humanities Research, Tianjin, China, 11–12 June 2016; pp. 15–19. [CrossRef]
32. Zhang, N.; Tan, L.; Li, F.; Han, B.; Xu, Y. Development and application of digital assistive teaching system for anatomy. *Virtual Real. Intell. Hardw.* **2021**, *3*, 315–335. [CrossRef]
33. Sartika, N.S.; Mauladaniyati, R. Analysis of Prospective Mathematics Teachers ' Reading Interest Through E-Book for Geometry Systems Course in New Normal Era. In Proceedings of the International Conference on Educational Studies in Mathematics (ICoESM 2021), Virtual, 13 August 2021; Volume 611, pp. 353–359.
34. Zhang, X.; Xu, J. Integration of Micro Lectures into the Blended Learning Discourse in Tertiary Education. *Asian Assoc. Open Univ. J.* **2015**, *10*, 13–28. [CrossRef]
35. Wijaya, T.T.; Cao, Y.; Weinhandl, R.; Yusron, E. Applying the UTAUT Model to Understand Factors Affecting Micro-Lecture Usage by Mathematics Teachers in China. *Mathematics* **2022**, *10*, 1008. [CrossRef]
36. Yang, Y.; Zhang, X.; Tian, D. Micro-Lecture Design and Practice In The Internet Plus Era. *Adv. Comput. Sci. Res.* **2017**, *74*, 673–678. [CrossRef]
37. Lavidas, K.; Komis, V.; Achriani, A. Explaining faculty members' behavioral intention to use learning management systems. *J. Comput. Educ.* **2022**, *5*, 143–155. [CrossRef]
38. Taamneh, A.; Alsaad, A.; Elrehail, H.; Al-Okaily, M.; Lutfi, A.; Sergio, R.P. University lecturers acceptance of moodle platform in the context of the COVID-19 pandemic. *Glob. Knowl. Mem. Commun.* **2022**. [CrossRef]
39. Liu, D.; Luo, J. College Learning from Classrooms to the Internet: Adoption of the YouTube as Supplementary Tool in COVID-19 Pandemic Environment. *Educ. Urban Soc.* **2021**. [CrossRef]
40. Fitriani, W.R.; Mulyono, A.B.; Hidayanto, A.N.; Munajat, Q. Reviewer's communication style in YouTube product-review videos: Does it affect channel loyalty? *Heliyon* **2020**, *6*, e04880. [CrossRef]
41. Wijaya, T.T.; Ying, Z.; Purnama, A. Using Hawgent Dynamic Mathematics Software in Teaching Trigonometry. *Int. J. Emerg. Technol. Learn.* **2020**, *15*, 215–222. [CrossRef]
42. Hermita, N.; Putra, Z.H.; Alim, J.A.; Tang, J.; Wijaya, T.T.; Li, L.; Pereira, J.; Tamur, M. The Hungry Ant: Development of Video-Based Learning on Polyhedron. *Int. J. Interact. Mob. Technol.* **2021**, *15*, 18–32. [CrossRef]
43. Han, J. Micro-lecture Teaching for Improving the Learning Effect of Non-English Majors at North China Electric Power University. *English Lang. Teach.* **2019**, *12*, 209–216. [CrossRef]
44. El Said, G.R. How Did the COVID-19 Pandemic Affect Higher Education Learning Experience? an Empirical Investigation of Learners' Academic Performance at a University in a Developing Country. *Adv. Hum.-Comput. Interact.* **2021**, *2021*, 6649524. [CrossRef]
45. Piramanayagam, S.; Seal, P.P. Hospitality students' adoption of e-Books during the COVID-19 pandemic: A developing country perspective. *Libr. Philos. Pract.* **2021**, *2021*, 1–17.
46. Wah, L.L.; Hashim, H. Determining pre-service teachers' intention of using technology for teaching english as a second language (Esl). *Sustainability* **2021**, *13*, 7568. [CrossRef]
47. Nikou, S.; Aavakare, M. An assessment of the interplay between literacy and digital Technology in Higher Education. *Educ. Inf. Technol.* **2021**, *26*, 3893–3915. [CrossRef]
48. Venkatesh, V.Y.L.; Thong, J.; Xu, X. Consumer acceptance and use of information Technology: Extending the unified theory of acceptance and use of technology. *MIS Q.* **2012**, *36*, 157–178. [CrossRef]
49. Zhao, W.; Mok, I.A.C.; Cao, Y. Factors influencing teachers' implementation of a reformed instructional model in China from the theory of planned behavior perspective: A multiple case study. *Sustainability* **2020**, *12*, 1. [CrossRef]
50. Moorthy, K.; Chun T'ing, L.; Ming, K.S.; Ping, C.C.; Ping, L.Y.; Joe, L.Q.; Jie, W.Y. Behavioral Intention to Adopt Digital Library by the Undergraduates. *Int. Inf. Libr. Rev.* **2019**, *51*, 128–144. [CrossRef]
51. Chávez Herting, D.; Cladellas Pros, R.; Castelló Tarrida, A. Habit and social influence as determinants of PowerPoint use in higher education: A study from a technology acceptance approach. *Interact. Learn. Environ.* **2020**, *8*, 1–17. [CrossRef]
52. Yu, C.W.; Chao, C.M.; Chang, C.F.; Chen, R.J.; Chen, P.C.; Liu, Y.X. Exploring Behavioral Intention to Use a Mobile Health Education Website: An Extension of the UTAUT 2 Model. *SAGE Open* **2021**, *11*, 21582440211055721. [CrossRef]
53. Venkatesh, M.; Morris, G.; Davis, G.B.; Davis, F.D. User acceptance of information technology: Toward a unified view. *Manag. Inf. Syst. Q.* **2003**, *27*, 425–478. [CrossRef]

54. Wijaya, T.T.; Zhou, Y.; Houghton, T.; Weinhandl, R.; Lavicza, Z.; Yusop, F.D. Factors affecting the use of digital mathematics textbooks in indonesia. *Mathematics* **2022**, *10*, 1808. [CrossRef]
55. Chatti, H.; Hadoussa, S. Factors Affecting the Adoption of E-Learning Technology by Students during the COVID-19 Quarantine Period: The Application of the UTAUT Model. *Eng. Technol. Appl. Sci. Res.* **2021**, *11*, 6993–7000. [CrossRef]
56. Balkaya, S.; Akkucuk, U. Adoption and use of learning management systems in education: The role of playfulness and self-management. *Sustainability* **2021**, *13*, 1127. [CrossRef]
57. Shao, D.; Lee, I.J. Acceptance and influencing factors of social virtual reality in the urban elderly. *Sustainability* **2020**, *12*, 9345. [CrossRef]
58. Alvi, I. College students' reception of social networking tools for learning in India: An extended UTAUT model. *Smart Learn. Environ.* **2021**, *8*, 18. [CrossRef]
59. Khalid, B.; Chaveesuk, S.; Chaiyasoonthorn, W. Moocs adoption in higher education: A management perspective. *Polish J. Manag. Stud.* **2021**, *23*, 239–256. [CrossRef]
60. Shen, C.W.; Ho, J.T.; Ly, P.T.; Kuo, T.C. Behavioural intentions of using virtual reality in learning: Perspectives of acceptance of information technology and learning style. *Virtual Real.* **2019**, *23*, 313–324. [CrossRef]
61. Alowayr, A. Determinants of mobile learning adoption: Extending the unified theory of acceptance and use of technology (UTAUT). *Int. J. Inf. Learn. Technol.* **2021**, *39*, 1–12. [CrossRef]
62. Nikolopoulou, K.; Gialamas, V.; Lavidas, K. Habit, hedonic motivation, performance expectancy and technological pedagogical knowledge affect teachers' intention to use mobile internet. *Comput. Educ. Open* **2021**, *2*, 100041. [CrossRef]
63. de Guinea, A.O.; Markus, M.L. Why break the habit of a lifetime? Rethinking the roles of intention, habit, and emotion in continuing information technology use. *MIS Q.* **2009**, *33*, 433. [CrossRef]
64. Garg, A. Investigating the Moderating Effects of Age and Gender on Customers' Use of Tablet Menu in Casual Dining Restaurants. *J. Qual. Assur. Hosp. Tour.* **2021**, *22*, 1–39. [CrossRef]
65. Abdul Rabu, S.N.; Hussin, H.; Bervell, B. QR code utilization in a large classroom: Higher education students' initial perceptions. *Educ. Inf. Technol.* **2019**, *24*, 359–384. [CrossRef]
66. Faqih, K.M.S.; Jaradat, M.I.R.M. Integrating TTF and UTAUT2 theories to investigate the adoption of augmented reality technology in education: Perspective from a developing country. *Technol. Soc.* **2021**, *67*, 101787. [CrossRef]
67. Hu, S.; Laxman, K.; Lee, K. Exploring factors affecting academics' adoption of emerging mobile technologies-an extended UTAUT perspective. *Educ. Inf. Technol.* **2020**, *25*, 4615–4635. [CrossRef]
68. Martins, M.; Farias, J.S.; Albuquerque, P.H.M.; Pereira, D.S. Adoption of technology for reading purposes: A study articles of e-books acceptance. *Brazilian Bus. Rev.* **2018**, *15*, 568–588. [CrossRef]
69. Kosiba, J.P.B.; Odoom, R.; Boateng, H.; Twum, K.K.; Abdul-Hamid, I.K. Examining students' satisfaction with online learning during the Covid-19 pandemic—An extended UTAUT2 approach. *J. Furth. High. Educ.* **2022**, *46*, 1–18. [CrossRef]
70. Alyoussef, I.Y. Massive open online course (Moocs) acceptance: The role of task-technology fit (ttf) for higher education sustainability. *Sustainability* **2021**, *13*, 7374. [CrossRef]
71. Alenezi, A.R. Modeling the Social Factors Affecting Students' Satisfaction with Online Learning: A Structural Equation Modeling Approach. *Educ. Res. Int.* **2022**, *2022*, 2594211. [CrossRef]
72. Yu, H.; Wang, S.; Li, J.; Shi, G.; Yang, J. Influence of Online Merging Offline Method on University Students' Active Learning through Learning Satisfaction. *Front. Psychol.* **2022**, *13*, 842322. [CrossRef]
73. Wu, B. Influence of MOOC learners discussion forum social interactions on online reviews of MOOC. *Educ. Inf. Technol.* **2021**, *26*, 3483–3496. [CrossRef]
74. Ly, H.T.; Khuong, N.V.; Son, T.H. Determinants Affect Mobile Wallet Continuous Usage in COVID 19 Pandemic: Evidence From Vietnam. *Cogent Bus. Manag.* **2022**, *9*, 2041792. [CrossRef]
75. Raffaghelli, J.E.; Rodríguez, M.E.; Guerrero-Roldán, A.-E.; Bañeres, D. Applying the UTAUT model to explain the students' acceptance of an early warning system in Higher Education. *Comput. Educ.* **2022**, *182*, 104468. [CrossRef]
76. Dahri, N.A.; Vighio, M.S.; Bather, J.D.; Arain, A.A. Factors influencing the acceptance of mobile collaborative learning for the continuous professional development of teachers. *Sustainability* **2021**, *13*, 13222. [CrossRef]
77. Acharjya, B.; Das, S. Adoption of E-Learning During the COVID-19 Pandemic. *Int. J. Web-Based Learn. Teach. Technol.* **2022**, *17*, 1–14. [CrossRef]
78. Pando-Garcia, J.; Periañez-Cañadillas, I.; Charterina, J. Business simulation games with and without supervision: An analysis based on the TAM model. *J. Bus. Res.* **2016**, *69*, 1731–1736. [CrossRef]
79. Taylor, S.; Todd, P.A. Understanding information technology usage: A test of competing models. *Inf. Syst. Res.* **1995**, *6*, 144–176. [CrossRef]
80. Alamri, M.M. Using blended project-based learning for students' behavioral intention to use and academic achievement in higher education. *Educ. Sci.* **2021**, *11*, 207. [CrossRef]
81. Cabero-Almenara, J.; Barroso-Osuna, J.; Martinez-Roig, R. Mixed, augmented and virtual, reality applied to the teaching of mathematics for architects. *Appl. Sci.* **2021**, *11*, 7125. [CrossRef]
82. McCarthy, J.W.; Cooley, J.A.; Benigno, J.P. Integrating music into small-group book reading activities increases the engagement of students with autism. *Evid. Based. Commun. Assess. Interv.* **2010**, *4*, 83–87. [CrossRef]

83. Anthony, B.; Kamaludin, A.; Romli, A. *Predicting Academic Staffs Behaviour Intention and Actual Use of Blended Learning in Higher Education: Model Development and Validation*; Springer: Dordrecht, The Netherlands, 2021. [CrossRef]
84. Almulla, M.A.; Alamri, M.M. Using conceptual mapping for learning to affect students' motivation and academic achievement. *Sustainability* **2021**, *13*, 4029. [CrossRef]
85. Kokkinou, E.; Kyriakides, L. Investigating differential teacher effectiveness: Searching for the impact of classroom context factors. *Sch. Eff. Sch. Improv.* **2022**, *33*, 1–28. [CrossRef]
86. Edumadze, J.K.E.; Barfi, K.A.; Arkorful, V.; Baffour, N.O., Jr. Undergraduate student's perception of using video conferencing tools under lockdown amidst COVID-19 pandemic in Ghana. *Interact. Learn. Environ.* **2022**, 1–12. [CrossRef]
87. Al-Maroof, R.S.; Alnazzawi, N.; Akour, I.A.; Ayoubi, K.; Alhumaid, K.; AlAhbabi, N.M.; Alnnaimi, M.; Thabit, S.; Alfaisal, R.; Aburayya, A.; et al. The effectiveness of online platforms after the pandemic: Will face-to-face classes affect students' perception of their behavioural intention (BIU) to use online platforms? *Informatics* **2021**, *8*, 83. [CrossRef]
88. Morchid, N. The Determinants of Use and Acceptance of Mobile Assisted Language Learning: The Case of EFL Students in Morocco. *Arab World Engl. J.* **2019**, *5*, 76–97. [CrossRef]
89. Hair, J.F.; Hult, G.T.M.; Ringle, C.; Sarstedt, M. *A Primer on Partial Least Squares Structural Equation Modeling (PLS-SEM)*; SAGE Publications: Thousand Oaks, CA, USA, 2016.
90. Castéra, J.; Marre, C.C.; Yok, M.C.; Sherab, K.; Impedovo, M.A.; Sarapuu, T.; Pedregosa, A.D.; Malik, S.K.; Armand, H. Self-reported TPACK of teacher educators across six countries in Asia and Europe. *Educ. Inf. Technol.* **2020**, *25*, 3003–3019. [CrossRef]
91. Wang, Q.; Zhao, G. ICT self-efficacy mediates most effects of university ICT support on preservice teachers' TPACK: Evidence from three normal universities in China. *Br. J. Educ. Technol.* **2021**, *52*, 2319–2339. [CrossRef]
92. Smeda, A.; Shiratuddin, M.F.; Wong, K.W. A structural equation modelling approach for adoption of e-book amongst mathematics and statistics (MAS) students at higher education institutions in Libya. *Int. J. Inf. Learn. Technol.* **2018**, *35*, 240–254. [CrossRef]
93. Akbulut, C.; Dindar, A.; Kucuk, S.; Sesen, B. Development and Validation of the ICT-TPACK-Science Scale. *J. Sci. Educ. Technol.* **2020**, *29*, 185–218. [CrossRef]
94. Elyazgi, M. Validating pupils' behaviour intention to use e-book technology in their learning. *Int. J. Eng. Technol.* **2018**, *7*, 511–518. [CrossRef]
95. Ahmad, F.A.R.O.B. The Effect of Augmented Reality in Improving Visual Thinking in Mathematics of 10th-Grade Students in Jordan. *Int. J. Adv. Comput. Sci. Appl.* **2021**, *12*, 352–360. [CrossRef]
96. Fornell, C.; Larcker, D.F. Evaluating structural equation models with unobservable variables and measurement error. *J. Mark. Res.* **1981**, *18*, 39–50. [CrossRef]
97. Kline, R.B. *Principles and Practice of Structural Equation Modeling*, 2nd ed.; Guilford Press: New York, NY, USA, 2005.
98. Hair, J.F.; Ringle, C.M.; Sarstedt, M. Partial least squares: The better approach to structural equation modeling? *Long Range Plann.* **2012**, *45*, 312–319. [CrossRef]
99. Doll, W.; Xia, W.; Torkzadeh, G. A confirmatory factor analysis of the end-user computing satisfaction instrument. *MIS Q.* **1994**, *18*, 357–369. [CrossRef]
100. Hair, J.; Black, B.; Babin, B.; Anderson, R.E.; Tatham, R.L. *Multivariate Data Analysis*, 6th ed.; Prentice Hall: New York, NY, USA, 2006.
101. Browne, M.W.; Cudeck, R. *Alternative Ways of Assessing Model Fit*; Sage: Beverly Hills, CA, USA, 1993.
102. Cohen, J. *Statistical Power Analysis for the Behavioral Sciences*; Lawrance Erlbaum Associates: Hillsdale, NJ, USA, 1988.
103. Lawson-Body, A.; Willoughby, L.; Lawson-Body, L.; Tamandja, E.M. Students' acceptance of E-books: An application of UTAUT. *J. Comput. Inf. Syst.* **2020**, *60*, 256–267. [CrossRef]
104. Abu-Al-Aish, A.; Love, S. Factors influencing students' acceptance of m-learning: An investigation in higher education. *Int. Rev. Res. Open Distance Learn.* **2013**, *14*, 82–107. [CrossRef]
105. David, A. Relative Effectiveness of Computer-based Instruction, Digital Video and Audio Packages on Students Achievement in ALgebra. *Int. J. Res. Sci. Technol. Math. Educ.* **2014**, *2*, 164–180.
106. Dajani, D.; Abu Hegleh, A.S. Behavior intention of animation usage among university students. *Heliyon* **2019**, *5*, e02536. [CrossRef] [PubMed]
107. Kala, D.; Chaubey, D.S. Examination of relationships among technology acceptance, student engagement, and perceived learning on tourism-related MOOCs. *J. Teach. Travel Tour.* **2022**, *22*, 1–18. [CrossRef]
108. Fu, Y.; Zhang, D.; Jiang, H. Students' Attitudes and Competences in Modeling Using 3D Cartoon Toy Design Maker. *Sustainability* **2022**, *14*, 2176. [CrossRef]
109. Teo, T. Examining the influence of subjective norm and facilitating conditions on the intention to use technology among pre-service teachers: A structural equation modeling of an extended technology acceptance model. *Asia Pac. Educ. Rev.* **2010**, *11*, 253–262. [CrossRef]
110. Choi, H.; Hoo-Jo, H. Influential Factors on Elementary School Teachers' Intention to Use MOOC. *Teach. Educ. Res.* **2018**, *57*, 16–29. [CrossRef]
111. Rohan, R.; Pal, D.; Funilkul, S.; Chutimaskul, W.; Eamsinvattana, W. How Gamification Leads to Continued Usage of MOOCs? A Theoretical Perspective. *IEEE Access* **2021**, *9*, 108144–108161. [CrossRef]

112. Arpaci, I.; Basol, G. The impact of preservice teachers' cognitive and technological perceptions on their continuous intention to use flipped classroom. *Educ. Inf. Technol.* **2020**, *25*, 3503–3514. [CrossRef]
113. Agustyaningrum, N.; Kepulauan, U.R.; Mahmudi, A. Dominant Factors that Cause Students ' Difficulties in Learning Abstract Algebra: A Case Study at a University in Indonesia. *Int. J. Instr.* **2021**, *14*, 847–866. [CrossRef]

electronics

Article

Is Initial Performance in a Course Informative? Machine Learning Algorithms as Aids for the Early Detection of At-Risk Students

Maura A. E. Pilotti [1,*], Emaan Nazeeruddin [2], Mohammad Nazeeruddin [2], Ibtisam Daqqa [1], Hanadi Abdelsalam [1] and Maryam Abdullah [1]

1 Department of Science and Human Studies, Prince Mohammad Bin Fahd University, Al Khobar 31952, Saudi Arabia; idaqqa@pmu.edu.sa (I.D.); habdelsalam@pmu.edu.sa (H.A.); mabdullah@pmu.edu.sa (M.A.)
2 Department of Engineering and Computer Science, Prince Mohammad Bin Fahd University, Al Khobar 31952, Saudi Arabia; 202000080@pmu.edu.sa (E.N.); nmohammad@pmu.edu.sa (M.N.)
* Correspondence: maura.pilotti@gmail.com

Abstract: The extent to which grades in the first few weeks of a course can predict overall performance can be quite valuable in identifying at-risk students, informing interventions for such students, and offering valuable feedback to educators on the impact of instruction on learning. Yet, research on the validity of such predictions that are made by machine learning algorithms is scarce at best. The present research examined two interrelated questions: To what extent can educators rely on early performance to predict students' poor course grades at the end of the semester? Are predictions sensitive to the mode of instruction adopted (online versus face-to-face) and the course taught by the educator? In our research, we selected a sample of courses that were representative of the general education curriculum to ensure the inclusion of students from a variety of academic majors. The grades on the first test and assignment (early formative assessment measures) were used to identify students whose course performance at the end of the semester would be considered poor. Overall, the predictive validity of the early assessment measures was found to be meager, particularly so for online courses. However, exceptions were uncovered, each reflecting a particular combination of instructional mode and course. These findings suggest that changes to some of the currently used formative assessment measures are warranted to enhance their sensitivity to course demands and thus their usefulness to both students and instructors as feedback tools. The feasibility of a grade prediction application in general education courses, which critically depends on the accuracy of such tools, is discussed, including the challenges and potential benefits.

Keywords: predictive validity; general education; learning algorithms; COVID-19; online learning; face-to-face learning

Citation: Pilotti, M.A.E.; Nazeeruddin, E.; Nazeeruddin, M.; Daqqa, I.; Abdelsalam, H.; Abdullah, M. Is Initial Performance in a Course Informative? Machine Learning Algorithms as Aids for the Early Detection of At-Risk Students. *Electronics* **2022**, *11*, 2057. https://doi.org/10.3390/electronics11132057

Academic Editor: Hung-Yu Chien

Received: 17 June 2022
Accepted: 28 June 2022
Published: 30 June 2022

1. Introduction

General education courses in undergraduate programs are assumed to ensure that all college students acquire the foundational interdisciplinary knowledge, as well as the analytical and communication skills that are necessary to address the demands of their selected majors and those of their chosen professions [1]. Failures in general education courses may ignite a cascade of undesirable effects, which can span from mild (the repetition of a course) to severe (e.g., academic dismissal, delayed degree attainment, and loss of financial aid eligibility). Thus, how students perform in such courses can be considered key in determining their academic success, including retention and graduation [2,3]. It is an accepted fact that the timing of the identification of at-risk students is a critical aspect of the effectiveness of the implementation of remedial interventions [4–6]. However, educators usually have very little information about students' performance during the first half of the semester, which can make the identification of at-risk students both challenging and broadly consequential if erroneous conclusions are reached. To wit, unrecognized difficulties (an event classified in signal detection theory as a miss) are likely to lead to course failure. Notwithstanding the need for valid early predictions of students' academic performance, which rely on limited information, most of the research on algorithms that are intended to assist educators' performance forecasts has relied upon much greater amounts of information collected within a much larger timeframe and has often involved discipline-specific subject matters [7–9]. Examples are predictions of final course grades in a particular subject matter based on students' grade point average (GPA), as well as grades in pre-requisite courses [10], or more simply, on students' academic history, as exemplified by their performance in past courses [11]. Yet, the algorithms that yield optimal results tend to vary considerably [12,13], along with a myriad of innovative stand-alone or hybrid solutions that appear as a regular stream in the extant literature [14]. As a result, the selection and subsequent use of a suitable technique for predicting at-risk students may become so challenging and overwhelming for an educator whose expertise is other than computer science that ignoring potentially viable technical solutions is the most likely course of action [15]. For such educators, continued reliance on personal intuition and conscious reasoning may seem preferable to the ordeal of understanding the technically dense machine learning literature. However, this comes at the cost of an increased likelihood of biases affecting the processing of students' information for assessment and decision-making [16,17], including personal preferences for the parameters to take into account and for the type and amount of data that are necessary to generate sensible predictions. Consider that, as the semester progresses, the amount of information about students that is available to an educator accumulates, but its utility decreases as remedial actions become harder to implement and their success becomes more uncertain [18,19]. Earlier predictions are unquestionably more valuable than later predictions, but at the beginning of a course, very little information is available to the educator, making predictions about a student's difficulties even more uncertain (e.g., is initial poor performance symptomatic of a momentary hurdle, perhaps linked to the idiosyncrasies of an assignment, or a reliable indication of serious issues?).

The COVID-19 pandemic has complicated the prediction matter by suddenly relocating students, most of whom were exclusively accustomed to face-to-face instruction, to online instruction. Although the synchronous online mode that is adopted by many institutions of higher learning has replicated many aspects of the face-to-face mode (e.g., real-time interactions in a virtual classroom), physical distance, technical idiosyncrasies, competencies, and other issues (e.g., students' degree and manner of adaptation to environmental changes) may have made learning in online courses different from that of face-to-face courses [20,21]. For instance, it has been proposed that the online mode has fostered the practice of a more continuous engagement in learning activities [20]. As evidence of change, studies have reported higher online performance (as measured by course grades) than pre-pandemic face-to-face performance [21–25]. However, other studies have reported declines or no change at all [4,25–27]. Thus, because uncertainty endures as to whether remote instruction during the pandemic has fostered relevant changes to students' learning, it remains unclear whether performance predictions that are made online and face-to-face can be considered equivalent.

In the present study, we examined whether algorithms that are commonly used for the predictions of final grades could be of assistance to educators in both face-to-face and online courses when the only information available to the educators is students' performance on the first test and assignment. Both assessment measures can be classified as formative assessment tools [28]. These are tools that are used by students to assess their learning in a course and by educators to determine the effectiveness of the instruction they deliver, thereby defining formative assessment as serving both diagnostic and feedback functions. In a course, formative assessment measures can be said to be particularly critical to students' academic success, since the information they provide has the potential to foster change in the way that students approach the curriculum and understand its demands, as well as in the way that educators teach. Thus, in principle, the earlier the assessment, the higher may be its impact on both students and educators. Formative assessment differs from summative assessment (i.e., final tests), whose primary aim is to measure learning comprehensively across the entire semester as an evaluation of the extent to which it meets pre-set learning outcomes. A summative assessment indicator is the final course grade that is given to each student at the end of the semester. The effectiveness of early formative assessment measures, each of which covers a portion of the curriculum to be acquired in a course, resides in their ability to adequately predict final course grades which reflect the student's learning of the entire course curriculum.

It is customary for institutions of higher education to demand that students meet a minimum performance requirement to gain access and remain enrolled in any degree program. At the institution that was selected for the present research, this requirement entails maintaining a GPA that is better than a C (greater than 79%). Thus, to ensure authenticity, we classified the final course grades into three performance categories: high (H—equal to or greater than 90%); medium (M—80–89.99%), and low (L—79% or below). This stringent classification scheme created categories of comparable size, while it minimized the impact of grade inflation and educators' grading idiosyncrasies, as well as reflected the standards of academic success at the selected institution.

At the outset, we recognized that the predictive validity of a forecast may refer to a variety of key parameters, such as accuracy [(hits + correct rejections)/all responses, including hits, correct rejections, misses, and false alarms)]; precision [hits/(hits + false alarms)]; and sensitivity [hits/(hits + miss)]. In the task of identifying at-risk students, however, correct rejections are not particularly relevant. Furthermore, false alarms are much less costly or even less relevant than misses. Namely, false alarms are likely to reflect cases of temporary difficulties experienced by individual students which are mistakenly identified as enduring and/or severe (a false alarm), thereby creating unnecessary but fleeting stress in such students. Thus, in the present study, we relied on sensitivity as a measure of the predictive validity of forecasts of at-risk students (i.e., learners receiving an L grade at the end of the semester). A sensitivity score for an L classification was conceptualized as a proportion, including the number of correctly classified L grades divided by the number of grades that were either misclassified as H and M or correctly classified as L.

The study involved female undergraduate students of a society that is in transition from a patriarchal order to one that is akin to gender equity in education and employment [29–32]. In such a society, of which a prototypical example is the Kingdom of Saudi Arabia (KSA), female students of college age are the main target of top-down gender-equity interventions. Decrees and massive financial investments aim to re-set the country's social structure to favor meritocracy for both sexes at the expense of tribal and patriarchal favoritism [33]. Thus, the academic success of female college students is a priority for the adequate development of the economic and social engine of KSA, making our research a window into the performance of this highly valued population, as well as into the utility of early performance assessment in the said population.

The current study tested several popular learning algorithm(s) to answer two interrelated questions:

a. Can at-risk students (defined as those with an end-of-the-semester score of L in a general education course) be effectively identified by very early performance indicators (i.e., grades on the first test and first assignment) through one of these algorithms?
b. Do predictions of at-risk students vary between face-to-face instruction and synchronous online instruction, as well as with the specific subject matter taught in a course?

We selected a sample of courses that are representative of the general education curriculum of a Saudi higher education institution that follows a curriculum of U.S. import and a student-centered pedagogy. The courses had been taught by the same instructor both online (during the pandemic) and face-to-face (before the pandemic) for at least three semesters in each mode. The acceptable sensitivity threshold for the selected algorithms was determined by a sample of educators who taught similar courses. We predicted that if early performance indicators cannot be relied upon to identify at-risk students, early predictions will exhibit a sensitivity score at the identified subjective threshold or below. This outcome is likely to be present if instructors are more lenient at the start of the semester, thereby making the results of the first assignment and test less representative of the demands that are placed upon students in the courses they teach. Alternatively, the higher a sensitivity score is above the threshold, the more the first assignment and test can be said to represent students' overall performance. The description of the specific algorithms that we selected and the rationale behind their selection are included in the Methods section.

2. Methods

2.1. Sample

The data set of the present study included the grades of 5158 female students that were enrolled in a general education course at a university in the Middle East (KSA). In the set, a random number uniquely identified each student and was associated with her grades on the first test and homework assignment, as well as her final course grade, all measured on a scale from 0 to 100. The data set included students who completed one of the following general education courses, which were carefully chosen to ensure an adequate representation of the general education curriculum that was adopted by the selected university and a minimal overlap of students: Arabic Cultural Studies ($n = 1314$); Introductory Psychology ($n = 847$); Statistics ($n = 612$); Wellness Education ($n = 1390$); and Written Communication ($n = 995$). The courses had been taught both face-to-face (before the pandemic; $n = 2614$) and synchronously online (during the pandemic; $n = 2544$) by the same faculty ($n = 10$). All faculty had at least 5 years of teaching experience in the sampled courses. If a student appeared in more than one course in our dataset, only the grades of one course would be included. Random selection dictated the course for which grades would be entered into the data set.

Both instructional modes relied on Blackboard for the posting of course materials, submission of assignments, and testing. Online classes also relied on Blackboard Collaborate, a platform that created a virtual classroom. In it, video, audio, and chat functions permitted participants to interact with each other in real-time. To ensure proper conduct during the tests that were administered in the virtual classroom, students were required to activate the video and microphone function of Blackboard Collaborate and rely on a lockdown browser application. In both online and face-to-face classes, anti-plagiarism software (i.e., Turnitin) was also used. Important to note is that the general education curriculum at the selected university followed a U.S. model in content and practice. Namely, the curriculum, whose content had been developed by the Texas International Education Consortium (TIEC) and approved by the Saudi Ministry of Education, required that English be used as the primary mode of communication and that the pedagogy adopted for instruction be student-centered. The university was known to rely on a standard-based grading system, according to which the performance of students in a course was defined by their attainment of the learning objectives that defined the curriculum of the course, regardless of how other students performed [34]. Faculty teaching any of the selected courses were required to comply with syllabi that were developed and approved by TIEC, thereby demanding activities (tests and assignments) that met an identical set of learning outcomes. In each of the selected courses, the first assignment and test generally covered at least 1/3 of the learning objectives that were specified by the entire curriculum. Since preliminary analyses yielded null differences for the variable "faculty" within the same course, this variable was not included in the data analyses described in the result section. The present research was conducted under the purview of the Deanship of Research of the selected institution.

2.2. Materials and Procedure

For convenience, students' grades (range: 0–100) were organized into 3 categories, high (H), medium (M), and low (L) performance. H stood for grades in the 90–100 range, M referred to grades in the 80–89 range, and L included grades between 79 and 0. Six different algorithms that are commonly used in educational research [35,36] were chosen to make predictions of students' final grades. These algorithms are K-Nearest Neighbor (KNN); Linear Regression (LR); Multi-Layer Perceptron (MLP); Naïve Bayes (NB); Random Forest (RF); and Support Vector Machine (SVM). Each algorithm is briefly described below. For each algorithm, an article from the pertinent literature describing its mathematical details is mentioned for the interested reader.

K-Nearest Neighbor (KNN) is a non-parametric classifier [37]. That is, it selects a letter grade as the final grade for a given student. KNN merely classifies a target object (e.g., a student's initial grades) by relying on a majority vote of its K neighbors (e.g., students with similar initial grades). The distance between the target object and each of its neighbors is what matters. In our research, to predict final grades (H, M, and L), the number of nearest neighbors used was 21. This number was chosen after fine-tuning the algorithm. The Euclidian distance between any two points was used to find the nearest neighbors.

Logistic regression [38] is a method used to predict a dependent variable (also called outcome variable) for a given set of independent variables (i.e., predictors). It is specifically used to calculate the probability of a categorical dependent variable (e.g., H, M, and L grades). For this study, multiclass logistic regression was used, as predictions needed to fall into more than two categories.

Multi-Layer Perceptron (MLP) is a type of Artificial Neural Network (ANN), which attempts to mimic how the human brain is organized and functions [39]. Namely, an ANN is a network of neurons (i.e., nodes) with the strength of their connections defined by their weight. The MLP that was used in this study was organized into an input layer, three hidden layers, and an output layer. It was a feed-forward network with a back-propagation algorithm for training. Adam optimizer, a stochastic gradient-based optimizer, was applied to optimize the log-loss function. A RELU activation function was used for the hidden layer.

The algorithm named Naïve Bayes (NB) works on the principles of Bayes' theorem, whose aim is to find the conditional probability of a given event [40]. NB assumes that each feature makes an individual and equal contribution to the outcome. For a dataset, NB first creates a frequency table for each attribute, which is then molded into likelihood tables. Lastly, the Bayesian theorem is used for calculating the posterior probability for each class. The class with the highest posterior probability is the prediction that is made by NB. Although in the grade prediction task, the assumption of independence for the variables that are used to make estimates is likely to be violated (e.g., first assignments and tests are unlikely to be independent measures of early performance), we included NB because of its recurrent presence in grade prediction research (see [35]).

The Random Forest (RF) algorithm relies on a tree-like structure consisting of nodes and leaves [41]. Each node corresponds to an input variable (e.g., a student's initial grades), whereas each leaf represents all the different outcomes that could be achieved (e.g., final grades illustrating H, M, and L performance). RF consists of many decision trees that are determined by the user. Each decision tree gives one prediction and the prediction with the highest number of votes becomes the model's final prediction. The main concept behind this algorithm is that many uncorrelated trees working as a community can outperform the working of a single tree. For the present research, we used 100 trees.

The core of a Support Vector Machine (SVM) is an algorithm where each data point is plotted in an n-dimensional space [42]. In such a space, the number of features is n and the value for each feature is a coordinate. Classification is achieved by finding a hyperplane, which is a line that separates groups of data. Finding the hyperplane with the maximum margin for all groups is an optimization problem. SVM is sensitive to the data points that are close to the border of any two groups. In our study, a multi-class SVM algorithm based on Gaussian Radial Basis Function (RBF) kernel was used to predict students' H, M, and L performance, as measured by final grades in the courses in which they were enrolled. The algorithm works by solving a single optimization problem that maximizes the margins between all the designed classes simultaneously.

For measurement purposes, the dataset was partitioned into 70% training and 30% testing. The parameter of sensitivity was used to illustrate the predictive validity of the estimates that were generated by each algorithm, since in grade prediction matters it is particularly costly to miss a student at risk. Sensitivity = [hits/(hits + miss)]. In this equation, a hit (true positive) is the number of students who are correctly identified as receiving a particular letter grade (e.g., the algorithm predicts an L for a particular student, which is indeed what the student has received at the end of the semester). A miss (false negative or type II error) is the number of students who are incorrectly identified as not receiving a particular letter grade (e.g., the algorithm predicts a grade other than an L for a particular student, but the student's grade is an L). We excluded other measures of performance as either irrelevant to the task of predicting at-risk students, such as a correct rejection, or minimally costly, such as a false alarm. In the grade prediction task, a false alarm (false positive or type I error) is the number of students who are incorrectly identified as receiving a particular letter grade (e.g., the algorithm predicts an L for a particular student, but the student has received a grade better than an L). A correct rejection (true negative), instead, is the number of students who are correctly identified as not receiving a particular letter grade (e.g., the algorithm predicts a grade other than an L for a particular student, which is indeed what the student has received).

3. Results

The results of the present study are organized into the following sections: a description of students' performance, and a description of the performance of the chosen algorithms in predicting the final course grades.

3.1. Students' Performance

To obtain a sample of grades that adequately reflected students' key performance levels in the courses in which they were enrolled, and bypassed grade inflation and instructors' grading idiosyncrasies, we classified the final course grades into three performance categories: High (equal to or greater than 90%); Moderate (80–89.99%); and Low (79% or below). The latter category included at-risk students. Table 1 displays the percentage of grades that were assigned to H, M, and L performance by course and instructional mode.

Table 1. Percentage of students with H, M, and L performance by course and instructional mode.

Course	Mode	High Performance	Medium Performance	Low Performance
ACS	Face-to-face	37.9%	37.6%	24.4%
	Online	65.6%	15.9%	18.5%
PSY	Face-to-face	25.1%	40.5%	34.4%
	Online	41.7%	36.4%	21.9%
STA	Face-to-face	16.7%	34.5%	48.8%
	Online	33.1%	50.4%	16.5%
WCO	Face-to-face	30.8%	33.6%	35.6%
	Online	41.9%	31.1%	27.0%
WED	Face-to-face	29.7%	31.7%	38.6%
	Online	57.3%	13.5%	29.2%

Note: ACS = Arabic Cultural Studies; PSY = Psychology; STA = Statistics; WCO = Written Communication; and WED = Wellness Education.

Overall, a greater percentage of students yielded L or M performance in face-to-face classes than in online classes, whereas a greater percentage of students yielded H performance online, χ^2 (2, n = 5158) = 285.34, $p < 0.001$. However, when we examined the frequency of grades H, M, and L in individual courses, a more nuanced pattern emerged about the relationship between performance level and instructional modality, χ^2 (2, n = 612–1390) $\geq$ 14.66, $p \leq 0.001$. In online classes, H was the most frequent score. The only exception was online STA for which M was the most frequent score. In face-to-face WCO and WED classes, there was a somewhat even distribution of L, M, and H scores. In face-to-face STAT classes, L was the most frequent score, whereas in face-to-face ACS classes most scores were either H or M. These patterns of frequency distribution were reflected in students' end-of-course feedback surveys for which STA was judged as a difficult course, but less so online. Although the other classes were reported not to be as difficult as STA, they were also seen as easier when they were online. However, at the start of the semester, STA was judged as more difficult when it was online.

3.2. Algorithms' Performance

The final course grades, labeled as H, M, or L, were used for estimation. We applied the selected algorithms to assess their ability to predict at-risk students in each of the selected performance categories. We relied on sensitivity scores as a measure of the quality of the estimation that was made. Table 2 displays the sensitivity scores of the L-performance category, which indexed at-risk students, as a function of instructional mode and type of course.

Table 2. Prediction of at-risk students (sensitivity scores) by instructional mode and course.

Algorithm	Modality	ACS	PSY	STA	WED	WCO	Mean
KNN	FtF	0.83	0.43	0.67	0.74	0.70	*0.70*
	Online	0.41	0.52	0.71	0.68	0.57	*0.50*
LR	FtF	0.86	0.43	0.67	0.69	0.74	*0.70*
	Online	0.41	0.58	1.00	0.63	0.76	*0.63*
MLP	FtF	0.86	0.55	0.71	0.69	0.70	*0.56*
	Online	0.41	0.52	0.90	0.62	0.68	*0.59*
NB	FtF	0.86	0.43	0.64	0.69	0.79	*0.73*
	Online	0.31	0.58	0.76	0.46	0.76	*0.55*
RF	FtF	0.76	0.43	0.76	0.74	0.70	*0.56*
	Online	0.41	0.48	0.71	0.69	0.57	*0.59*
SVM	FtF	0.75	0.40	0.67	0.69	0.63	*0.48*
	Online	0.25	0.48	1.00	0.61	0.61	*0.52*
Mean		*0.59*	*0.49*	*0.77*	*0.66*	*0.68*	

Note: KNN = K-Nearest Neighbor; LR = Logistic Regression; MLP = Multi-Layer Perceptron; NB = Naive Bayes; RF = Random Forest; SVM = Support Vector Machine. ACS = Arabic Cultural Studies; PSY = Psychology; STA = Statistics; WCO = Written Communication; and WED = Wellness Education.

To determine the extent to which actual instructors in actual classrooms would tolerate misclassifications of at-risk students as likely to do well in their courses, we presented 10 faculty who had experience in at least one of the selected courses with the following scenario: "Imagine that after a couple of weeks into a semester, you are asked to take over a class from another instructor who was abruptly granted a leave of absence for health reasons. A colleague offers you an algorithm that can help you identify students who will not do well in this class. Imagine that at present, unbeknownst to you, 10 students will not do well in this class without some sort of early intervention. What is the maximum number of students out of 10 that the algorithm can fail to correctly identify as being at-risk for you to deem the algorithm unlikely to be useful? Alternatively, what is the minimum number of students out of 10 that the algorithm should correctly identify as being at-risk for you to deem the algorithm useful? Keep in mind that the early identification of at-risk students in real life is a complex task, which may lead educators to misclassify some students as likely to do well (misses or false negatives). Thus, provide a realistic number that would apply to you as representative of your teaching experience". The answers included sensitivity rates ranging from 0.9 to 0.6, leading to an average of 0.7 (the value the arrow points to).

Thus, in our study, we considered a sensitivity score of 0.7 or above as acceptable, which was treated as the threshold of subjective effectiveness.

Was there an algorithm that could be described as superior in both face-to-face (FtF) and online instruction? Figure 1, which plots sensitivity as a function of the type of algorithm and mode of instruction, shows that KNN, LR, and NB consistently performed more effectively face-to-face than online. SVM, MLP, and RF yielded poor performance both online and face-to-face. Interestingly, no algorithm performed adequately (i.e., above the threshold of subjective effectiveness) in online classes.

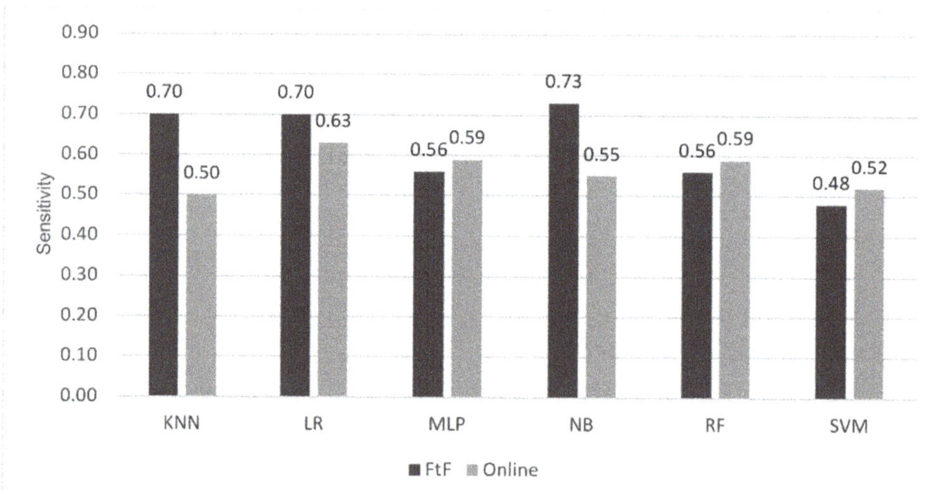

Figure 1. Prediction of at-risk students (sensitivity scores) by instructional modality and algorithm.

Across algorithms, did sensitivity change as a function of the type of course and mode of instruction? Figure 2, which plots sensitivity scores as a function of course and mode of instruction, shows two dissimilar patterns: ACS, WCD, and WED yielded more effective predictions face-to-face than online. Instead, STA yielded more effective predictions online. However, except for ACS and less so for STA, the differences between the modes of instruction were minor.

To better understand the pattern that was yielded by ACS, WCO, WD, STA, and PSY, we examined whether specific algorithms contributed to it. Figures 3–7, which plot the sensitivity scores as a function of the algorithm and mode of instruction in each of these courses, illustrate that the effectiveness of algorithms in making predictions depended on both the type of course and the mode through which the instruction was delivered. Thus, educators would be well advised to consider both variables in selecting algorithms for predicting at-risk students in the classes they teach. For instance, although ACS, WCO, and WED showed greater effectiveness in the prediction task (as measured by the threshold of subjective effectiveness of 0.7) in the face-to-face mode, not all algorithms did so across all courses. To illustrate, ACS was an exception, as all algorithms made effective predictions in face-to-face classes and none in online classes. However, such a clear pattern was not obtained in WCO and WED. Specifically, only KNN and RF made predictions that were above the threshold of effectiveness in face-to-face WCO classes, whereas all algorithms, except SVM, made predictions that were above the threshold of subjective effectiveness in face-to-face WED classes. Furthermore, only LR and NB made effective predictions in online WED classes. In contrast to the checkered pattern of WCO and WED, STAT exhibited a pattern that was largely the opposite of ACS. Namely, all algorithms yielded predictions that were above the threshold of subjective effectiveness online, whereas only MLP and RF made predictions that were above said threshold face-to-face. Irrespective of whether

PSY was delivered face-to-face or online, the predictions of all the algorithms were poor, all well below the threshold of subjective effectiveness.

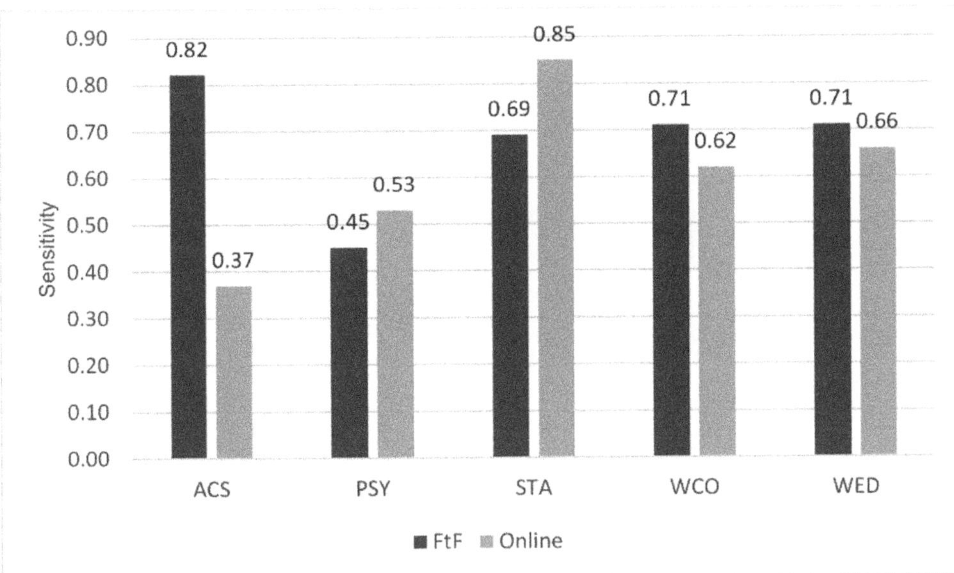

Figure 2. Prediction of at-risk students (sensitivity scores) by instructional modality and course type.

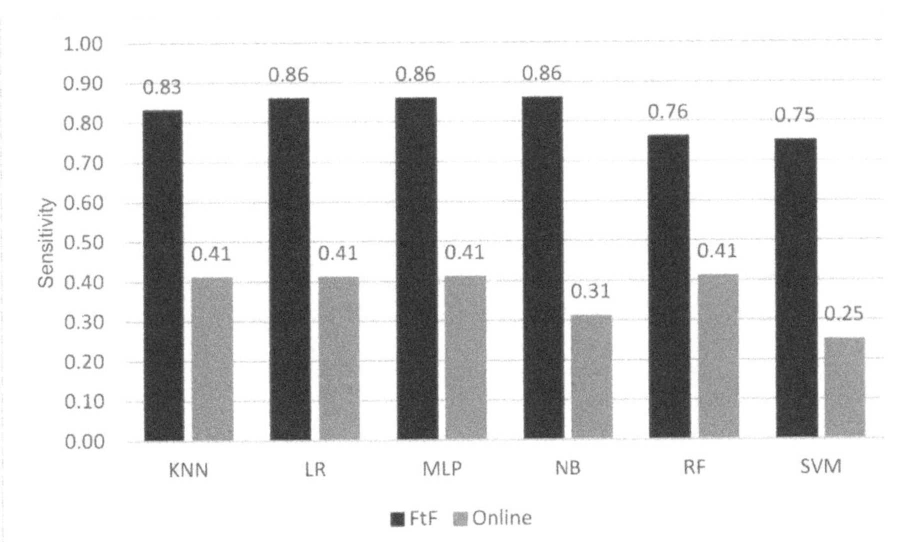

Figure 3. ACS course: prediction of at-risk students (sensitivity scores) by instructional mode.

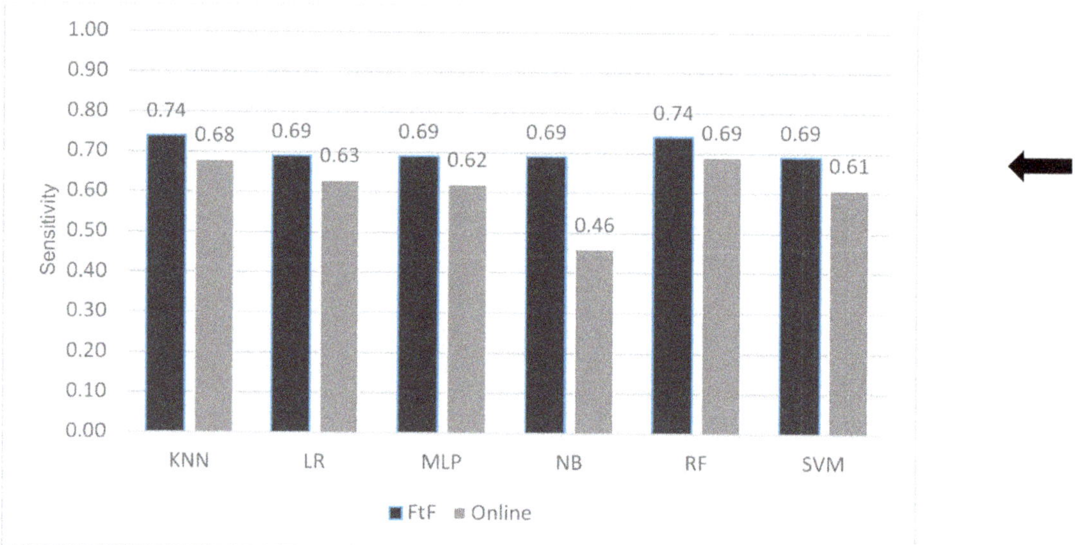

Figure 4. WCO course: prediction of at-risk students (sensitivity scores) by instructional mode.

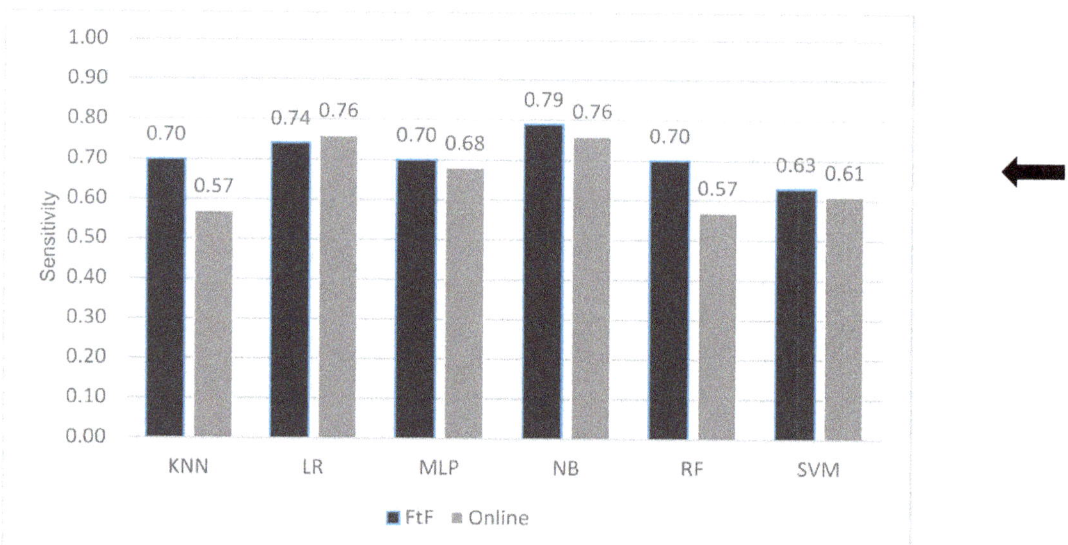

Figure 5. WED course: prediction of at-risk students (sensitivity scores) by instructional mode.

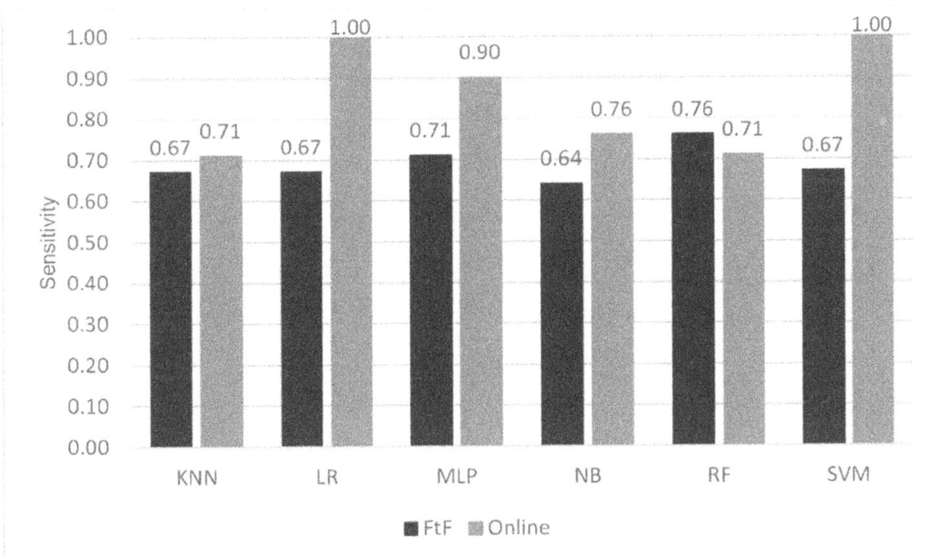

Figure 6. STA course: prediction of at-risk students (sensitivity scores) by instructional mode.

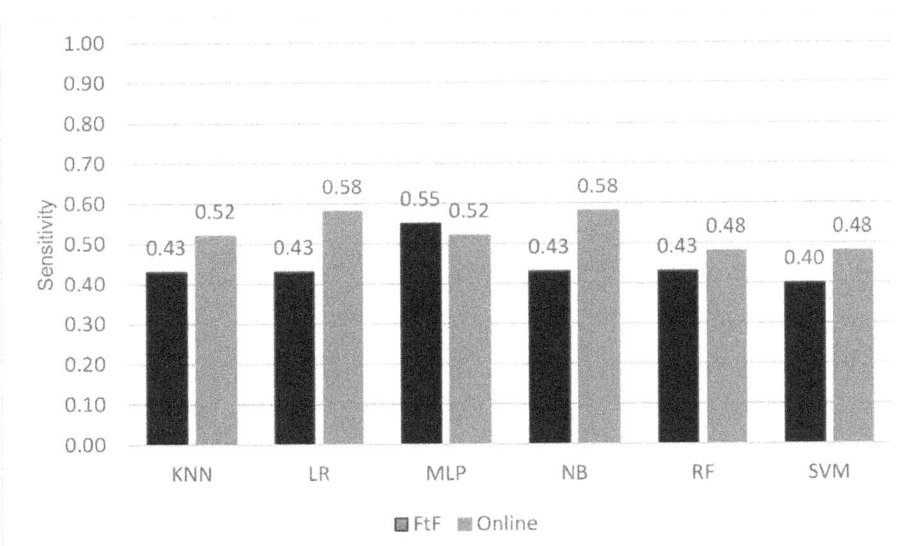

Figure 7. PSY course: prediction of at-risk students (sensitivity scores) by instructional mode.

4. Discussion

The findings of the present research can be summarized in two points. First, if the overall predictive validity of machine learning algorithms is of interest, our evidence suggests that they tend to yield a higher predictive validity (as indexed by sensitivity scores) in face-to-face than in online classes. Is this higher predictive validity due to changes in the way that students approach the curriculum of a course? Is it due to changes in the way that educators deliver content and/or assess learning? We explicitly selected courses that were taught by the same experienced educators and followed the same curriculum requirements online and face-to-face. Educators' self-reports did not illustrate that the standards of early formative assessment (i.e., the first assignment and test) were changed between face-to-face and online courses. Yet, the lower predictive validity of early formative assessment online indicated that these measures were less useful to both educators and students when embedded in online classes. Inquiries through focus groups and informal exchanges with both students and faculty did not clarify this puzzle, mostly leading to the acknowledgment by both parties that the adaptation to online courses was more challenging for students than the adaptation to face-to-face courses. The following themes were frequently mentioned by students and corroborated by faculty: more time devoted to understanding how to navigate materials posted online (Blackboard) and how to use them; feelings of isolation and perceived distance from the instructor; and fewer opportunities for informal interactions with the instructor and classmates. Instructors reported more initial inquiries regarding course contents and requirements online than face-to-face, often noting that students who were accustomed to on-campus classes required more time to navigate and feel comfortable with the online mode. Thus, qualitative evidence seemed to point to educators who, aware of students' adaptation challenges, might have become tacitly more lenient when assessing performance, even though they purported not to have changed their standards of assessment. However, this pattern may offer a misleading picture since the variables course type and mode of instruction interacted.

Second, algorithms such as KNN and RF were consistently better predictors of at-risk students in face-to-face courses in the humanities and social sciences, such as ACS, WCO, and WED, whereas they were better predictors online when the course covered mathematical knowledge and skills (STA). The flexibility of KNN and RF may be particularly useful if the SARS-CoV-2 virus, which causes the COVID-19 disease, persists in affecting people's lives, thereby forcing university administrators to continue relying on the online mode or adopt hybrid modes for courses that are offered at their institutions.

As for the differences in algorithms' predictive validity between online and face-to-face, the interaction of course type and mode of instruction was not entirely clarified by the self-reports that were produced by students and educators; however, a consistent theme emerged. At the very beginning of the semester, some classes were reported by students as likely to be more difficult online (e.g., STA), whereas others (e.g., ACS and WED) were seen as potentially easier online. These biases could be thought of as capable of shaping students' behavior inside and outside the virtual classroom. For instance, consider that the anxiety that was experienced by female students towards Math courses increased considerably when the courses were delivered online. Increased anxiety might have led students to pay more attention in class and devote more time to class activities across the entire semester, thereby potentially leading to three interrelated outcomes: (a) enhancing their performance across the entire semester; (b) rendering a view, at the end of the semester, of the online STAT course as easier to manage than expected; and (c) making even initial formative assessment measures more likely to reflect overall course performance online than face-to-face. Instead, the initial expectation of easier online courses, coupled with educators' purported leniency that was driven by the opposite expectation, might have had a quite different impact. Namely, expectations might have unnecessarily lessened students' effort towards class activities, and relaxed educators' grading standards to ease students' adaptation to online courses, thereby making initial formative assessment less likely to reflect overall course performance online.

Our research suggests that particular machine learning algorithms can be used to make informed predictions regarding students' performance attainment, but the predictive validity of each algorithm has to be first assessed as a function of two important variables: course type and instructional mode. Our study adds to the growing body of grade prediction studies that rely on machine learning algorithms [8,10,43–45] by pointing to the relevance of such variables to interventions that are intended to foster academic success in an understudied student population. In our research, the latter is represented by young women of college age from a society that has only recently implemented and enforced gender equity guidelines. Our research also contributes to the extant literature by relying on a subjective criterion of effectiveness that is produced by faculty with direct experience in teaching the courses that are included in our sample. Too often, studies examining the predictive validity of different algorithms have focused on relative comparisons but have failed to give readers an idea of how to conceptualize a desirable outcome for the actual situations/conditions they face.

5. Conclusions

We believe that research in educational settings should be motivated by the intention to improve participants' existing conditions [46–48]. As such, we subscribe to the main tenets of action research according to which the aim of a research project is practical. Namely, it is to identify a problem, condition, or situation; propose and implement a solution that is intended to bring improvement to the very people who participate in the research; assess the effectiveness of the solution; and either (a) start from the beginning if the outcome is unsatisfactory, or (b) broaden the reach of the purported solution if the outcome is within the expected parameters [49]. Thus, our goal is to rely on machine learning algorithms as feedback tools for students to assess their learning, and for faculty to assess their teaching. If improvement is needed, such tools can also inform the nature of the changes to be implemented in a university's curriculum and instruction.

To this end, we recognize that too often, algorithms and related data mining techniques are mainly accessible to educators who possess a background in computer science, and, more precisely, in artificial intelligence [15]. Educators with diverse backgrounds are frequently unable to access data mining techniques, thereby preventing their application to a much wider educational field. Our goal at the selected institution is to offer faculty with backgrounds outside computer science access to such techniques via workshops and mentorship efforts. Specifically, we plan to develop an easy-to-use early-warning system that relies on KNN to identify students at risk in particular courses, depending on the mode of instruction that is used to deliver their content. Currently, we have data supporting the effectiveness of an early-warning system using KNN in face-to-face ACS, WCD, and WED courses, as well as in online STA. The choice of KNN is based on its yielding the best relative performance for different modes of instruction in the courses that are selected for our study. However, the dismal performance of KNN, along with that of all the other algorithms in PSY, suggests that the early formative assessment measures of this course need to be examined closely to determine whether they indeed fit the learning outcomes of the course. A similar examination of the early formative assessment measures of the other courses that were selected for the present examination may also be warranted to improve their predictive validity. Of course, the predictive validity of KNN for at-risk students in general education courses that were not included in the present research, especially those involving natural sciences and math, will also need to be examined.

Existing early warning systems for the identification of at-risk students may be too general and thus become poor predictors of academic difficulties in particular courses. They also often rely on norm-referenced scores, which take into account how other students perform, instead of criterion-referenced scores, which consider how students perform relative to the learning outcomes that are set by the curriculum of different courses (as illustrated by summative assessment measures). Thus, an algorithm, such as KNN, that uses criterion-based scores to predict at-risk students may be seen as particularly helpful by educators. Indeed, it has the ability not only to effortlessly identify students who experience difficulties, but also to inform revisions of the curriculum and assessment protocol to ensure adequate coverage of the learning outcomes of a course. The attainment of such learning outcomes is particularly critical to general education courses, which lay the foundations for academic success in major-specific courses [50–53].

Two important lessons that are learned from our investigation and the pertinent extant literature are reminders of the limitations of our study. First, machine learning solutions for grade prediction, albeit most useful early in a course, may require adaptation to the particular student population and the academic environment that an educator or administrator has selected for assessment and intervention. Namely, each educational setting may have features that are common to other educational settings (e.g., reliance on the synchronous online mode), thereby allowing a technical solution to be transferred, and features that may be unique to it. Unique features introduce uncertainties by questioning the transfer of the solution to other settings. For instance, the students selected for the present investigation are exposed to undergraduate courses guided by the principles of student-centered instruction intended to promote deep learning at the expense of rote learning. It is unclear whether our findings generalize to students who are exposed to a different type of instructional principles. Thus, uncertainties, defined by their statistical properties (e.g., parametric or non-parametric factors), and origin (e.g., internal to the learner or environmental, etc.), are likely to depend on the student populations that educators select for assessment and intervention, and the specific factors they deem relevant. Second, a lesson that is learned from the extant literature is that, in a vast array of problem domains, computational models are relentlessly evolving and are often so complex that educators without knowledge in computer science are left out. Innovative algorithmic solutions may be applied to the grade prediction needs of an institution and its faculty [54,55] if the unique properties of the grade-prediction conundrum in any given setting (including students and academic environment) are integrated and computing resources for training and inference purposes are made available e.g., [56]. However, such models also need to become more transparent and user-friendly for non-experts to ensure broad and reliable adoption [15]. Our paper is a modest call for action that specifically targets non-expert educators and administrators to approach the field of machine learning due to its potential benefits to the quality of learning and teaching.

Author Contributions: All authors M.A.E.P., E.N., M.N., I.D., H.A. and M.A., contributed equally to the research, including conceptualization, methodology, formal analysis, data curation, writing—original draft preparation, writing—review and editing, and project administration. All authors have read and agreed to the published version of the manuscript.

Funding: This research received no external funding.

Data Availability Statement: Data are available upon request.

Acknowledgments: We thank the members of the Cognitive Science Cluster for their feedback.

Conflicts of Interest: The authors have no conflict of interest to declare that are relevant to the content of this article.

References

1. Lei, S.A.; Lei, S.Y. General education curricula affecting satisfaction and retention of undergraduate students: A review of the literature. *Education* **2019**, *139*, 197–202.
2. Warner, D.B.; Koeppel, K. General education requirements: A comparative analysis. *J. Gen. Educ.* **2009**, *58*, 241–258. [CrossRef]
3. Millea, M.; Wills, R.; Elder, A.; Molina, D. What matters in college student success? Determinants of college retention and graduation rates. *Education* **2018**, *138*, 309–322.
4. AbdelSalam, H.M.; Pilotti, M.A.E.; El-Moussa, O.J. Sustainable math education of female students during a pandemic: Online versus face-to-face instruction. *Sustainability* **2021**, *13*, 12248. [CrossRef]
5. Daniel, A.M. Identification of skill-appropriate courses to improve retention of at-risk college freshmen. *J. Coll. Stud. Retent. Res. Theory Pract.* **2020**, *24*, 126–143. [CrossRef]
6. Hannafin, K.M. Technology and the support of at-risk students. *J. Gen. Educ.* **1991**, *40*, 163–179. Available online: https://www.jstor.org/stable/27797135 (accessed on 10 December 2021).
7. Alyahyan, E.; Düştegör, D. Predicting academic success in higher education: Literature review and best practices. *Int. J. Educ. Technol. High. Educ.* **2020**, *17*, 3. [CrossRef]
8. Bujang, S.D.A.; Selamat, A.; Ibrahim, R.; Krejcar, O.; Herrera-Viedma, E.; Fujita, H.; Ghani, N.A.M. Multiclass prediction model for student grade prediction using machine learning. *IEEE Access* **2021**, *9*, 95608–95621. [CrossRef]
9. Rastrollo-Guerrero, J.L.; Gomez-Pulido, J.A.; Durán-Domínguez, A. Analyzing and predicting students' performance by means of machine learning: A review. *Appl. Sci.* **2020**, *10*, 1042. [CrossRef]
10. Huang, S.; Fang, N. Predicting student academic performance in an engineering dynamics course: A comparison of four types of predictive mathematical models. *Comput. Educ.* **2013**, *61*, 133–145. [CrossRef]

11. Bydžovská, H. Are collaborative filtering methods suitable for student performance prediction? In *Progress in Artificial Intelligence: Lecture Notes in Computer Science*; Pereira, F., Machado, P., Costa, E., Cardoso, A., Eds.; Springer: New York, NY, USA, 2015; Volume 9273, pp. 230–425. [CrossRef]
12. Fahd, K.; Venkatraman, S.; Miah, S.J.; Ahmed, K. Application of machine learning in higher education to assess student academic performance, at-risk, and attrition: A meta-analysis of literature. *Educ. Inf. Technol.* **2021**, *27*, 3743–3775. [CrossRef]
13. Albreiki, B.; Zaki, N.; Alashwal, H. A systematic literature review of student' performance prediction using machine learning techniques. *Educ. Sci.* **2021**, *11*, 552. [CrossRef]
14. Kanetaki, Z.; Stergiou, C.; Bekas, G.; Jacques, S.; Troussas, C.; Sgouropoulou, C.; Ouahabi, A. Grade Prediction Modeling in Hybrid Learning Environments for Sustainable Engineering Education. *Sustainability* **2022**, *14*, 5205. [CrossRef]
15. Lee, K.M.; Yoo, J.; Kim, S.W.; Lee, J.H.; Hong, J. Autonomic machine learning platform. *Int. J. Inf. Manag.* **2019**, *49*, 491–501. [CrossRef]
16. Gilovich, T.; Griffin, D.; Kahneman, D. *Heuristics and Biases: The Psychology of Intuitive Judgment*; Cambridge University Press: Cambridge, UK, 2002.
17. Meegan, D.V. Zero-sum bias: Perceived competition despite unlimited resources. *Front. Psychol.* **2010**, *1*, 191. [CrossRef]
18. Lu, O.H.T.; Huang, A.Y.Q.; Huang, J.C.; Lin, A.J.Q.; Ogata, H.; Yang, S.J.H. Applying learning analytics for the early prediction of students' academic performance in blended learning. *Educ. Technol. Soc.* **2018**, *21*, 220–232.
19. Quinn, R.J.; Gray, G. Prediction of student academic performance using Moodle data from a Further Education setting. *Ir. J. Technol. Enhanc. Learn.* **2020**, *5*, 1–19. [CrossRef]
20. Gonzalez, T.; de la Rubia, M.A.; Hincz, K.P.; Comas-Lopez, M.; Subirats, L.; Fort, S.; Sacha, G.M. Influence of COVID-19 confinement on students' performance in higher education. *PLoS ONE* **2020**, *15*, e0239490. [CrossRef]
21. Iglesias-Pradas, S.; Hernández-García, Á.; Chaparro-Peláez, J.; Prieto, J.L. Emergency remote teaching and students' academic performance in higher education during the COVID-19 pandemic: A case study. *Comput. Hum. Behav.* **2021**, *119*, 106713. [CrossRef]
22. Engelhardt, B.; Johnson, M.; Meder, M.E. Learning in the time of Covid-19: Some preliminary findings. *Int. Rev. Econ. Educ.* **2021**, *37*, 100215. [CrossRef]
23. El Said, G.R. How Did the COVID-19 Pandemic affect higher education learning experience? An empirical investigation of learners' academic performance at a university in a developing country. *Adv. Hum. Comput. Interact.* **2021**, *2021*, 6649524. [CrossRef]
24. Elzainy, A.; El Sadik, A.; Al Abdulmonem, W. Experience of e-learning and online assessment during the COVID-19 pandemic at the College of Medicine, Qassim University. *J. Taibah Univ. Med. Sci.* **2020**, *15*, 456–462. [CrossRef] [PubMed]
25. Zheng, M.; Bender, D.; Lyon, C. Online learning during COVID-19 produced equivalent or better student course performance as compared with pre-pandemic: Empirical evidence from a school-wide comparative study. *BMC Med. Educ.* **2021**, *21*, 495. [CrossRef] [PubMed]
26. Foo, C.C.; Cheung, B.; Chu, K.M. A comparative study regarding distance learning and the conventional face-to-face approach conducted problem-based learning tutorial during the COVID-19 pandemic. *BMC Med. Educ.* **2021**, *21*, 141. [CrossRef]
27. Hussain, A.; Chau, J.; Bang, H.; Meyer, L.; Islam, M. Readiness, reception, and performance of students in a communications course delivered amid the pandemic. *Am. J. Pharm. Educ.* **2021**, *85*, 8617. [CrossRef]
28. Boston, C. The concept of formative assessment. *Pract. Assess. Res. Eval.* **2002**, *8*, 9. [CrossRef]
29. Ahmed, W. Women empowerment in Saudi Arabia: An analysis from education policy perspective. *Middle East Int. J. Soc. Sci.* **2020**, *2*, 93–98.
30. Barry, A. Gender differences in academic achievement in Saudi Arabia: A wake-up call to educational leaders. *Int. J. Educ. Policy Leadersh.* **2019**, *15*, 17. [CrossRef]
31. Pilotti, M.A.E. What lies beneath sustainable education? Predicting and tackling gender differences in STEM academic success. *Sustainability* **2021**, *13*, 1671. [CrossRef]
32. Syed, J.; Ali, F.; Hennekam, S. Gender equality in employment in Saudi Arabia: A relational perspective. *Career Dev. Int.* **2018**, *23*, 163–177. [CrossRef]
33. Hvidt, M. Economic diversification and job creation in the Arab Gulf countries: Applying a value chain perspective. In *When Can Oil Economies be Deemed Sustainable?* Luciani, G., Moerenhout, T., Eds.; Palgrave Macmillan: London, UK, 2021; pp. 281–300.
34. Sadler, D.R. Interpretations of criteria-based assessment and grading in higher education. *Assess. Eval. High. Educ.* **2005**, *30*, 175–194. [CrossRef]
35. Marbouti, F.; Diefes-Dux, H.A.; Madhavan, K. Models for early prediction of at-risk students in a course using standards-based grading. *Comput. Educ.* **2016**, *103*, 1–15. [CrossRef]
36. Romero, C.; Ventura, S. Educational data mining: A review of the state of the art. *IEEE Trans. Syst. Man Cybern. Part C Appl. Rev.* **2010**, *40*, 601–618. [CrossRef]
37. Zhang, Z. Introduction to machine learning: K-nearest neighbors. *Ann. Transl. Med.* **2016**, *4*, 136. [CrossRef] [PubMed]
38. Kleinbaum, D.G.; Dietz, K.; Gail, M.; Klein, M.; Klein, M. *Logistic Regression*; Springer: New York, NY, USA, 2002.
39. Kruse, R.; Mostaghim, S.; Borgelt, C.; Braune, C.; Steinbrecher, M. (Eds.) Multi-layer perceptrons. In *Computational Intelligence*; Springer: New York, NY, USA, 2022; pp. 53–124.
40. Xu, S. Bayesian Naïve Bayes classifiers to text classification. *J. Inf. Sci.* **2018**, *44*, 48–59. [CrossRef]

41. Boulesteix, A.L.; Janitza, S.; Kruppa, J.; König, I.R. Overview of random forest methodology and practical guidance with emphasis on computational biology and bioinformatics. *Wiley Interdiscip. Rev. Data Min. Knowl. Discov.* **2012**, *2*, 493–507. [CrossRef]
42. Bhavsar, H.; Panchal, M.H. A review on support vector machine for data classification. *Int. J. Adv. Res. Comput. Eng. Technol.* **2012**, *1*, 185–189.
43. Pereira, F.D.; Oliveira, E.H.; Fernandes, D.; Cristea, A. Early performance prediction for CS1 course students using a combination of machine learning and an evolutionary algorithm. In Proceedings of the 2019 IEEE 19th International Conference on Advanced Learning Technologies (ICALT), Maceio, Brazil, 15–18 July 2019; Volume 2161, pp. 183–184. [CrossRef]
44. Friedman, J.H.; Bentley, J.L.; Finkel, R.A. An algorithm for finding best matches in logarithmic expected time. *ACM Trans. Math. Softw. TOMS* **1977**, *3*, 209e226. [CrossRef]
45. Hung, J.L.; Wang, M.C.; Wang, S.; Abdelrasoul, M.; Li, Y.; He, W. Identifying at-risk students for early interventions—A time-series clustering approach. *IEEE Trans. Emerg. Top. Comput.* **2015**, *5*, 45–55. [CrossRef]
46. Sáez Bondía, M.J.; Cortés Gracia, A.L. Action research in education: A set of case studies? *Educ. Action Res.* 2021; *latest articles*. [CrossRef]
47. Ali, A.D. Implementing action research in EFL/ESL classrooms: A systematic review of literature 2010–2019. *Syst. Pract. Action Res.* **2020**, *33*, 341–362. [CrossRef]
48. Lufungulo, E.S.; Mambwe, R.; Kalinde, B. The meaning and role of action research in education. *Multidiscip. J. Lang. Soc. Sci. Educ.* **2021**, *4*, 115–128.
49. Jacobs, S.D. A history and analysis of the evolution of action and participatory action research. *Can. J. Action Res.* **2018**, *19*, 34–52. [CrossRef]
50. Benander, R.; Lightner, R. Promoting transfer of learning: Connecting general education courses. *J. Gen. Educ.* **2005**, *54*, 199–208. [CrossRef]
51. Aloi, S.L.; Gardner, W.S.; Lusher, A.L. A framework for assessing general education outcomes within the majors. *J. Gen. Educ.* **2003**, *52*, 237–252. [CrossRef]
52. Jeske, J. Nurturing rich general education courses. *J. Gen. Educ.* **2002**, *51*, 103–114. [CrossRef]
53. Landon-Hays, M.; Peterson-Ahmad, M.B.; Frazier, A.D. Learning to Teach: How a Simulated Learning Environment Can Connect Theory to Practice in General and Special Education Educator Preparation Programs. *Educ. Sci.* **2020**, *10*, 184. [CrossRef]
54. Tutsoy, O.; Polat, A.; Çolak, Ş.; Balikci, K. Development of a multi-dimensional parametric model with non-pharmacological policies for predicting the COVID-19 pandemic casualties. *IEEE Access* **2020**, *8*, 225272–225283. [CrossRef]
55. Tutsoy, O. COVID-19 epidemic and opening of the schools: Artificial intelligence-based long-term adaptive policy making to control the pandemic diseases. *IEEE Access* **2021**, *9*, 68461–68471. [CrossRef]
56. Kang, S.J.; Lee, S.Y.; Lee, K.M. Performance comparison of OpenMP, MPI, and MapReduce in practical problems. *Adv. Multimed.* **2015**, *2015*, 7. [CrossRef]

electronics

MDPI

Article

ICT Usage for Cross-Curricular Connections in Music and Visual Arts during Emergency Remote Teaching in Slovenia

Eda Birsa [1], Matjaž Kljun [2] and Barbara Kopačin [1,*]

[1] Faculty of Education, University of Primorska, 6000 Koper, Slovenia; eda.birsa@pef.upr.si or eda.birsa@upr.si
[2] Faculty of Mathematics, Natural Sciences and Information Technologies, University of Primorska, 6000 Koper, Slovenia; matjaz.kljun@upr.si
* Correspondence: barbara.kopacin@upr.si; Tel.: +386-05-6631-269

Abstract: Due to the COVID-19 pandemic, the entire process of teaching and learning moved online. This forced teachers and pupils to heavily rely on information and communications technology (ICT) and make adjustments to the new mode of teaching and learning in educational institutions. We conducted a qualitative case study by interviewing 24 teachers from Slovene primary schools focusing on the implementation of cross-curricular connections in music and visual arts content with the support of ICT during the period of emergency remote teaching. We found that when planning and implementing the cross-curricular learning process, teachers insufficiently took advantage of possibilities offered by modern ICT. The manner of implementing cross-curricular connections showed uncertainties in terms of understanding their specifics, resulting in the inefficient transfer of concepts taught, the results of which were seen in pupils' work. This might additionally show the negative influence of parental supervision on the creative thinking and expression of pupils. The present study emphasizes the lack of ICT competences on the part of all participants in the educational process. Our findings show the need to educate teachers by eliminating the uncertainties related to the implementation of distant cross-curricular connections while meaningfully applying ICT adapted to pupils' competences.

Keywords: cross-curricular integration; music art; visual art; remote teaching; ICT; interdisciplinary assignments; elementary education; learning process; music subjects; visual art subjects

Citation: Birsa, E.; Kljun, M.; Kopačin, B. ICT Usage for Cross-Curricular Connections in Music and Visual Arts during Emergency Remote Teaching in Slovenia. *Electronics* **2022**, *11*, 2090. https://doi.org/10.3390/electronics11132090

Academic Editor: Cheng Siong Chin

Received: 6 June 2022
Accepted: 27 June 2022
Published: 3 July 2022

Publisher's Note: MDPI stays neutral with regard to jurisdictional claims in published maps and institutional affiliations.

1. Introduction

In Slovenia, as in much of the world, the entire process of teaching and learning moved online in spring 2020 due to COVID-19 pandemic. This forced teachers and pupils to heavily rely on information and communications technology (ICT) and make immediate adjustments to the new mode of teaching and learning. When incorporating ICT in the planning and implementing of the remote teaching–learning process in the first and second period of education (pupils aged six to eleven), it is necessary to think about how ICT can contribute to integrated consideration of course content, and consequently to the efficient acquisition of knowledge. However, this is not as straightforward as might be thought, as the ICT equipment in schools and that available to individuals in their homes are often very different, as are the digital competences of all involved.

In order to ensure pupils' progress in creative subjects such as music and visual arts during remote teaching and distance learning, it is even more necessary to incorporate modern teaching strategies that promote holistic learning of content with the support of suitable ICT. Modern teaching approaches, including cross-curricular connections and ICT, have already been used by teachers in the traditional implementation of the teaching–learning process. However, several ambiguities are known to exist, and inefficient connections can often be implemented in practice which do not contribute to holistic knowledge acquisition [1]. As these findings are from 2011, and focused on year five teachers as well as on visual arts only, we wanted to expand the study by including all teachers from the first and second period of education and then to investigate whether remote teaching and increased use of ICT led to any changes in this respect. This is particularly important, as the use of modern ICT can contribute to the sustainable development of pupils' individual and technological literacy [2–4]. To this end, we conducted a study with 24 teachers who teach both music and visual arts courses to pupils in key stage 1 and 2 in order to investigate cross-curricular connections in music and visual arts content with the support of ICT during the period of emergency remote teaching.

The paper is structured as follows. The next section presents related works. Section 3 covers the materials and methods used. Section 4 presents the results, and the final section focuses on discussion.

2. Related Work

In the related work section, we first look at ICT in education, ICT in remote teaching, and ICT in cross-curricular connections, with a focus on Music and Visual Arts courses in the Slovenian education system. The section concludes with a discussion of the gaps in knowledge, the aims of this study, and the research questions it addresses.

2.1. ICT in Music and Visual Arts Education in Slovenia

ICT incorporates hardware and software that enable the functioning of video, audio, and data transfers over computer networks. All of this is possible with desktop and laptop computers, tablets, smartphones, and a plethora of applications. In the field of education, ICT can be used either in the formation of the teaching–learning process or as technical aid to deliver the content [2]. The application largely depends on the technological equipment of the environment where classes take place and the ICT competences of teachers, parents, and pupils. ICT competences determine an individual's skills regarding the use of computers and associated software, communication tools, and the internet, involving skills for incorporating technology into everyday processes, designing information, and information retrieval and evaluation [2,3,5].

Toselli [6] emphasises that ICT should never be forced, and should be gradually applied only when pupils know how to use it. Additionally, Toselli mentions synchronous (real time) technologies such as video conferencing, instant messaging, and telephone as well as asynchronous methods and forms of remote teaching and learning such as social networking services, e-mail, shared cloud documents (Google Drive https://www.google.com/drive/ (accessed on 17 November 2021) or Office 365 https://www.office.com/ (accessed on 17 November 2021)), audio and video recordings, video channels, and the web. Teachers can choose asynchronous tools to forward teaching material prepared in advance, which pupils then cover on their own, or can use synchronous video conferencing to provide live classes [7,8].

The national curriculum for music art contains recommendations for gradual and planned introduction of ICT, including television, radio, video projector, CD, MP3 and DVD players, computers, and interactive tables [9]. It advises planning the use of ICT during the preparation of the learning content and that ICT must support learning objectives while taking pupils' experiences with ICT into consideration. It states that ICT should be used as an aid for the presentation of different music-related information, while the teacher's role should not be diminished. With the transition into the second period of education it is assumed that pupils have gained experiences and understanding of elementary music theory and expression, which should be considered while embedding ICT in music activities. The advantage of using ICT is seen especially in the "transfer, storage and organization of music contents and audio records, research of sound and sound polyphony, in the transfer of acoustic performances into musical recordings, search of information and in setting up new forms of socialisation and (music) communication among users, and between users and source of (music) information" [9].

The national curriculum for visual art specifies methods of using ICT in the planning and implementation of visual art assignments. Examples include drawing in graphics editors, creating photo montages, photography, animations, or animated films, graphics in combined techniques, etc. It is up to teachers to select appropriate computer programs for pupils to efficiently and creatively take advantage of the opportunities they offer. Even software that is easy to master can support creative artistic expression and stimulate general creativity while offering the possibility of experimenting, as seen in Figure 1 [5,10]. Recently, tablet computers with stylus pens have allowed for more genuine graphic expression, while teachers can use them to motivate pupils as well as for demonstration purposes. Digital cameras are another technology that has become ubiquitous, and can be used in combination with computers to support pupils' creativity [5].

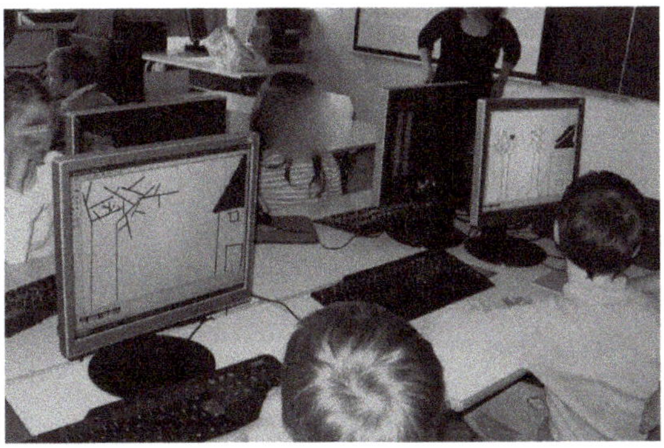

Figure 1. Possibilities of using ICT in the process of learning visual art in class for drawing.

2.2. *ICT and Emergency Remote Teaching in Slovenia*

Here, it is appropriate to differentiate distance learning and emergency remote teaching [11]. Distance learning is planned well in advance for when pupils and teachers are separated by time and space. Emergency remote teaching replaces traditional forms of teaching in a time of crisis, and focuses on remote delivery of content with only space being a separator. At the beginning of the COVID-19 pandemic in March 2020, emergency remote teaching became the reality, and teachers' ICT competences suddenly became very important. Both in Slovenia and elsewhere in the world, as Blahušiaková and others [12] mention, teachers and students have encountered completely new challenges. All primary school teachers and pupils in Slovenia have access to Arnes http://www.arnes.si/zavod-arnes/ (accessed on 17 November 2021) (Slovene Academic Research Network) and eAsistent https://www.easistent.com/ (accessed on 17 November 2021) services. Arnes provides the video conferencing systems Zoom https://zoom.us/ (accessed on 17 November 2021) and MS Teams https://www.microsoft.com/sl-si/microsoft-365/microsoft-teams/group-chat-software (accessed on 17 November 2021) for remote teaching as well as the Moodle Learning Management System https://moodle.org/ (accessed on 17 November 2021) (often referred to as an e-classroom) for designing and uploading teaching materials, quizzes, crosswords, and essay questions [2,13].

When planning and implementing a remote teaching and/or distance learning process, teachers must think of an efficient application of modern ICT to support various creative activities and assess what additional value ICT can provide to music and visual art subjects. For example, through creative activities, ICT can support integrated understanding of artistic content and optimal knowledge transfer involving various senses [2]. In addition, Toselli [6] argues that teachers in remote teaching have never been closer to their pupils, because they can reach each pupil via headphones or speakers, meaning that pupils have a sense of being directly addressed. As a consequence, the individualisation of learning and teaching comes even more to the forefront.

Until 2020, digital competences of pupils in the first and second period of education covered only basic skills. The use of word processors (e.g., MS Word https://www.microsoft.com/sl-si/microsoft-365/p/word/ (accessed on 17 November 2021)) and email were usually not needed in the teaching–learning process, despite the fact that these generations are called 'digital generations' [14]. The introduction of emergency remote teaching, however, required pupils and their parents/guardians to master various new ICT skills and competences [2,11,14,15].

2.3. ICT and Cross-Curricular Connections of Content in Music and Visual Arts

In the first and second period (key stages 1 and 2) of education in Slovenia, music and visual arts are taught by the same teacher, as the transfer of knowledge and skills can be achieved with more comprehensive consideration of concepts from both subjects using cross-curricular connections. This has been shown to have beneficial effects on the cognitive aspects of pupils [16–19]. As in individual subjects, teachers have full autonomy in the implementation of the cross-curricular teaching–learning process with music and visual art content, including choosing methods and forms of work and in planning activities with the available ICT. Cross-curricular connections need to be planned in a way that pupils are able to connect concepts already learned with new ones learned during the class [19,20].

Special attention must be paid to the planning of clear and concrete goals that can indicate the knowledge that pupils are to acquire in each particular art subject, including particular concepts of music and visual arts, concepts related to creativity, skills developed in the use of various art materials, instruments and aids, and increased aesthetic, emotional, and social qualities. Moreover, teachers must not overlook the aspect of the learning sequence, and must carefully consider the appropriateness of selected content they are planning to connect as well as appropriate learning and teaching strategies with modern ICT in order to effectively transfer knowledge at a distance. It is necessary to envisage the learning sequence of content through individual ICT, as this enables students to more easily and comprehensively understand novel musical and visual art concepts [21].

2.4. Purpose of the Study

Teaching methods in music and visual arts are constantly changing with offerings from novel ICT. Due to the COVID-19 pandemic in 2020, these changes were much faster than they would have been otherwise. Teachers were forced to replace even established teaching methods that did not rely on ICT with ones using ICT in order to support remote teaching and distance learning. In such a situation, cross-curricular integration represents a way of facilitating the cognitive process through the effective delivery of learning content, and can enable holistic understanding of various subjects.

When planning a cross-curricular teaching–learning process for music and visual arts, it is necessary to carefully choose the content in order to achieve the planned goals of both subjects. In traditionally-conducted teaching–learning, processes inefficient implementation of cross-curricular connections is the norm [1,22]. Irregularities and lack of knowledge of the specifics of planning and implementing connections are often issues when a musical or artistic activity is used only as a diversification of another subject instead of establishing mental strategies and knowledge transfer with equal consideration of the concepts or content of the exposed subject areas. Our goal was therefore to determine whether the 'new digital reality' encouraged teachers to consider learning content holistically and how they implemented cross-curricular connections remotely with the support of ICT. In particular, we wanted to investigate which ICT teachers used for planning and implementing their lectures and which ICT they planned for pupils to use in order to complete their assignments. We wanted to uncover whether the ICT intended to be used in the implementation of the lectures and ICT in completing the assignments was indeed used as planned or whether there were discrepancies between plans and implementation where cross-curricular connections were concerned. Based on the purpose and goals of this research, we formulated several research questions; however, in this paper we focus on the following two:

RQ1: Which ICT do teachers use, and how do they using them in planning and implementing cross-curricular connections involving music and visual arts content during remote teaching?

RQ2: Which ICT do teachers plan for pupils to use in completing their cross-curricular assignments during remote teaching?

In the following paper, we present the course and results of our qualitative case study.

3. Materials and Methods

To carry out the research, which took place in the 2020/2021 school year, we used the descriptive method for empirical pedagogical research. We conducted a qualitative case study, in which we included teachers from selected primary schools. We chose an approach that allowed us to gain an in-depth understanding of the subjective aspects involved in the use of ICT in learning and in teaching interdisciplinary content in music and fine arts.

3.1. Participants

The research involved 24 purposefully selected teachers (three men) employed in the first (from 6 to 9 years) and second (from 10 to 12 years) period of education in primary schools in the southwestern part of Slovenia. Their average age was 33.8 years (±9.08), and their average number of teaching years was 11.2 (±10.5).

3.2. Instrument and Procedure

For the purpose of the research, we conducted semi-structured interviews with the participating teachers in October 2020. We conducted these interviews individually through the Zoom online platform. When conducting the interviews, we used the funnel technique, starting the conversation with open-ended questions in the context of the research questions and continuing with more concrete and content-specific questions which were formed on the spot during the interview. During the conversation, we let teachers explain the strategies they used without additional sub-questions, thus avoiding any influence on the answers. Results of these interviews are quoted directly in this paper.

We reviewed the written conversations and determined the coding units based on the research objectives and the principles of grounded theory [23]. Content-related codes were connected, and defined relevant concepts from which we derived categories related to insights into how teachers used ICT in distance teaching of music and art content, namely, using ICT for planning the remote teaching–learning process, and strategies for using ICT in the implementation of cross-curricular connections in music and visual arts. Both categories were documented in the obtained teacher statements, which were left unchanged and are denoted below by italics.

Based on the presented results, we performed consensual validation with teachers [24] to confirmed that what was written was accurate and credible data and findings that reflected the actual view of the research participants.

4. Results

The responses were grouped into the use of ICT in (i) planning the remote teaching–learning process; (ii) implementing cross-curricular connections in teaching music and visual arts; and (iii) planning cross-curricular assignments. The following subsections cover these three groups.

4.1. Using ICT in Planning the Remote Teaching–Learning Process

When the teaching–learning process first moved online in March 2020, most teachers did not plan to use video conferencing tools, as they were not familiar with them. They mainly followed the now mostly outdated recommendations for the use of ICT from the previously mentioned national curricula [9,10]. Teachers commonly used word processors and presentation programs to prepare cross-curricular materials (Figure 2), while using their own photos to explain various concepts. Several teachers prepared detailed written instructions of the teaching–learning process available to both pupils and parents. These materials were uploaded to the e-classroom (Google classrooms http://classroom.google.com (accessed on 22 November 2021), Microsoft Teams, Moodle, and eAsistent Xooltime) together with assignments and sometimes quizzes to consolidate and assess the knowledge acquired. The selection of the e-classroom platform was at first left to teachers, and it was only later that the management of individual schools decided on a unified platform.

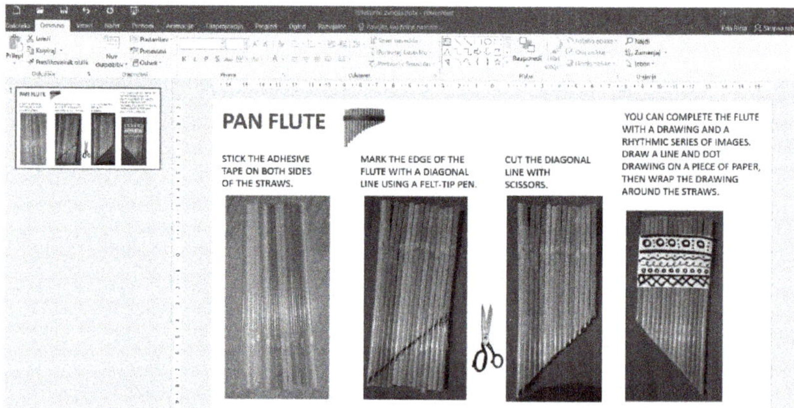

Figure 2. Using PowerPoint in the planning of a remote cross-curricular assignment for building a rattle, a lesson including both music and visual arts concepts.

Teachers used social networking sites and the web in general for searching and exchanging information and materials with other teachers. Multimedia materials were found on the websites of textbook publishing houses (Državna založba Slovenije (https://www.dzs.si/katalog/10237338/osnovna-sola (accessed on 2 July 2022), Mladinska knjiga https://www.ucimte.com/ (accessed on 2 July 2022) and RokusKlett https://www.rokus-klett.si/za-ucitelje/ (accessed on 2 July 2022)), on YouTube, and on teachers' own CDs. Sometimes, teachers filmed their own singing. All materials were primarily made available in the e-classroom. Teachers stressed that this preparation was based on their own knowledge of various computer programs and other ICT. Most of them pointed out a lack of ICT competences: "*... that I had too little knowledge in the field of ICT. We helped each other a lot among the teachers. I was afraid that I would not be successful in my work and that the students would not achieve the goals of the subject.*" Several teachers initially needed help from family members to prepare these materials: "*I am lucky that my older children at home are much more ICT aware and willing to help when I didn't know how to use all the options that ICT offers us.*"

Although offered by the Ministry of Education for decades, teachers mentioned that they had not attended ICT training for a long time, and were mainly self-educated for basic use. This is in line with a report from 2016 mentioning the lack of ICT skills among teachers in primary schools [14]. Despite the fact that various measures have been prepared at the national level to ensure and improve the digital competences of the population, especially by providing equal access to the internet and equal opportunities for all people, regardless of their abilities and acquired e-skills [2,14], a shortage and obsolescence of ICT equipment, mostly among primary school children and especially in families with many primary school children, was present in the autumn of 2020 [25].

Teachers who participated in training in the past said they had already forgotten the skills they acquired, as they did not use them in class. In addition, these training sessions did not provide the specific knowledge needed for emergency remote teaching. Additional ICT training was organised on school days after the epidemic was declared in the country.

Because grasping, understanding, and remembering everything in these short training sessions was not possible, teachers supported and helped each other, and parents came to their aid as well. With increasing familiarity teachers started to use modern ICT, which they became acquainted with through training, peer support, and the recommendation guidelines offered by [21]. Several teachers started to use ICT to identify prior knowledge and assess newly acquired knowledge; they used assignments from e-materials and e-textbooks https://eucbeniki.sio.si/ (accessed on 22 November 2021), employed services such as Socrativ https://b.socrative.com/ (accessed on 22 November 2021) and Kliker https://kliker.sio.si/ (accessed on 22 November 2021) (classroom response systems), and used more advanced functions of the available software, such as the 'immersion reader' in Microsoft Teams, which helped year 1 and year 2 pupils to listen to written instructions (Figure 3, allowing them to work more independently and flexibly [26,27].

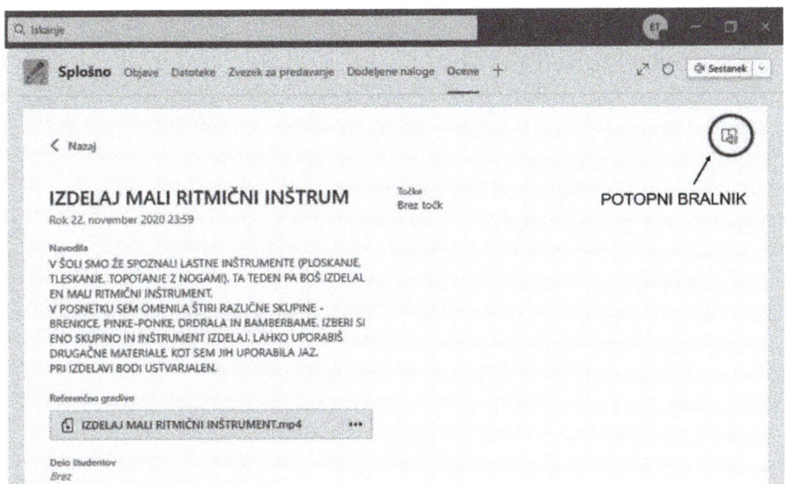

Figure 3. Use of Microsoft Teams e-classroom in the planning of a remote cross-curricular assignment including music and visual art.

When presenting and mentioning the possible usage of audio and video editing software (e.g., Audacity https://www.audacityteam.org/ (accessed on 22 November 2021), OpenShot https://www.openshot.org/ (accessed on 22 November 2021)), mind or concept maps (e.g., SimpleMind https://simplemind.eu/ (accessed on 22 November 2021), Creately https://creately.com/ (accessed on 22 November 2021), Coggle https://coggle.it/ (accessed on 22 November 2021), e-listovnik https://listovnik.sio.si/ (accessed on 22 November 2021)), music composition and notation software (e.g., MuseScore https://musescore.org/ (accessed on 22 November 2021) seen in Figure 4), graphics editors (Microsoft Paint, Gimp, Inkscape), and other services from the web (e.g., Artsteps https://www.artsteps.com/ (accessed on 22 November 2021) virtual galleries) as well as advanced software recommended by Breznik and Eyer [21], most participants had never heard of nor tried the majority of them. As already mentioned in the literature [2,28,29] the willingness of teachers to introduce modern approaches as well as ICT to teaching music and visual arts remains very low, despite the fact that using ICT in the process of learning can offer pupils opportunities for active and holistic learning of musical and artistic concepts [2,30].

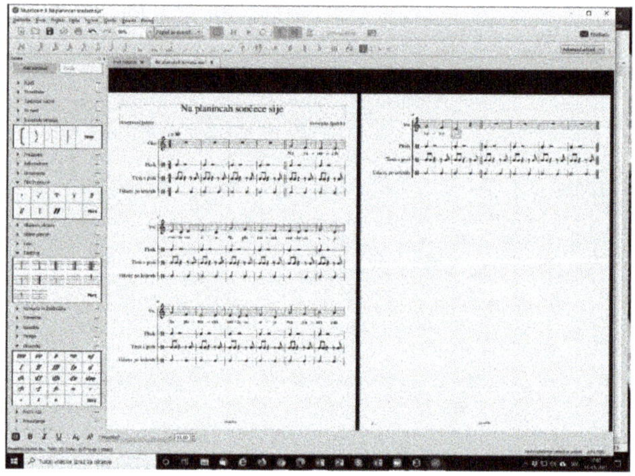

Figure 4. Use of MuseScore software during remote teaching.

4.2. Strategies for Using ICT in the Implementation of Cross-Curricular Connections in Teaching Music and Visual Arts

Teachers mostly used two ways of presenting content: (i) through the e-classroom of each particular subject by uploading text documents, presentations and audiovisual recordings, and (ii) through video conference systems (ZOOM, and to a lesser extent Cisco Webex, Microsoft Teams, Skype, and Jitsi Meet). For music art, teachers mostly presented and explained the selected musical content (concepts) via PowerPoint presentations that included audio and/or video clips (online content and local content (files and CDs) played in Windows Media Player, Microsoft Groove Music, iTunes). Several used interactive materials prepared by publishing houses. Teachers complained that they had problems with creating music as a group, as different internet connections caused delays in sound and pupils were deprived of fully experiencing sound images, that is, sound performances created with one or more instruments simultaneously [30]. In order to consolidate and test the acquired knowledge of musical contents, several teachers offered interactive quizzes. Visual art content (concepts) was mainly discussed over PowerPoint presentations. Only one of the interviewees encouraged the understanding of artistic concepts by analysing artwork found online that he included in the PowerPoint presentation.

Teachers participating in the survey mostly carried out cross-curricular connections in which music art played the leading role and visual art had a supportive role, in a similar way as presented in [4]. Our interviews revealed inadequate understanding of the leading and supporting role. For example, many teachers did not take advantage of the visual art subject for integrated transfer of knowledge because they only used it for artistic expression of the music motif, which in practice is too often understood as an appropriate implementation of cross-curricular connections; Birsa came to similar conclusions in [1].

4.3. Strategies for Planning and Using ICT for Pupils to Use in Completing Their Cross-Curricular Assignments

The sudden move to remote teaching in March 2020 required pupils to quickly adopt particular ICT skills, such as using e-classrooms, joining and participating in video conferences, opening, creating and editing documents, and listening to and watching audio-visual recordings, etc. on their smartphones and computers. Initially, pupils had difficulties in creating the planned assignments due to inadequate or non-existent ICT, however, these were gradually eliminated. Teachers and schools then had the possibility to better prepare for remote teaching for the school year starting in September 2020 [25].

Teachers described their pupils as often sending in recordings of assignments such as singing and playing small rhythmic instruments (Figures 5 and 6) via e-mail, the web file sharing service we-transfer (https://wetransfer.com/) (accessed on 22 November 2021), the document sharing system Google Drive, the instant messaging system Viber https://www.viber.com/ (accessed on 22 November 2021), and Microsoft Teams. Teachers mentioned that pupils wanted to show their products, both music and visual artworks, live via video conference in order to obtain live feedback from teachers. An important observation from teachers was that the artistic motifs in these recordings were not depicted in accordance with the artistic development of pupils, which indicates the influence of parents on the implementation of artistic assignments. Their works revealed uncreative thinking and the use of art techniques in pupils' works that are not typical of children in the first and second period of education. Several teachers said: *"Because I know the students and their abilities I was surprised that some interdisciplinary assignments were unexpectedly and surprisingly accomplished."*

Figure 5. A clip from a video of a ten-year-old pupil performing a rhythm task.

Figure 6. A clip from a video of a six-year-old pupil performing a rhythm task on her home-made rhythmic instrument.

Several teachers converted their lecture presentations to videos and uploaded them to YouTube in order to enable convenient viewing, which was welcomed by both parents and pupils. While pre-recorded explanations recommended by Kustec et al. [31] allow pupils to watch the content multiple times, it has been argued that this method is not particularly suitable for music and visual arts pupils in the first and second period of education due to the specifics of the subjects [9,10]. The teaching–learning process of both art subjects emphasises teacher–pupil interaction, and one-way communication fully supports a process in which pupils are focused on creative thinking and making as well as on cross-curricular integration [19].

This was observed in pupils' creations following cross-curricular assignments. Figure 7 shows an example of inefficient cross-curricular connection implemented with respect to the concept of rhythm being taught during a particular lecture. Pupils did not have the opportunity to learn the concept holistically through a pre-recorded video. While the goals of the music art lecture were achieved, the visual artwork served mainly for decorating the instrument. The emphasis of the latter was on depicting the motif with stereotypical images rather than on becoming familiar with and connecting a new visual art concept, characterised by different but constantly repeated rhythmic sequences. Teachers made insufficient use of ICT to support a holistic consideration of the concept being taught and to provide a variety of options for visual expression with computer tools to create different rhythmic sequences (multiplying images and shapes, changing colours, mirroring) using software such as MS Paint.

Figure 7. 'Pan's flute' created by a year one pupil, with rhythmically distributed lines and points.

5. Discussion

The following discussion section is divided into subsections based on the two research questions presented in Section 2 and the summary of the results presented in Figure 8. The third subsection highlights limitations and future directions.

5.1. Which ICT Do Teachers Use, and How Do They Use Them in Planning and Implementing Cross-Curricular Connections of Music and Visual Arts Content during Remote Teaching?

The two main problems that we identified in our study are lack of knowledge about how ICT can be used in the teaching and learning process, and understanding of cross-curricular connections between music and visual arts during remote teaching.

Despite the increased use of computers, the internet, and various computer software programs in education, this does not mean that they are utilised effectively or that teachers use them to their full potential. The lack of in-depth ICT knowledge is evident from the teachers' answers, which show that the ICT skills of most respondents did not exceed the use of basic computer software. According to teachers' statements about the diversity of choice in computer tools and programs, they lacked knowledge of the possibilities that modern ICT offers for creating efficient cross-curricular connections, which is in line with other studies [2,3,10,32]. Bohak Adam and Metljak [33] investigated Slovenian teachers' competences and use of ICT in the field of music during the COVID-19 pandemic. They investigated whether teachers had any problems in transferring to online music teaching and how or if they had improved their knowledge of ICT. They found no statistically significant differences between teachers with different professional experience with respect to having problems with ICT. In addition, they showed that teachers had technical problems and problems with communication with students and emphasised other problems, such as students' unresponsiveness and ignorance or lack of knowledge concerning ICT.

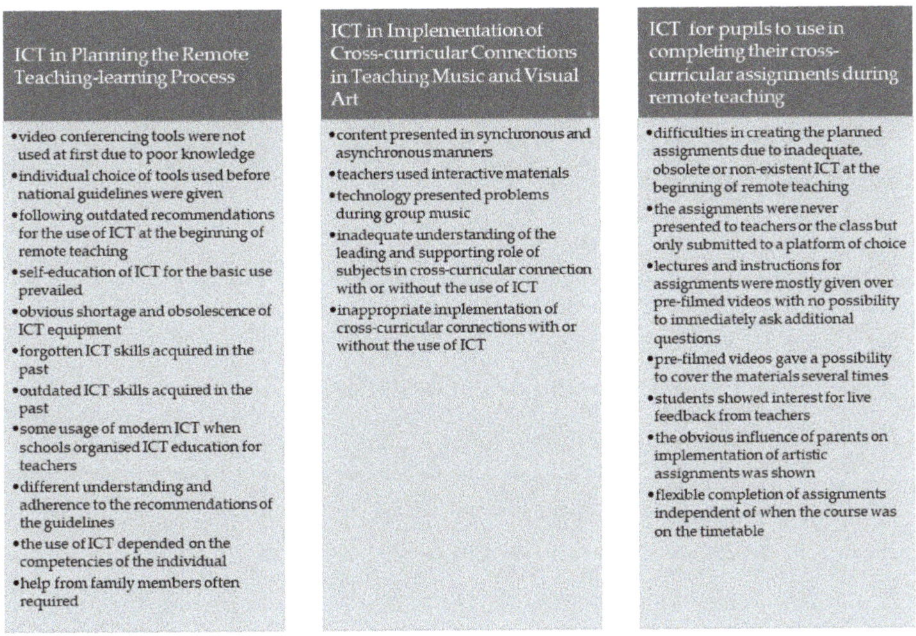

Figure 8. The main findings summarised and divided based on the research questions.

Despite the fact that the curricula of both art subjects from 2011 and the guidelines for the use of ICT which were provided to teachers in 2020 contain recommendations for ICT usage, the interviewees mostly did not follow them. The interviews revealed that teachers did not always take advantage of all the possibilities of cross-curricular connections available with ICT to implement these connections. One said: "… *when we switched to distance learning, my thoughts were occupied by problems related to the use of ICT, so I probably often neglected the possibilities of interdisciplinary links between music and fine arts.*" It is true, however, that implementation of cross-curricular connections with ICT depend on both teachers' ICT competences and on the guidelines available in curricula, guidelines for the use of ICT [5,21], and other teaching materials such as interactive materials offered by publishing houses, which may or may not promote cross-curricular connections.

As already mentioned, the authors of both national curricula [9,10] and guidelines for ICT use [5,21] for music and visual arts highlight the use of ICT for knowledge enrichment, skills development, and personality formation, as well as key competences for lifelong learning. The guidelines for ICT use for music education [21] emphasize that students develop their use of digital technology, independent learning, and critical judgement. They develop digital competences in the areas of collecting, storing, sharing, and communicating information, e.g., about music, composers, etc. They learn in depth about different digital tools for online communication. The guidelines for visual art education [5] do not highlight how ICT can enrich the teaching and learning process. Rather, it mentions that pupils can develop different art skills and understanding of theoretical underpinnings with or without the help of ICT. Our study reveals that the guidelines for the use of ICT for visual arts are outdated. While the national curricula for music education include detailed cross-curricular connections, these are barely mentioned for visual arts [34]. However, neither national curricula nor the newer national guidance for the use of ICT provide concrete information on how to implement cross-curricular lessons using ICT.

One should not overlook the fact that lack of knowledge of the content of different subject areas and of cross-curricular connection strategies can lead to inefficient and inadequate cross-curricular connections, even if teachers are highly digitally literate [14,35]. Therefore, it is very important to use the planning stage to consider strategies for transferring music and visual arts knowledge and skills in the remote educational process through meaningful use of ICT. This makes the implementation of cross-curricular connections much more efficient, as Birsa and Kopačin [29] have previously pointed out.

The experiences reported by our participants indicate that with appropriate guidelines and additional training in both ICT and cross-curricular connections, these challenges can become opportunities to learn about new themes, strategies, and ways of communicating.

5.2. Which ICT Do Teachers Plan for Pupils to Use in Completing Their Cross-Curricular Assignments during Remote Teaching?

An analysis of the ICT used by the pupils to complete cross-curricular assignments during distance learning showed that they used both hardware and software that they were already familiar with, as well as hardware and software that required the help of an adult. The assignments showed that teachers wanted to encourage creative expression in both music and visual arts subjects. However, especially in the visual art assignments, the problem of adults interfering with pupils' work and thus influencing their creative thinking was highlighted. Duh and Vrlič [36] emphasised that the creative learning process should encourage pupils to develop a variety of ideas in order to find new possibilities and solutions to assignments, to experiment with art techniques and materials, to enquiry various sounds on instruments, etc. This was made impossible by parents' intervention in pupils' creative cognitive processes. The fact that pupils often required help from their parents to use ICT, which is understandable considering the pupils' age, contributed to this phenomenon. Other research [37], however, has shown positive effects on the realization of the learning process when the parents of students were involved in it.

One of the reasons that parents helped students more than they would have done otherwise could be that teachers provided students with too many or too complex tasks, which they were not able to complete in the time allotted. This is precisely the problem highlighted by Toseli [6], who notes that the first reaction of teachers when adopting distance learning was to give students too many tasks, which turned out to be counterproductive and a wrong didactic decision. He says that while fewer tasks for students requires better organization and more work for the teacher, it encourages greater collaboration between teachers in different subject areas and the development of distance learning that is not just about homework and revision without the student being present. This leads to the proposal that the implementation of both synchronous and asynchronous distance learning should be oriented towards more project-based, creative, interdisciplinary, and flexible tasks. In this way, students can be more motivated and creative in the artistic field, and are able to be successful without the help of an adult.

5.3. Limitations and Future Work

This qualitative research is limited to participants in the Slovenian educational environment. In subsequent research, the size and the heterogeneity of the included sample in the Slovenian learning area could be increased, and perhaps teachers from other countries could be added in order to expand the research to an international context. Nevertheless, the conducted research addressed several of the practical issues that arise for teachers in connection with the implementation of interdisciplinary links between music and art content in the new digital reality. Further research could additionally examine the situation in the field of teachers' digital competences in connection with the implementation of the learning process of art subjects and the integration of new ICT skills in the return to the school environment.

6. Conclusions

The aim of this qualitative research was to uncover the use of ICT by teachers in the first and second period of education (key stages 1 and 2) for planning and implementing cross-curricular connections of music and visual arts content during remote teaching. Teachers initially used ICT which they had already used before emergency remote teaching began. Later, they were forced to adapt and become acquainted with the use of modern hardware, software, and online services, and had to adjust their assignments to their pupils' ICT competences. Teachers' ICT competences were not influenced by previous professional training offered nationally by the Ministry of Education, as it turned out that teachers did not attend them often, or that if they did the acquired knowledge proved to be useless for remote teaching, as it was outdated. Furthermore, teachers did not know how to take advantage of all the opportunities offered by ICT, as previously pointed out by Duh and colleagues [28].

A particular focus of this study was on using ICT in the implementation of cross-curricular connections for the remote teaching of music and visual arts. This study shows that teachers used both asynchronous (e.g., e-classroom) and synchronous (e.g., video conference) delivery of learning contents to implement the remote teaching–learning process. Teachers used interactive materials to explain novel concepts discussed in both arts subjects on certain occasions, however, they did not encourage efficient cross-curricular connections. This study revealed a lack of knowledge about the specifics of cross-curricular connections, and as such, this situation has not changed over the past ten years. Most teachers used one subject only as a motif [1], while others used one subject only as a support for the leading subject and its learning goals [36]. Cross-curricular assignments were modestly performed, and ICT was insufficiently applied.

Another objective of this study was to learn more about the planning and use of ICT for cross-curricular assignments. It turned out that parents helped their children with both ICT usage and with their assignments. Parents' influence on their child's creative thinking and expression can hinder their creative and cognitive development. This was mainly visible in pupils' final products (e.g., artwork), which did not show solutions typical of the developmental stage of children aged between six and eleven. This study emphasises the importance of teachers' live feedback on the works submitted by pupils. Students' pride in showing their creations and receiving live response is very important also in remote teaching.

In order to improve remote teaching and render cross-curricular connections effective with the application of ICT, changes and updates to the curriculum of both art subjects should be considered, at least at the primary school level. New methods and forms of work should be suggested with innovative and interactive digital learning environments, including recommendations for the reasonable application of modern ICT in the teaching–learning processes of music and visual arts. To an extent, this has already been noted in the literature [38]; however, the period of emergency remote teaching intensively stressed the importance of it. Furthermore, it is necessary to update the interactive teaching materials in the art subjects; while these are currently available on the websites of Slovenian publishing houses, they have not been updated for years. This could promote effective cross-curricular connections of both subjects [28]. In addition, it is necessary to investigate the ICT competences of pupils under the age of 16 in Slovenia, as research thus far has focused on people aged between 16 and 74 [14]. This would allow teachers to adjust the use of the appropriate ICT to pupils' age and skills, supporting the further leveraging of ICT competences.

Author Contributions: Conceptualization, E.B. and B.K.; methodology, E.B. and B.K.; validation, E.B., B.K. and M.K.; formal analysis, E.B. and B.K.; research, E.B., B.K. and M.K.; data curation, E.B., B.K. and M.K.; writing—original draft preparation, E.B., B.K. and M.K.; writing—review and editing, E.B., B.K. and M.K. All authors have read and agreed to the published version of the manuscript.

Funding: This research received no external funding.

Institutional Review Board Statement: The study was conducted in accordance with the Declaration of Helsinki, and approved by the Institutional Review Board of the University of Primorska (application number 2020-12 approved on 14 May 2020).

Informed Consent Statement: Informed consent was obtained from all subjects involved in the study.

Data Availability Statement: In this study we obtained qualitative data through interviews. We assured the participants that the data obtained would only be used for research purposes, not be disseminated, and be securely stored and appropriately discarded after the research is finished.

Acknowledgments: We wish to thank all participating primary school teachers for their cooperation in the study.

Conflicts of Interest: The authors declare no conflict of interest

References

1. Birsa, E. Transfer likovnega znanja in izkušenj v novo učenje. In *Sučasni Perspektyvy Osvity*; Omelčenko, S., Ed.; Institute for Foreign Languages: Horlivka, Ukraine, 2016; pp. 283–294, 432–433.
2. Istenič Starčič, A. *Izobraževalna Tehnologija in Izgradnja Avtentičnega Učnega Okolja*; Fakulteta za Gradbeništvo in Geodezijo v Ljubljani: Ljubljana, Slovenia, 2020; Available online: https://repozitorij.uni-lj.si/IzpisGradiva.php?lang=slv&id=116682 (accessed on 15 November 2021).
3. Krek, J.; Metljak, M. *Bela Knjiga o Vzgoji in Izobraževanju v Republiki Sloveniji 2011*; Zavod Republike Slovenije za Šolstvo: Ljubljana, Slovenia, 2011.
4. Rutar Ilc, Z.; Pavlič Škerjanc, K. *Medpredmetne in Kurikularne Povezave: Priročnik za Učitelje*; Zavod Republike Slovenije za Šolstvo: Ljubljana, Slovenia, 2010.
5. Kocjančič, F.N. *Smernice za Uporabo IKT pri Predmetu Likovna Umetnost—Delovna Verzija*; Zavod Republike Slovenije za Šolstvo: Ljubljana, Slovenia, 2016; Available online: https://www.zrss.si/digitalnaknjiznica/smernice-ikt-lum/files/assets/common/downloads/publication.pdf (accessed on 12 December 2021).
6. Toselli, L. *La Didattica a Distanza—Funziona, se sai Come Farla*; Edizioni Sonda Srl.: Milan, Italy, 2020.
7. Hrastinski, S. Asynchronous and synchronous e-learning. *Educ. Q.* **2008**, *31*, 51–55.
8. Zmazek, B.; Pesek, I.; Lipovec, A. Edupedia.si oziroma razlagamo.si. *UMniverzum Interna Rev. Univ. Mariboru* **2020**, *12*, 12.
9. Holcar, A.; Borota, B.; Breznik, I.; Jošt, J.; Kerin, M.; Kovačič, A.; Lango, J.; Novak, T.M.; Kafol, B.S. *Program Osnovna Šola—Glasbena Vzgoja—Učni Načrt*; Zavod RS za Šolstvo in Ministrstvo za Šolstvo in Šport: Ljubljana, Slovenia, 2011. Available online: https://www.gov.si/assets/ministrstva/MIZS/Dokumenti/Osnovna-sola/Ucni-nacrti/obvezni/UN_glasbena_vzgoja.pdf (accessed on 15 January 2022).
10. Kocjančič, F.N.; Karim, S.; Kosec, M.; Opačak, Ž.; Prevodnik, M.; Rojc, J.; Velikonja, A.; Zupančič, T.; Kepec, M.; Prevodnik, M. *Program Osnovna Šola—Likovna Vzgoja—Učni Načrt*; Zavod RS za Šolstvo in Ministrstvo za Šolstvo in Šport: Ljubljana, Slovenia, 2011. Available online: https://www.gov.si/assets/ministrstva/MIZS/Dokumenti/Osnovna-sola/Ucni-nacrti/obvezni/UN_likovna_vzgoja.pdf (accessed on 15 January 2022).
11. Hodges, C.; Moore, S.; Lockee, B.; Trust, T.; Bond, A. The Difference Between Emergency Remote Teaching and Online Learning. WY IT Matters to Hihher Education—Educause Review 2020. Available online: https://er.educause.edu/articles/2020/3/the-difference-between-emergency-remote-teaching-and-online-learning (accessed on 19 October 2021).
12. Blahušiaková, M.; Mokošová, D.; Šoltés, E. Education in online environment from students' and teachers' perspective. *Int. J. Cogn. Res. Sci. Eng. Educ. IJCRSEE* **2021**, *9*, 203–226. [CrossRef]
13. Ministrstvo za Izobraževanje, Znanost in Šport. Priporočila Osnovnim in Srednjim Šolam za Izvajanje Izobraževanja na Daljavo. 2020. Available online: https://www.gov.si/novice/2020-03-13-priporocila-osnovnim-in-srednjim-solam-za-izvajanje-izobrazevanja-na-daljavo/ (accessed on 15 January 2022).
14. Digitalna Slovenija 2020, Strategija Razvoja Informacijske Družbe do Leta 2020. Digitalizacija Slovenije z Intenzivno in Inovativno Uporabo IKT in Interneta v Vseh Segmentih Družbe. 2016. Available online: https://www.gov.si/assets/ministrstva/MJU/DID/Strategija-razvoja-informacijske-druzbe-2020.pdf (accessed on 7 February 2022).
15. Baloh, B. Kurikularno področje jezik v vrtcih v čezmejnem prostoru v slovenski Istri in na Tržaškem. In *Slovenski Javni Govor in Jezikovno-Kulturna (Samo)zavest—Obdobja 38*; Tivadar, H., Ed.; Znanstvena Založba Filozofske Fakultete: Ljubljana, Slovenia, 2019; pp. 449–456. Available online: https://centerslo.si/wp-content/uploads/2019/10/Obdobja-38_Baloh.pdf (accessed on 19 October 2021).
16. Fogarty, R. *How to Integrate the Curricula*, 3rd ed.; Corwin A SAGE Company: Thousand Oaks, CA, USA, 2009.
17. Huat See, B.; Kokotsaki, D. Impact of arts education on children's learning and wider outcomes. *Rev. Educ.* **2016**, *4*, 234–262.
18. Kim, J.H. Music teachers' understanding of blended learning in Korean elementary music classes. *Music Educ. Res.* **2021**, *23*, 311–320. [CrossRef]
19. Tacol, T. *Creative Lessons of Visual Art Education*; Digit Ltd.: Bratislava, Slovakia, 2007.
20. Barnes, J. *Cross-Curricular Learning 3-14*; Sage Publications: Los Angeles, CA, USA; London, UK; New Delhi, India; Singapore; Washington, DC, USA, 2015; Available online: https://books.google.si/books?hl=sl&lr=&id=piKJCwAAQBAJ&oi=fnd&pg=PP1

&dq=cross-curricular+connection+between+music+and+art&ots=Ve1LHEXmBq&sig=9WZaH6YInyNYcNNgIIvSz5cRE8g&redir_esc=y#v=onepage&q=cross-curricularconnectionbetweenmusicandart&f=false (accessed on 12 August 2021).

21. Breznik, I.; Eyer, D. *Smernice za Uporabo IKT Glasbena Umetnost v Osnovni Šoli in Glasba v Gimnaziji*; Zavod Republike Slovenije za Šolstvo: Ljubljana, Slovenia, 2020; Available online: https://skupnost.sio.si/pluginfile.php/919516/mod_folder/content/0/Posodobljene%20IKT%20smernice%20GUM%20_GLA_16.7.2020.pdf?forcedownload=1 (accessed on 9 September 2021).
22. Birsa, E. Teaching strategies and the holistic acquisition of knowledge of the visual arts. *CEPS J. Cent. Educ. Policy Stud. J.* **2018**, *8*, 187–206. [CrossRef]
23. Mažgon, J. *Razvoj Akcijskega Raziskovanja na Temeljnih Predpostavkah Kvalitativne Metodologije*; Filozofska Fakulteta: Ljubljana, Slovenia, 2008.
24. Mesec, B. *Uvod v Kvalitativno Metodologijo*; Visoka Šola za Socialno Delo: Ljubljana, Slovenia, 1998.
25. Pouk na Daljavo: Nekateri Primorski Osnovnošolci še Vedno Brez Računalnikov, Novice, Radio Koper, 6 November 2020, Slovenija. Available online: https://www.rtvslo.si/radiokoper/novice/pouk-na-daljavo-nekateri-primorski-osnovnosolci-se-vedno-brez-racunalnikov/541415 (accessed on 5 October 2021).
26. Veletsianos, G.; Houlden, S. An analysis of flexible learning and flexibility over the last 40 years of Distance Education. *Distance Educ.* **2019**, *40*, 454–468. [CrossRef]
27. Ajtnik, M. Poučevanje glasbe kot ustvarjalna naloga. *Vzgoja Izobr.* **2001**, *32*, 7–8.
28. Duh, M.; Bratina, T.; Krašna, M. Elementary teacher's competences for multimedia learning materials production = Kompetencije učitelja u osnovnim školama za pripremu materijala za multimedijsko učenje. *Informatologia* **2013**, *46*, 333–342.
29. Birsa, E.; Kopačin, B. Medpredmetno povezovanje glasbenih in likovnih vsebin v zamejskem šolstvu. In *Medpredmetno Povezovanje: Pot do Uresničevanja Vzgojno-Izobraževalnih Ciljev = Cross-Curricular Integration: The Path to the Realisation of Educational Goals*; Volk, M., Štemberger, T., Sila, A., Kovač, N., Eds.; Založba Univerze na Primorskem: Koper, Slovenia, 2020; pp. 133–146.
30. Oblak, B. *Glasbena Slikanica 1: Učbenik za Prvi Razred Devetletne Osnovne Šole—Priročnik za Učitelje*; Državna Založba Slovenije: Ljubljana, Slovenia, 1999.
31. Kustec, S.; Logar, V.; Krek, M.; Flogje, A.; Truden Dobrin, P.; Ivanuš Grmek, M. *Vzgoja in Izobraževanje v Republiki Sloveniji v Razmerah, Povezanih s COVID-19. Modeli in Priporočila*; Ministrstvo za Izobraževanje, Znanost in Šport ter Zavod Republike Slovenije za Šolstvo: Ljubljana, Slovenia, 2020.
32. Eynon, R. Becoming digitally literate: Reinstating an educational lens to digital skills policies for adults. *Br. Educ. Res. J.* **2021**, *47*, 146–162. [CrossRef]
33. Bohak Adam, T.; Metljak, M. Experiences in distance education and practical use of ICT during the COVID-19 epidemic of Slovenian primary school music teachers with different professional experiences. *Soc. Sci. Humanit. Open* **2022**, *5*, 100246. [CrossRef] [PubMed]
34. Birsa, E. Primerjava Učnih Načrtov in Vključenost Priporočil za Medpredmetno Povezovanje Likovnih Vsebin = Comparison of Syllabi and Inclusion of Recommendations for Interdisciplinary Integration of Visual Arts Contents'. *Rev. Za Elem. Izobr. J. Elem. Educ.* **2017**, *10*, 163–180.
35. Svalina, V.; Sukop, I. Listening to music as a teaching area in Croatian primary schools: The teacher's perspective. *Music Educ. Res.* **2021**, *23*, 321–334. [CrossRef]
36. Duh, M.; Vrlič, T. *Likovna Vzgoja v Prvi Triadi Devetletne Osnove Šole—Priročnik za Učitelje Razrednega Pouka*; Rokus Klett: Ljubljana, Slovenia, 2003.
37. Abakumova, I.; Bakaeva, I.; Grishina, A.; Dyakova, E. Active learning technologies in distance education of gifted students. *Int. J. Cogn. Res. Sci. Eng. Educ. IJCRSEE* **2019**, *7*, 85–94. [CrossRef]
38. Manches, A.; Plowman, L. Computing education in children's early years: A call for debate. *Br. J. Educ. Technol.* **2017**, *48*, 191–201. [CrossRef]

MDPI
St. Alban-Anlage 66
4052 Basel
Switzerland
Tel. +41 61 683 77 34
Fax +41 61 302 89 18
www.mdpi.com

Electronics Editorial Office
E-mail: electronics@mdpi.com
www.mdpi.com/journal/electronics

www.ingramcontent.com/pod-product-compliance
Lightning Source LLC
Chambersburg PA
CBHW061823210926
51850CB00006B/787
* 9 7 8 3 0 3 6 5 6 4 6 0 9 *